Sociology of the Sciences Yearbook

Volume 31

More information about this series at http://www.springer.com/series/6566

Karen Kastenhofer • Susan Molyneux-Hodgson
Editors

Community and Identity in Contemporary Technosciences

 Springer

Editors
Karen Kastenhofer
Institute of Technology Assessment
Austrian Academy of Sciences
Vienna, Austria

Susan Molyneux-Hodgson
Department of Sociology, Philosophy
and Anthropology
University of Exeter
Exeter, United Kingdom

ISSN 0167-2320   ISSN 2215-1796   (electronic)
Sociology of the Sciences Yearbook
ISBN 978-3-030-61727-1   ISBN 978-3-030-61728-8   (eBook)
https://doi.org/10.1007/978-3-030-61728-8

This Springer imprint is published by the registered company Springer Nature Switzerland AG
The registered company address is: Gewerbestrasse 11, 6330 Cham, Switzerland

Preface

This volume resulted from two conference panels and one dedicated workshop. In September 2014, Susan Molyneux-Hodgson (then University of Sheffield) and Morgan Meyer (then Agro ParisTech) organised a panel on '*Synthesising futures: Analysing the socio-technical production of knowledge and communities*' at the biennial conference of the European Association for the Study of Science and Technology (EASST) in Torun, Poland. Two years later, in September 2016, Karen Kastenhofer (Austrian Academy of Sciences), Sarah Schönbauer (then University of Vienna), and Niki Vermeulen (University of Edinburgh) organised a track on a related theme, '*(Techno)science by other means of communality and identity*', at the EASST conference in Barcelona, Spain. Then, in February 2017, Karen Kastenhofer (Austrian Academy of Sciences), Martina Merz (Alpen Adria University Klagenfurt), Ulrike Felt, Max Fochler, Anna Pichelstorfer (University of Vienna), and Niki Vermeulen (University of Edinburgh) organised a three day workshop on '*Community and identity in contemporary technosciences*' in Vienna, Austria,[1] co-funded by EASST, STS Austria, and the Austrian Science Fund (via project V 383-G15). This final event sought to bring the various strands of the almost 3-year discussion into dialogue with each other and establish a statement on the state of the art on community and identity in technoscience.

The 2017 Vienna workshop saw 16 papers presented and discussed. Nine of those papers have now been revised and written up for this volume, and four further contributions have been solicited from the two previous events. We want to thank all the participants involved in the conferences and workshop – whether speaking or in the audiences – for their contributions to this collective endeavour!

The contributions in this volume benefited from an extensive review process that included not only the editorial board of the Sociology of the Sciences Yearbook series and the guest editors of this volume but also a large number of external

---

[1]http://www.sts-austria.org/wp-content/uploads/2017/02/Programm_CIT_HP.pdf. Accessed 24 April 2019.

reviewers that provided helpful feedback and thereby supported this project funda-mentally. Among the latter feature Sandra Beaufays (University of Duisburg-Essen), Bernadette-Bensaude-Vincent (professor emeritus, Université Paris I Panthéon-Sorbonne), Alexander Bogner (Austrian Academy of Sciences), Jane Calvert (University of Edinburgh), Ana Delgado (University of Oslo), Max Fochler (University of Vienna), Scott Frickel (Brown University), Stephen Hilgartner (Cornell University), Sabina Leonelli (University of Exeter), Dagmar Lorenz-Meyer (Charles University, Prague), Sabine Maasen (Technical University of Munich/Munich Center for Technology in Society), Robert Meckin (University of Manchester), Morgan Meyer (Mines ParisTech, PSL University), Monika Nerland (University of Oslo), Bart Penders (Maastricht University), Simone Rödder (University of Hamburg), Philip Shapira (University of Manchester and Georgia Institute of Technology), Lisa Sigl (University of Vienna), Esther Turnhout (Wageningen University), Niki Vermeulen (University of Edinburgh), Caroline Wagner (Ohio State University), Bridgette Wessels (University of Glasgow), Matthias Wienroth (Newcastle University), and Sally Wyatt (Maastricht University) as reviewers of individual chapters and four anonymous reviewers of the entire volume.

We hope you find the volume as interesting and useful as we have found the process of discussing and collating contributors' ideas!

Institute of Technology Assessment   Karen Kastenhofer
Austrian Academy of Sciences
Vienna, Austria

Department of Sociology, Philosophy   Susan Molyneux-Hodgson
and Anthropology
University of Exeter, UK
Exeter, UK

May 2020

Contents

Contributors

Clemens Blümel Research area 'Research Systems and Science Dynamics', German Centre for Higher Education Research and Science Studies (DZHW), Berlin, Germany

Béatrice Cointe CNRS, Centre de Sociologie de l'Innovation, i3, Mines ParisTech, PSL, Paris, France

Carlos Cuevas-Garcia Munich Center for Technology and Society (MCTS), Technical University of Munich, Munich, Germany

Sarah R. Davies Department of Science and Technology Studies, University of Vienna, Vienna, Austria

Alexander Degelsegger-Márquez Executive Unit of Digital Health and Innovation, Gesundheit Österreich GmbH, Vienna, Austria

Juliane Jarke Institute for Information Management Bremen (ifib) & Centre for Media, Communication and Information Research (ZeMKI), University of Bremen, Bremen, Germany

Pierre-Benoît Joly Laboratoire Interdisciplinaire Sciences Innovations Sociétés (LISIS), CNRS, INRAE, Université Gustave Eiffel, Marne-la-Vallée, France

Karen Kastenhofer Institute of Technology Assessment, Austrian Academy of Sciences, Vienna, Austria

Susan Molyneux-Hodgson Department of Sociology, Philosophy and Anthropology, University of Exeter, Devon, UK

Marianne Noël Laboratoire Interdisciplinaire Sciences Innovations Sociétés (LISIS), CNRS, INRAE, Université Gustave Eiffel, Marne-la-Vallée, France

Benjamin Raimbault Laboratoire Interdisciplinaire Sciences Innovations Sociétés (LISIS), CNRS, INRAE, Université Gustave Eiffel, Marne-la-Vallée, France

Andrea Schikowitz Friedrich Schiedel Endowed Chair for Sociology of Science, Munich Center for Technology in Society (MCTS), Technical University Munich, Munich, Germany

Sarah M. Schönbauer Munich Center for Technology in Society (MCTS), Munich, Germany

Inga Ulnicane Centre for Computing and Social Responsibility, De Montfort University, Leicester, UK

Bettina Bock von Wülfingen Department of Cultural History and Theory, Humboldt University, Berlin, Germany

Caitlin D. Wylie Program in Science, Technology, and Society, University of Virginia, Charlottesville, VA, USA

List of Figures

List of Tables

Chapter 1
Making Sense of Community and Identity in Twenty-First Century Technoscience

Karen Kastenhofer and Susan Molyneux-Hodgson

1.1 New Wine in Old Bottles?

Modern societies are changing and the sciences' character, institutions and functions change with them. From Price's (1963) 'Big Science', via Gibbons and colleagues' (1994) 'mode 2 knowledge production' to Haraway's (1996) technoscience or Nordmann and colleagues' (2011) discussion of the epochal break thesis, from accounts of the universities' new entrepreneurialism and critical discussions of an ongoing projectification of scientific work and its repercussions on the various dimensions of 'epistemic living spaces' (Felt 2009; Felt 2016) to the diagnosis of a 'medialization' of science (Weingart 2012), the sociology of science literature points at an ongoing qualitative change in the scientific system at large, linked to wider societal changes through processes of co-production (Jasanoff 2006). With these accounts of fundamental change comes the necessity to re-evaluate classical conceptions of scientific sociality and identity as they have been promulgated with the emergence of the sociology of science over the past century. Such a re-evaluation is very likely not only faced with shifts in its empirical attention, but also with persisting conceptual weaknesses, ambiguities, even incommensurabilities, and several theoretical turns and diversifications that the field has undergone. In addition, some formerly less contested concepts may have become the battleground of far more fundamental and politically laden conflicts within the wider context of societies at large. One may thus be tempted to simply omit former horizons of analysis

K. Kastenhofer (✉)
Institute of Technology Assessment, Austrian Academy of Sciences, Vienna, Austria
e-mail: karen.kastenhofer@oeaw.ac.at

S. Molyneux-Hodgson
Department of Sociology, Philosophy and Anthropology, University of Exeter, Exeter, UK
e-mail: s.hodgson@exeter.ac.uk

© The Author(s) 2021
K. Kastenhofer, S. Molyneux-Hodgson (eds.), *Community and Identity in Contemporary Technosciences*, Sociology of the Sciences Yearbook 31,
https://doi.org/10.1007/978-3-030-61728-8_1

1

such as those of (scientific) community and identity, to shelve them as outdated themes in view of these many complicating facets.

This book is dedicated to pursuing another route. Admittedly, the concepts of both 'community' and 'identity' have complex histories, they span multiple social sciences and seem to hit a nerve in contemporary society in a way that complicates their scientific discussion. This situation is not improved by their initial transfer from general sociology to the sociology of science having happened in a seemingly haphazard manner. Nevertheless, our premise is that the study of scientific communities and identities is of enduring importance, evidenced by ongoing, lively research interest as well as science policy initiatives that explicitly target these two dimensions of science and our ambition is to not shy away from a complex and complicated theme. We also hold that to discuss both concepts in relation to each other allows for a deeper understanding.

Our starting point here is to explore new work that addresses community and identity constellations within contemporary techno-scientific environments. On this basis, we ask how we can make sense of conceptions of community and identity in the rapidly shifting contexts in which scientific and technical actors work. What do, or can, 'community' and 'identity' mean in these times of strategic science, transdisciplinarity and identity politicking? What can we gain from discussing both, community *and* identity, together? Unavoidably, we will thus touch on theoretical weaknesses, unsolved puzzles and societal nerves linked to these concepts. This work can hence only be an effort in initiating—or better—reviving a research programme that is as old as the sociology of science itself. Our central thesis holds that scientific identity and community still matter in many respects, but that they have changed fundamentally during the past decades, in their character, qualities, roles and accomplishments. They have done so alongside shifts in the discourse of science-society relations and with some major modifications in scientific governance regimes. Moreover, we must explicitly acknowledge that changes in scientific communality impinge on options for, and the significance of, identity; and changes in scientific identity constellations impinge on options for, and characteristics of, community.

In this chapter, we begin with an empirical case that we have both encountered, independently, and that has challenged our own thinking, provoking us to re-open the black boxes of 'community' and 'identity' and to re-address their conceptual basis. We then move on to a short delineation of the conceptualisations of community and identity in past sociologies of science. Consecutively, we present the chapters of this volume, their takes on community and identity constellations and effects on the contemporary technosciences as institutions, practices and living spaces. We do so with a focus on common themes that we have pulled to the fore from the various contributions. In a final discussion, we take stock of where our attempt at re-addressing community and identity in contemporary technoscientific contexts has brought us, which ambiguities could not be resolved and which questions seem promising starting points for further conceptual and empirical endeavour. The final assessment of whether the task of filling new wine into old bottles is worth the effort, is left to the reader and to the next generation of sociologists of science.

1.2 Staged Communities, Manufactured Disciplines, and Strategic Identities

In 2003, the Massachusetts Institute of Technology (MIT) launched a new social format for undergraduate science students to gather, collaborate, and compete; to learn; and to construct, reflect on, and sell their own visionary products. All this took place in the context of a fairly new research area labelled 'synthetic biology', depicted as the rational engineering of biological systems at all levels of hierarchy—from individual molecules to whole organisms (cp. e.g. Serrano 2007). Nowadays, around 6000 people each year—primarily university students—gather in multidisciplinary teams, work on a self-defined project 'to design, build, test, and measure a system of their own design using interchangeable biological parts and standard molecular biology techniques', and present their results at an annual Jamboree—the International Genetically Engineered Machine (iGEM) competition. This annual process is depicted as 'instrumental in the building of the discipline of synthetic biology', 'appeal[ing] to young minds', and 'captur[ing] the attention of industry academics and governments'. In 2017, the iGEM Foundation added an 'After iGEM program', supporting a 'global network' of former participants, including a wider network of advisors and staff that support the student endeavour. In its own depiction, '[t]his global network is leading the field, taking what they learned in the competition and expanding it to continue to build a better world.'[1]

The phenomenon of iGEM has been addressed within science (e.g. Goodman 2008; Smolke 2009; Dixon and Kuldell 2011; Kelwick et al. 2015; Tsui and Meyer 2016) as well as among STS scholars (e.g., Balmer and Bulpin 2013; Damm et al. 2013; Frow and Calvert 2013; Mercer 2015), as has the phenomenon of synthetic biology more broadly. In these works, iGEM is discussed in terms such as: an educational experiment; a medium for the development and sharing of a normative ethos; an extended reflexivity within science; and an essential (or illustrative) part of an ongoing scientific revolution. It is safe to say that the phenomena of iGEM and synthetic biology have become foci of reflection and analysis in both technical and sociological camps, leading to interestingly similar topics and threads of discussion (varying mostly in the amount of critique applied and in the mode of 'reality' attributed to them). Yet, a resolution of what to make of these phenomena is not yet in sight. Do competitions signify an entertaining diversion from the seriousness of the day-to-day business of science (just like the ever-present Gary Larson cartoons in the life sciences' laboratories around the world), a strategy to better cope with sometimes boring routines? Are competitions primarily strategic instruments aimed

---

[1] All quotations are from: http://igem.org/Main_Page. Accessed 7 July 2018.

at boosting new hype cycles via new buzzword, mobilising new generations of young researchers and triggering new funding (thus denoting specific constellations in the current science governance regime)? Or are fields such as synthetic biology and their novel interaction spaces emblematic of fundamental changes within identity constellations and the way communality in science is organised and effectuated? And, if so, how will the ways that collectivity and identity are conceived of and organised within science writ large be affected?

iGEM and other new formats of interaction (mobile science festivals, science slams or 'dance your PhD' competitions) within and beyond science do—in sometimes very explicit and strategic ways—address aspects of identity and communality, combining a social engineering attitude with a revolutionary emphasis. iGEM is explicitly advertised as instrumental in 'the building of a discipline' as well as in the training and moulding of a new generation of scientists. It does so in a way that differs in fundamental ways from a teacher–disciple-based inter-generational interaction model (Bulpin and Molyneux-Hodgson 2013) or a 'purely academic' educational vision, without denying its didactic and socialising purpose. But rather than only educating future synthetic biologists and establishing such identification with a field, iGEM also aims at producing a global network of former participants. iGEM hence constitutes a consciously designed and staged temporal interaction among peers that elicits lasting identification with the respective event, its participants, and its culture. Moreover it focuses on peer-to-peer interaction rather than inter-generational encounters; on innovation based upon a specific template of parts, actions and interaction patterns rather than on slowly changing disciplinary traditions (with which it aims to break); and on an innovation culture inspired by an idealised atmosphere of friendship-based and fun-fuelled small business incubators rather than a sincere, and serious, tradition-laden ivory tower attitude.

Research fields such as synthetic biology, systems biology, and nanotechnology have each been presented as 'communities to be manufactured' by funding initiatives and science lobbyists alike; by the 'European Systems Biology Community' initiative promoted by the 'Infrastructure for Systems Biology in Europe' consortium[2] as well as by diverse national funding initiatives that specifically target networking, collaborative projects, or the establishment of temporal, dedicated research centres. But these initiatives do not develop necessarily in ways that lead to a strict identification of individual scientists with these manufactured communities. Attachments to older labels remain strong even when a scientist ventures into new waters (Molyneux-Hodgson and Meyer 2009), often due to concerns about the transitory nature of funding and research policies. Indeed, the next buzzword is likely to be already waiting in the pipeline of science lobbying and funding networks, so that over-identification with an emerging field presents a risk that the label one buys into might not deliver or simply go out of fashion, succeeded by an ever-newer label. What can we make of such 'staged' or 'provisional' communities or (labelled) 'communities without members' (Kastenhofer 2013)? Recent examples

---

[2]http://community.isbe.eu/content/what-site-0, accessed 30 September 2019.

such as the iGEM competition, or synthetic biology more generally, challenge our conceptualisation of what the character of communality and identity in contemporary science is and how it is negotiated, organised, and made use of. The concepts invoke or explicitly build upon traditional categories and understandings of 'discipline', 'community', and 'identity', but at the same time call into question the traditional patterns, modes, and reference points of sociality. It is this context in which the analyses presented here are situated.

To date, empirical case studies have addressed such new configurations (cp. for identity, e.g. Calvert 2010 for systems biologists, Felt et al. 2013 for sustainability researchers), while theoretical discussions have highlighted changing conceptions and approaches to communality and identity within sociology and social psychology (Wetherell 2010). Overall, these studies and theoretical debates seem to put forward more questions than answers. They point not only to the idea that the institutional, social, and cultural conditions and conceptions of doing and being in science are shifting (Gläser et al. 2016) but also to as-yet unsolved conceptual ambiguities, inconsistencies, and gaps relating to forms of scientific communality and identity.

With this in mind, our scope and objective for the book must be outlined in the humblest of ways: it does not claim to interrogate, let alone integrate, all existing conceptual approaches nor to solve the many puzzles accompanying current questions of identity and community in science. Rather, the work is dedicated to (1) illuminating selected new analyses of recent empirical phenomena and contemporary heterogeneities relating to community and identity with an emphasis on potential changes within the underlying academic milieu; (2) addressing some of the ways community and identity relate to each other in these contemporary contexts; and (3) indicating how these empirical observations relate to some long-standing theoretical ambiguities and debates. To do so, we have gathered empirical and conceptual studies that can serve as exemplars of the specificity of contemporary constellations within the technosciences and that provide discussion of potential conceptual ramifications.

1.3 From Communality to Communities—The Socio-Cultural Organisation and Differentiation of Science

Within the sociology of the sciences, two modes of referring to scientific community are clearly discernible: early scholars like Hagstrom (Hagstrom 1965) explicitly refer to science as *a* community in the singular—thereby highlighting the communitarian (and thereby social) aspect of science. Such work is 'concerned with the operation of social control within the scientific community, with the problem of discovering the social influences that produce conformity to scientific norms and values.' (Ibid, p. 1) Its result is a 'social turn' in the conception of science, fuelling the emergence of a sociology of science. Later analyses such as Mullins' (1972) famous reconstruction of molecular biology's development from a small group of phage specialists to a scientific specialty can be linked to this social conception of science. From there, it seems but a small step to differentiate sub-communities with distinct social and cultural characteristics *in the plural*—preparing what has much later been labelled as the 'cultural turn' within science studies.

With Hagstrom, the step towards the plural is explained in reference to 'the subcommunities of colleagues within which recognition is awarded' (ibid, p. 2). Its methodological counterpart consists of a multi-disciplinary empirical sample, including 'established' and 'new' disciplines, disciplines in which the 'exigencies of research require some formal organisation' with disciplines without such exigencies (ibid, p. 4)—a methodological choice that hints at an *a priori* idea about a socio-cultural differentiation of science into disciplinary (sub)communities and (sub)-cultures. But other than the reference to science as a community in the singular (or Merton's science as a social institution and Polanyi's science as a republic, to add just two further renowned examples), the reference to science as a collection of subcommunities or subcultures not only comes with a plural, it also comes with less conceptual consideration about the character of the social entity these (sub)commu-nities or (sub)cultures represent.[3]

As with Hagstrom (1965), the later spike in interest on cultural aspects of science within STS was based upon the theoretical assumption as well as the methodological *sine qua non* that different cultures co-exist and can be compared on empirical grounds, necessitating the existence of specific social entities that exhibited this cultural plurality and could be studied case-by-case, without theorising the underly-ing socio-cultural differentiation in detail. Fleck (1983 [1974])—as one prominent forerunner of the cultural turn within science studies—introduced the notions of thought style and thought collective, building on a psychology of perception rather than a socio-epistemological perspective and leaving some essential questions to mere statements, such as the assumption that one can be a member of plural thought collectives. Other early science studies scholars focused on national scientific styles (Jamison 1997) with a focus on the history of scientific ideas. Scholars like Harwood (1993) or Knorr-Cetina (1999) have reconstructed isochronic 'styles of scientific thought' or 'epistemic cultures', whereas scholars like Pickstone (2001) or Hacking (2002) have delineated 'ways of knowing' or 'styles of scientific reasoning' in an attempt to develop a historical epistemology (Kusch 2010), but have stayed fuzzy enough about the related social dimensions and entities.[4] With an edited volume on the 'regional and national configurations of research fields', Merz and Sormani (2016a) provided fundamentally new insights into the socio-political dimensions of localised styles.

---

[3]The specificity of scientific communities in comparison to other communities has been highlighted and extemporised in scholarly work on epistemic communities (Meyer and Molyneux-Hodgson 2010) or 'wissenschaftliche Produktionsgemeinschaften' (Gläser 2006).

[4]Historical accounts of (fundamental) change thus culminated in a seemingly unsolvable debate about the validity of the accompanying epochal break theses (e.g. Nordmann et al. 2011)—maybe because the institutional dimension had been unaddressed.

Thus, scientific communities as socio-cultural phenomena are constantly, if indirectly, addressed as reference points for a socio-cultural diversification in the scientific production of knowledge. How the social unit of these cultures should be conceived of and how the individual relates to such styles, collectives, and cultures often remains implicit. Specialities, scientific fields (Whitley 1984), disciplines, university departments (Becher and Trowler 2001) or novel types of science centres (Hackett and Parker 2016), invisible colleges (Crane 1972), or transepistemic arenas (Knorr-Cetina 1982), opposing politicised networks (Haas 1992, 1994; Bonneuil 2006; Böschen et al. 2010) or Scientific/Intellectual Movements (Frickel and Gross 2005; Parker and Hackett 2012) figure as potential suspects to locate and generate diversity.[5] The relation between culture, style, or collective and the individual scientist has also been resolutely fuzzy. Over a certain period, studies of socialisation, enculturation, career tracks, collaboration, and membership aimed to link individuals and their respective collectives. Communities were depicted as effectuated by academic socialisation, academic institutions and research practice, while at the same time being constitutive of the latter (Becker et al. 1961; Liebau and Huber 1985; Traweek 1988; Wenger 1998; Becher and Trowler 2001; Beaufays 2003; Arnold and Fischer 2004).

But the more science studies lost interest in the individual as a central site of epistemic action and focussed more on conditions, contexts, networks, actor-networks, and hinterlands, the less salient these themes became. Questions traditionally raised by social psychologists or anthropologists lost momentum. With more recent empirical examples of technoscientific convergence and transformation in the realms of nano-, bio-, info- and cogno-engineering, the disciplinary nature of these phenomena has been called into question on empirical rather than theoretical grounds, rendering disciplinary categories as proxy even less attractive than previously and putting the mode of communality they represent up for discussion (Frickel 2004 on genetic toxicology; Kastenhofer 2013 on systems biology; Lewis and Bartlett 2013 on bioinformatics; Balmer et al. 2016 on synthetic biology; for a general discussion, see also Weber 2010 and Merz and Sormani 2016b). Waves of digitalisation, projectification, massification, mediatisation and new public management seem to overrun scientific institutions, and the significance of scientific community and specialty communities is once again called into question. Concurrently, calls for the active building and shaping of specific kinds of communities arise

---

[5]In Merton's case, it is the disciplinary layer that is referred to as proxy for a relating social entity. Many other scholars go down the same road without really explicating this choice of proxy: Fleck's (1947/1983) thought collectives are illustrated in reference to disciplines; Knorr-Cetina in her work on epistemic cultures (1999) chooses her empirical cases on a (sub)disciplinary basis—comparing 'molecular biology' with 'experimental high energy physics'—and bases her argumentation heavily on these categories. While she does aim at going beyond a disciplinary ontology, 'replacing notions such as discipline or specialty with that of an epistemic culture' (ibid, p. 3), the (sub)disciplinary categories remain central points of reference in her reconstructive work. Many ensuing empirical studies have aligned with this pattern of taking disciplinary categories as a proxy for social units of cultural differentiation in their methodological approach and argumentation.

within science policy, taking for granted that we need communities, and 'functional', 'interdisciplinary', 'ethical', and 'responsible' ones at that. While analysts proclaim an era of 'communities of promise' (Brown 2003) that coalesce around the promotion of distinct technoscientific visions or utopias, science policy seems to fully buy into the 'promise of communities'.[6]

1.4   The Scientific Persona and Identity

Against the background of general shifts within science and with the more specific shifts adhering to the dimension of scientific institutions and communalities, the theme of scientific identity also comes into play. Any conception of scientific community speaks to a certain conception of identity: an identity that comes with being a member of a community, an identity that is constructed, provided and experienced within a community, or as an identity shared, a communality that forms the basis of a community. Questions of identity thereby stretch from the identity of science at large (e.g., Weingart 2003), to questions related to the scientific identity of its personae (Daston 2003, see below), and to questions of disciplinary or other more specifying identities of scientists. They also encompass questions about local differences or historical shifts in the quality and momentum of identification, denoted by terms such as 'provisional identities' (Ibarra 1999) or 'liquid identity' (Bauman 2004).

If the theoretical as well as empirical feasibility of community as a concept is being drawn into discussion in our work here, the challenge posed is even greater for the conceptualisation of identity. Margaret Wetherell in her elegant introduction to the field of identity studies thus summarised:

> Nearly every scholar who works on identity complains about its slippery, blurred and confusing nature. Identity is notoriously elusive and difficult to define and nearly every generation of scholars since the 1950s has included some keen to dismiss it as a consequence, concluding it has no analytic value or purchase. (Wetherell 2010, p. 3).

She goes on to delineate the various paths identity research has taken, from a focus on identity as a subjective individual achievement or as social subjectivity, to identity as linked to group membership and belonging or identity as an ethical and political category. She outlines deconstructions and critique of the concept and category on theoretical, political, and empirical grounds. Still, Wetherell comes to the conclusion that identity should not be discarded as a concept within the social sciences:

> In my view, identity continues to be good to think with precisely because of the features which some have found difficult and irritating: the long chronology, the accretion of many layers of meanings, the rich sweep, the heated debates, the constitutive ideologies, fantasies and fictions, the politics, and the very many ambiguities. (ibid, pp. 23–24)

---

[6]Originally put forward within the analytical as well as managerial framework of 'communities of practice' (Wenger 1998).

There are rich seams of research on scientific identity in sociological and anthropological traditions, alongside the (social) psychological take of Wetherell and colleagues (e.g. Jenkins 2014). Yet identity has been less frequently referred to in explicit terms in science studies, although some scholars have linked issues of identity with issues of disciplinarity and disciplinary socialisation.[7] The questioning of the disciplinarity of contemporary sciences has come with a perceptible rise of studies scrutinising and problematising disciplinary identity (see, for instance, Kurath and Maasen 2006, Kurath 2010 or Chari et al. 2012 on disciplinary identity within nanoscience and toxicology; Calvert 2010 or Osbeck and Nersessian 2017 on disciplinary identity within systems biology). Further effects of such a problematisation of disciplinary identity on the individual as well as the systems' level are scarcely discussed.

Another route to addressing identity in science studies has been linked to an interest in the scientific persona. Lorraine Daston coined this field of interest (Daston 2003; Daston and Sibum 2003; Daston and Galison 2007), depicting the scientific persona as a link between individual biography, scientific institutions, and society at large. The field has recently developed with a primarily historical focus (e.g. Paul 2016) and with a focus on the moral dimensions of being a scientist in various socio-historical contexts (e.g. Shapin 2008). While disciplinary differentiations gain less attention, an aim to better understand contemporary conditions of being in science is a clear motivation of this historical enterprise. It is closely related to analyses of 'academic identity schisms' that go with contemporary shifts in the organisational culture(s) of higher education institutions (Henkel 2000), such as those between an academic identity and a managerial identity (Winter 2009). Also in this corpus of literature, values and norms play an important point of reference.

Furthermore, the strategic dimension of staging one's own disciplinary identity— a dimension introduced already by Gieryn's (1983) account of the instrumentalisation of a scientific identity within strategic boundary work—has been re-invoked against contemporary post-disciplinarity or new disciplinarity (Marcovich and Shinn 2011). In recent studies, identity is once again framed as a medium of strategic positioning, but also as a medium of experimentation, highlighting the discursive construction of identity, rendering identities provisional (Ibarra 1999), contextual and multiple. The strategic account is thus combined with an account of fluidity, resonating with what Bauman has put forward as aspects of a 'liquid modernity' (Bauman 2004).

When it comes to discussing how scientific community and identity are linked to each other, how they are co-produced and how they reverberate, the sociology of science literature does not provide a lot of material. The mostly implicit take that both scientific community and scientific identity are heavily based on disciplinary categories does not further this matter and every conception of community, every take on identity will come with its own explicit or—in most cases—implicit take on

---

[7]Again, similarly to how community has in many cases been linked to disciplinarity in empirical studies.

how these two aspects of scientific life are (or should be) linked to each other. With fundamental shifts in science and society, both aspects seem to change gradually and simultaneously in many respects, in their quality and in the ways they come to matter. Both terms seem to now denote an outdated era of 'the scientific life', an era that maybe has only existed as an ideal type all along, when at the same time these very terms are rediscovered as levers for socially engineering science and—more importantly—its outputs in quantitative and qualitative terms.

In some respects uniting community and identity spheres, studies of academic socialisation (enlisted exemplarily in the previous section) and 'community of practice' thinking (Wenger 1998) can be described as a middle ground, escaping overly deterministic structural frameworks and moving beyond narrowly localised moments of interaction as a means to describe and explain. They take practice-based theories as a point of departure in understanding how communities coalesce around *what people do together* and the common resources and meanings they (re-)produce in the process. An important element is that learning and collaboration are the key mechanisms through which community membership, belonging, and identities are gained and engendered. Communities, their representatives and practitioners are thus co-produced. Identity formation is positioned as an important element in learning and collaborative practice,

> [b]ecause learning transforms who we are and what we can do, it is an experience of identity. It is not just an accumulation of skills and information, but a process of becoming—to become a certain person or, conversely to avoid becoming a certain person (Wenger 1998, p. 215).

The emerging sociological work on 'epistemic communities' is also instructive in bringing together distributed, relational, and 'networked' aspects of community formation with a focus on the experiences of participation and the formation of individual identities. The notion of epistemic communities draws attention to the practices through which communities are produced together with their practitioners as 'site[s] where knowledge standards and practices are being negotiated, and a sense of belonging and researcher identity is being defined' (Lorenz-Meyer 2010, p. 8). Similarly, Meyer and Molyneux-Hodgson (2010, p. 5) propose probing how the connections established in building new epistemic communities 'shape, demarcate and articulate identities of present and future knowledge producers, and the individual and collective trajectories on which the latter navigate'. The prevalent politically-framed understandings of scientific communities and identities remain overly consensual. The policy-directed formation of new communities and identities remains conceptually too simplistic, leaving relating interventions ineffective at best. Thus, 'how a community comes to be assembled...the work, the politics, the materialities, the identities, the uncertainties that go into the formation and maintenance of a community clearly deserve our attention' (ibid, p. 1). The papers in this volume clearly respond to this invocation.

1.5   Synthesising Communities: Processes of Qualitative Change, Heterogeneity, Theoretical Demarcation and Social Engineering

The chapters in the first section illustrate a range of conceptualisations of the phenomenon of collectivity and do so from four directions: (1) they focus on processes of qualitative change of collectivity over time and thereby elucidate specific stages; (2) they zoom in on the heterogeneity of members, roles, and identities of communities; (3) they raise the question of which terms and concepts best enable empirical analysis of collectivity, moving towards a demarcation of community from other social formations and for a distinction in specific cases of community; and (4) they relate to avenues for socially engineering science and its output. Although these four themes are closely connected, the individual chapters provide a specific empirical case and a specific primary focus that elaborate on each theme. The empirical cases cover the transnational emergence of synthetic biology (Raimbault and Joly, Degelsegger-Márquez, Blümel), the local emergence of a field (supramolecular chemistry) (Noël), a regional cluster on microbial biofuel production (Cointe), international collaborations in nanotechnology (Ulnicane), engineering laboratories in the United States (Wylie), and communities of European eGovernment practitioners (Jarke).

1.5.1   Processes of Qualitative Change

Mullins' (1972) analysis of the development of molecular biology as an institutionalised scientific specialty between 1935 and 1972 along a four-stage process, starting with a loose paradigm group, and undergoing intermediate stages of a rapidly expanding communication network and a dogmatic cluster, still serves as a point of reference when processes of field emergence are addressed by sociologists of science, especially so in the empirical context of the biotechnosciences. Others may have criticised, refined, or replaced Mullins' model, but the general ambition to combine a focus on historical stages with a characterisation of corresponding social entities has not lost its momentum. It has only recently been revived by the analysis of the emergence of new technoscientific fields as Scientific/Intellectual Movements (Frickel and Gross 2005; Parker and Hackett 2012); an approach applied by scholars like Vermeulen (2018) to phenomena such as systems biology, 'discuss[ing] three different movements in the emergence of systems biology: aggregation, circulation and oscillation' (ibid, p. 1766).

Marianne Noël, in her historical reconstruction of the emergence of supramolecular chemistry in Strasbourg in the second half of the twentieth century, draws on Mullins' approach. Like Mullins, she combines an analysis of intellectual and social characteristics. Based on the publication profile of a central figure, the French chemist Jean-Marie Lehn, she develops three periods that resonate with, but do not totally conform to, Mullins' stages. They comprise the emergence of the paradigm, the emergence of the specialty, and the institutionalisation of supramolecular chemistry. Her argument also hints at a broader paradigm shift within chemistry that not only inspired the emergence of supramolecular chemistry (in line with Mullins' account) but also became normalised parallel to its establishment. Noël mentions that Lehn at the height of his career founded a journal. Rather than calling it 'Journal for Supramolecular Chemistry', he decided on 'Chemistry: A European Journal'. This choice can be interpreted as denoting at least two moves: one to frame the supramolecular paradigm not as an addition to existing chemical stances but as a now ubiquitous standpoint in all of chemistry; and another one to posit this ubiquity of the supramolecular paradigm in chemistry as a potentially European approach— creating space to envisage a 'European school of chemistry'—that could be populated by Lehn, his network, and his legacy (we will return to the concept of 'research schools' and 'local differentiation' later). This story of institutionalisation culminates with the creation of the 'Institut de Sciences et d'Ingénierie Supramoléculaire' (ISIS) and the design and realisation of a dedicated building. The institutional and architectural organisation, expressed by the label of a 'research hotel' (*hotel à projets*), becomes emblematic of a specific style of organising science, of envisioning a specific notion in scientific career patterns, and of engineering sociality in science.

An interesting difference between the formative analyses of Mullins' generation of scholars (including 'finalisation theory' put forward by the Starnberg group, Böhme et al. 1973) and current work within the sociology of science is that the former emphasised a final, 'mature' climax whereas today scholars are more fascinated by the seemingly frozen states of emergence of the contemporary technosciences. In line with this contemporary take on the theme of historical change, processes of community formation are mostly studied as phenomena of (ongoing) emergence in this volume's further case studies. The idea of a potential halt to this process at a future 'mature stage' may lurk in the background of some analyses, but as Molyneux-Hodgson and Meyer noted (2009) the question of whether a community *ever* stabilises is firmly on the table.

Clemens Blümel's contribution to this volume illustrates this stance well. Based on an in-depth, multi-method analysis of review articles in synthetic biology published between 2002 and 2012, he reconstructs three differing types of reviews that concur with different historic periods. The first set are highly cited and closely linked via cross-citation. This type of article is then replaced by narrative accounts of the history of synthetic biology from 2007 onwards that provide a larger structure within which events are located. These articles put 'heroic objects' at centre stage rather than foundational fathers or spokesmen (c.f Abir-Am 1985 on molecular biology). The third, evaluative, type of review is most frequent post-2010 and highlights application of the science. Overall, Blümel sees a shift from authoritative to narrative forms of constructing a field that aligns with a narrative normalisation achieved and represented by joint histories. One could also depict this shift as one from prophets to apostles. Some congruence with Mullins' paradigm group, communication network, cluster, and institutional stages or with Stent's (1968) romantic, dogmatic, and academic phase can be postulated perhaps. In contrast to earlier publication analyses (especially Bastide et al. 1989), Blümel highlights that a large quantitative share of review articles signals field emergence rather than field maturity in his case; field maturity is signalled by a distinctive (narrative or evaluative) type of review article.

Benjamin Raimbault and *Pierre-Benoît Joly* provide a further reconstruction of synthetic biology's emergence but focus on specificities of this emergence phase rather than on its (inevitable) obsolescence and hence point towards a conceptualisation of and interest in emergence as a 'perpetuated state' or 'end in itself'. Contributing to a new political sociology of science (cf Frickel and Moore 2005), the authors reconstruct this process by means of a multi-sited ethnography as well as scientometric accounts, zooming in on actor constellations (with reference to Latour's Actor Network Theory) and field characteristics (with reference to Bourdieu's field theory and to strategic action fields, Fligstein and MacAdam 2012). The approach puts power constellations and rules governing legitimate action at centre stage and acknowledges not only internal factors such as scientific credibility but also external ones like societal relevance. The authors thus frame 'the emergence of a techno-scientific field as a multiscalar and progressive establishment of a new set of epistemic and social rules' (p. 102). They define three determinant dimensions of this emergence: heterogeneity across the emerging community (of disciplines, research questions, visions, and social norms), which they see as constitutive of the process of emergence; hierarchy (between a core-group and peripheral members or temporary visitors) with strategic alliances; and autonomy, which in the case of synthetic biology is closely tied to its links to industry (and potentially also to science policy). We will return to all three dimensions below when we address the idea of heterogeneities.

In the context of reconstructing qualitative change at the level of research projects, *Béatrice Cointe* provides an enlightening analysis of how a distinct research project—an interdisciplinary project on potential avenues for biofuel production in microorganisms, funded in the form of a regional excellence initiative and excellence cluster—fares in and impacts on the (local, existing) research community. She thus re-constructs a 'project-ed community' as a local community that undergoes an instance of projectification just as much as its making or intensification is the envisioned outcome of a project. What the project 'really is' and what processes it entails, is revealed to be ambiguous, multifold, and fuzzy in the author's analysis, based on an in-depth participatory ethnographic study. In her final diagnosis, she depicts the project as 'an experiment in community-making: a nutrient broth. . .in which some collaborations may thrive while others barely catch on' (p. 140).

A similar point is addressed by *Inga Ulnicane* in her contextual comparative analysis of two international research collaborations in nanotechnology; one a 'bottom-up' initiative emerging from informal relations between individual scientists, and the other initiated by a funded European Commission framework programme. Her empirical material comprises publications; citations; organisational, project, and CV data; interviews; and protocols from site visits. In her resulting 'tale of two scientific communities' she reflects on differences and similarities in the dynamics of academic self-organisation versus external steering of research collaborations (thereby implicitly upholding an ontology in which science funding is external to science). To describe the dynamics of research collaborations she refers not to a linear evolution of the emergence of scientific fields or communities but rather to a circular movement: her model of collaborative processes features recurring phases of emergence and renewal, formal collaboration, informal collaboration, and scientific output.

Marianne Noël discusses some aspects of the emergence of supramolecular chemistry in reference to Marcovich and Shinn's (2011) 'new disciplinarity'. She thereby refers to recent discussions about the explanatory power of the category of disciplines within the sociology of the sciences. On the one hand, this category has been historised lately, reframed as an emerging property of eighteenth and nineteenth century continental European science only, and hence not necessarily valid for other local or historical contexts. On the other hand, scholars challenged the assumption that disciplines constitute a primary ordering category for *all* phenomena or all practical contexts relevant for the sociology of the sciences. Notwithstanding all this critique and relativisation, it was not only the first generation of analysts of emergent social entities in science that was dedicated to the emergence of disciplines (cp. Lemaine 1976); the reference to disciplines as an ordering category within the science system still abounds today, in scientists' accounts as well as in sociological literature. Many of the contemporary alternative conceptions such as interdiscipline (Frickel 2004, Weber 2010), post-normal science (Funtowicz and Ravetz 1992), or Mode 2 (Gibbons et al. 1994) are based on the idea of a precursor situation in which disciplines were central and would not work without this reference (just as postmodernity is tied to the preceding conception of modernity or technoscience builds on a preceding conception of modern science). In many instances, the reference to disciplines is also under-theorised, with the category being used as a proxy for any social or institutional differentiation related to scientific content; a fate very similar to that of the category 'community'. *Carlos Cuevas-Garcia*, in the second section of this volume, provides another sketch of the fate of the disciplinary category within the sciences, depicting disciplinarity as a 'widely taken for granted ideal from the Nineteenth century onwards', with disciplinary boundaries being to some extent illusionary, 'but real in their consences.' (p. 151)

As noted above, *Béatrice Cointe* embeds her analysis in a general observation of an ongoing projectification in science. According to Felt (2016), projects are nowadays a 'key organizing principle into science'. They co-organise day-to-day scientific work and international collaboration as well as the practice of science policy. Just as many other contemporary organising parameters (like the local and regional clusters as described by *Bettina Bock von Wülfingen* and *Marianne Noël* or the emerging fields of synthetic biology and nanotechnology addressed by *Benjamin Raimbault* and *Pierre-Benoît Joly, Clemens Blümel* and *Alexander Degelsegger-Márquez* in this volume), projects bring together actors across teams, countries, disciplines, and actor fields (such as science, policy, industry, and society). They thereby 'often embed research in extra-scientific financial, societal and political concerns, echoing (and maybe institutionalising) Knorr-Cetina's "transepistemic arenas of research" (1982) introduced as a critique of analyses focusing on "specialty communities"' (Cointe this volume). The author also refers to 'umbrella terms' (Rip and Voss 2013), another observational category demarcating fundamental shifts in the contemporary scientific system. And just as with projectification, the phenomenon of the 'umbrella term' is closely linked to changes in the quality and influence of the funding regime with which the science system is currently confronted. Accounts of 'perpetuated emergence' (Kastenhofer 2016) or of implementations and effects of new public management approaches at universities add up to such diagnoses of post-modernity within academia (Rip 2004). And just as with the discussion surrounding the 'new disciplinarity', the thesis of an 'epochal break' (Nordmann et al. 2011) lurks in the background without being decisively answered. Addressing this would be a perfect contribution to 'a more deeply historicised sociology of scientific knowledge' (Hess 2005). In this volume, a related question is: are there fundamental shifts in the conditions, qualities, and relevance of communality and collectivity in science? While no single contribution in this volume will be able to answer this question, all of them contribute new ideas to the discussion.

In all the analyses, change is configured as something that needs extra force, while stability is treated as the likely default historical avenue. It is all the more important to mention here that a focus on practices brings to the fore the constant need for enactment in both change and stability. Such a practice-oriented approach is followed not only by some of the chapters focussing on identity but is also central in analyses of performed communities, probably most visibly in *Juliane Jarke*'s account of 'communities by template'.

1.5.2 Heterogeneities

Leaving aside the proximal, short, medium, and long-term historical foci of the chapters, accounts of heterogeneity can also be found at the isochronic level. These discussions confront a naïve conception of scientific communities in which such communities are made up of identical members. The naïve position can result implicitly from studies that only compare the 'comparable', e.g. focussing on one subgroup of a community: students throughout the socialisation process; Principle Investigator accounts in a series of interviews; or core figures as authors of influential articles. Rarely are such implicit conceptions of community confronted with a theoretical relativisation. The chapters of this volume, however, point towards several kinds of heterogeneity: communities are heterogenous with respect to their members' career stages (*Wylie*); the roles their members exhibit (*Raimbault and Joly*); and prevalent paradigms and associated sub-communities (*Degelsegger-Márquez*). Moreover, scientific collectivities like those realised by research projects encompass diverse alternative realities (*Cointe*).

Caitlin Donahue Wylie's contribution to this volume on 'how undergraduate students contribute to engineering laboratory communities' puts heterogeneity and the related issue of a community's ontology centre stage: who do we as sociologists of science configure as part of a community? Who takes part in co-shaping it? Who is indispensable in sustaining it? These questions touch on issues of membership, be it of symbolic or practical nature, from an internalist or analytical perspective. Wylie's account focuses on the—often downplayed—role of students within a laboratory research group and acknowledges their centrality in day-to-day research practice and in the shaping of research projects and research programmes. The two laboratories studied are involved with the engineering of material properties and electronic sensor systems at a 'medium-sized public research university in the United States'. Beyond classical socialisation studies, the author frames students not only as future members of a community but also highlights that such 'novices bring important aspects to a community' such as open-mindedness, multi-disciplinary skills, and knowledge. Her account gives a sense of the adventurousness that students can contribute to a research laboratory ecosystem and goes on to show how students influence research and thereby 'actively contribute to the construction of knowledge and communities.' Interestingly, she also notes that novices can fulfil this function precisely because they are 'situated at the periphery', providing 'a wisdom of peripherality' (Wenger 1998, p. 146, this volume).

The focus taken by *Wylie* on student positioning confronts a challenging theoretical situation, and the conceptual aspects are far from being fully realised. Nevertheless, the centrality of the theme—especially within contemporary technoscientific contexts—is difficult to ignore, especially when carefully reading through the other chapters of the volume. Whereas Mullins (1972) refers to students only indirectly, as 'being a student of' someone, in some of this volume's chapters, the functions that graduate and PhD students fulfil—even before they are acknowledged members of academia—occur again and again. In *Noël*'s account of supramolecular chemistry's emergence, interviewees stress how much they 'tried to interest graduate students' in a specific phase of field formation and how much they 'insisted that several post-doctoral co-workers enter the field'. In *Ulnicane*'s analysis of two nanotechnology collaborations, joint PhD students are central in launching and stabilising the international networks. The PhD phase entails career risks and precariousness as well as distinct institutional demands (namely to enrol at a university accredited for the supervision of PhDs) and norms (namely to demonstrate international mobility), as does the postdoc phase (namely the institutional demand to apply for an open, paid position and to demonstrate expertise and success on an international scale), all of which demand or at least encourage international mobility in the cases she studied. As a result, she observes that 'both collaborations were launched by internationally mobile early career researchers' (p. 113). Conversely, a collaboration can also come to an end when people move to other sites during their highly mobile early career stages. It was the novices that strengthened inter-group ties via short visits 'to learn techniques and join experiments' and ensuing fellowships. Just like Wylie, Ulnicane notes that PhD students ask questions and thereby foster interdisciplinary dialogue, changing the quality of the interaction. Moreover, the motivation 'to give PhD researchers an opportunity to present their work, learn and find opportunities' for future jobs was depicted as a central element of the projects.

The project level was also seen as especially important to these novices as it 'lends visibility and group identity to young scientists'. Projects allow for 'communication and collaboration across all levels of hierarchy', while getting support from the organisations' top levels.

The focus on novices demonstrates that communities are stratified, not homogenous; they consist of diverse members and roles. Heterogeneity of roles and fields is addressed in exemplary fashion by *Raimbault and Joly*. With reference to Fligstein and McAdam's conception of strategic action fields (2012) as

> a constructed mesolevel social order in which actors interact with one another on the basis of shared understandings about the purposes of the field, relationships to others in the field (including who has power and why), and the rules governing legitimate action in the field (ibid, p. 9)

they stress the importance of distinct roles within a field. With reference to Mullins' (1972) model, Crane's 'invisible college' (1972), Collin's 'core set' (1981), Frickel and Gross' (2005) role differentiation of high status scientists, networks and young scholars, and Fligstein and McAdam's (2012) depiction of incumbents and challengers, they set the scene for their own discussion of various roles within the emergence of synthetic biology and relate this to the dimension of 'hierarchy'. They come close to devising roles *distinct* to the subfields, clusters, or discourses within synthetic biology that can be adopted. Their scientometric analysis hence results in a differentiation of the field into four clusters: biobrick engineering, protocell creation, genome engineering, and metabolic engineering (cf O'Malley et al. 2008). These four clusters are positioned in different ways, fulfil different roles, and face different fates throughout the history of the field. Their characterisation evokes a visionary role for biobrick engineering, whereas metabolic engineering fulfils the role of delivering applications. With reference to a further dimension, namely 'autonomy', they discuss roles exhibited by non-scientific fields, as emerging fields 'have a Russian doll structure', intertwining the micro-, meso-, and macro level and rendering seemingly internal/external interactions key components of emergence processes. This is a depiction that comes close to Knorr-Cetina's conception of transepistemic arenas—a notion also strengthened in Cointe's contribution. 'Autonomy' thus refers to autonomy from other fields rather than autonomy from extra-scientific contexts. An analysis of the core-group of synthetic biologists reveals that most of these are affiliated with the biobricks approach, based in the US, and have close ties to business development—depicting remarkable homogeneity within a heterogenous and changing context.

Heterogeneity of roles is also addressed in *Blümel*'s contribution: the shift from authoritative to narrative forms could also be depicted as a shift from prophets to apostles. Mullins (1972) referred to an earlier role differentiation put forward by Ben-David and Randall Collins in the context of a field's emergence, namely: forerunners, founders, and followers (Ben-David and Collins 1966). Mullins' own model devises roles only indirectly, such as the charismatic (or dogmatic) role detrimental in the cluster phase. Indeed, according to Mullins (1972, p. 79), 'leadership and charisma may be the most important factors [during the cluster phase], much more important, for example, than accuracy in intellectual judgement.' *Blümel* replaces the charismatic or heroic founding figures central to Mullins and others'

(Abir-Am 1985) earlier accounts with 'inaugural devices as heroic objects'. He thus relocates the (narrated) centre of activity and actor-ship within contemporary phenomena of field emergence, again raising the question whether the fundamental logics of the empirical phenomena we observe in science have changed or whether it is primarily our interpretive sociological frame that undergoes changes over time.

Heterogeneity of paradigms resulting in sub-differentiation in the emerging field of synthetic biology is addressed by *Raimbault and Joly* and by *Degelsegger-Márquez*. The latter points to a differentiation into an orientation towards understanding along an epistemic paradigm on the one hand and an orientation towards construction along an engineering paradigm on the other hand with references to earlier work by other scholars (Kastenhofer 2013; Ramibault et al. 2016). As a result, he doubts that the field will ever represent a disciplinary community—a question first posed by Molyneux-Hodgson and Meyer (2009) a decade ago—and characterises it rather as an interdiscipline with 'porous epistemic boundaries and temporal forms of institutionalisation', perhaps even limited to specific national environments like the UK.

Heterogeneity in the isochronic perception or construction of one and (arguably) the same object—in this case a project—is also a main concern in *Cointe*'s case study. What the project really is about, what its boundaries are, who and what belongs to it, is processed differently and with different results in three parallel 'versions' of the project and its collective. The author thus analyses three versions of a 'project-ed community': as an argumentative device in its documents; as a strategic venture in its institutional arrangements; and as an arena for scientific work in its daily research activity. These avenues 'provide different, but coherent pictures of the project-ed community', though the version of the project as an arena of research appears fuzzier than the other versions—'it turned out to be quite elusive as a whole in practice' as it was not always clear who or what was (an enduring) part of it and who or what was not. Thus, 'projects are more than temporary arenas of research, they are also argumentative devices that justify and display excellence and relevance, they serve to imagine future research communities and to start building them', they represent 'strategic entities that integrate scientific practices into coherent narratives to further the interests and ambitions of various parties', they are 'an experiment in community-making: a nutrient broth in which some collaborations may thrive while others barely catch on' (as also illustrated by Ulnicane). In short: projects exhibit versatile functions. In a similar vein, *Degelsegger-Márquez* characterises synthetic biology not only as a (heterogenous) field or an interdiscipline but even more so as 'a set of community-making devices' (a term originally developed in Molyneux-Hodgson and Meyer 2009), offering two alternative understandings and co-existing in distinct dimensions.

1.5.3   Theoretical Demarcations

The accounts we have provided above of qualitative change over time and isochronic heterogeneities in scientific collectives touch on the conceptual question of what kind of social entities we speak when referring to 'scientific communities'. Although all chapters refer to this label, evidently not all chapters address the same phenomenon. This becomes visible in alternative denominations such as (paradigm) group, (project) team, collaboration, network, cluster, field (as in 'hot field' or 'emerging field'), specialty, and discipline or by further specification such as 'community of practice' (*Wylie* and *Jarke* in line with Wenger 1998), 'project-ed community' (*Cointe*), or 'community of knowledge application' (*Degelsegger-Márquez*). Some authors content themselves with a loose or implicit definition of collectivity, others confront the theoretical discomfort head-on. The latter approach displays an interesting diversity, as we now elucidate.

Alexander Degelsegger-Márquez' contribution explicitly refers to early classical definitions of community within the sociological literature, that is, to authors like Ferdinand Tönnies, Max Weber, and Talcott Parsons. Within this context, 'community' is defined in contrast to 'society'. In this kind of conception, individuals are born into communities and realise all their actions and social relations within them. Strong social ties building on (diverse) personal interactions, roles, values, beliefs, and subjective feelings are emphasised. Performing community is characterised by affectivity, functional diffuseness, particularism, valuation by ascription, and an orientation towards the collective. It is opposed to affective neutrality, functional specificity, universalism, valuation by achievement, and orientation towards oneself. The tensions of this classical understanding of community with the modern framing of science are more than obvious; a tension that later spurs the radicality of Hagstrom's or Kuhn's thesis that the science system also exhibits characteristics of a community. Merton's scientific ethos (1993 [1942]) can be understood as presenting an intermediate position: science does follow an ethos (which points towards the quality of a community), but this very ethos is to mostly not follow the classical logics of a community (especially with the norms of universalism and organised scepticism). Degelsegger-Márquez sketches recent conceptualisations that frame communal aspects of science as a distinct kind of communality, namely one that is oriented towards the production of knowledge (Gläser 2006). As with Merton, the resulting conception of community is bounded: such scientific 'production communities' are tied to a function; they represent 'the mode of social organisation that allows scientists to produce … scientific knowledge'. In a further step, the author introduces theses of qualitative heterogeneity and/or change in science, referring to Elzinga's (1993) hybrid epistemic communities driven by rationales from policy-making or commercial application, Frickel's (2004) interdisciplines characterised by porous boundaries and a collective interest in problem-solving, and the proliferation of the engineering approach in the sciences along a technoscientific paradigm leading to the hybrid constellation of 'techno-epistemic communities' and a stratification along techno-epistemic visions rather than disciplines ('communities of vision', Kastenhofer 2013, p. 133). From this, the author devises his own concept in close reference to the aforementioned literature, namely that of a 'community of knowledge application', based on his own empirical material on the case of synthetic biology undertaken between 2008 and 2016). This concept explicitly acknowledges the formative impact of a strong practical and/or visionary focus on the utilisation of research output within non-epistemic contexts. Degelsegger-Márquez relates this to an observable difference between communities of knowledge production and communities of knowledge application, and sees only the latter realised within the field of synthetic biology: 'Under the umbrella term synthetic biology, we observe researchers sharing views on knowledge application while differing in their outlook on knowledge production (p. 178).'

Thus, in *Degelsegger-Márquez*' contribution the tension of a proclaimed community of synthetic biology against the background of epistemic heterogeneity is dissolved by the delineation of a community of knowledge application. The outlined tension between early sociological conceptions of community and later conceptions of scientific communities within the sociology of the sciences—or more broadly speaking, science studies—stays unresolved. Other chapters add further theoretical foundations to discussion of communities in science. *Cointe* refers to various contemporary scholars exhibiting the common trait 'that they are not primarily based on disciplines and often embed research in extra-scientific financial, societal and political concerns'; an argument that comes close to *Raimbault and Joly's* account of synthetic biology, albeit not explicated by these authors. They enlist Latour's Actor Network Theory as well as Bourdieu's classical field theory and Fligstein and McAdam's theory of strategic action fields. Thereby, a focus on arenas and networks stands in some tension to a focus on (the emergence of) distinct social entities, such as Bourdieu's scientific fields. *Noël's* strong reference to Mullins' four stage model also builds on different patterns and roles of communication, co-authorship, apprenticeship, and colleagueship but leads to the characterisation of a distinct specialty, just as Stichweh's disciplinarity and Marcovich and Shinn's 'new disciplinarity' emphasise 'disciplines as stable referent of practitioners, providing the intellectual and material resources, language and universe of questions'. Both takes on the theme can also be found in *Ulnicane's* account of nanotechnology collaborations and her reference to Crane's 'invisible colleges'. On the one hand, this account refers to more or less stable entities like disciplines or research organisations, on the other hand it emphasises collaboration on the basis of complementary expertise and heterogenous roles, as well as influences from an extra-scientific arena. *Wylie* refers to Fleck's 'thought collectives' and to Kuhn, but also evokes ethnographies like the one by Traweek (1988), and refers to Lave and Wenger's 'communities of practice'.

Lave and Wenger's 'communities of practice' (Lave and Wenger 1991; Wenger 1998) play a major role in *Jarke's* account of two half manufactured, half emerging European communities of eGovernment practitioners. She provides a very different case to, and scholarly references from, Degelsegger-Márquez, yet with just as much theoretical emphasis when establishing her concept of a 'community by template'. She delineates the career of Lave and Wenger's concept in various bodies of literature, from the initial ambition 'to investigate learning as a social and situated practice', to first attempts at nurturing such communities so as to enhance organisational productivity, and finally to contemporary initiatives to establish or build trans-local, virtual 'communities of practice'. Along this observed development in theoretical and practical contexts, the relation of distributed practice, exchange relations, objects of exchange, community, and membership plays a crucial but ambiguous role. These are obviously co-created in some way, while the strict uni-directional causalities along a one-dimensional ontology searched for by management scholars somehow evades the analyst's gaze. As an answer to the analytical aspect of this twofold problem, the author opts for a practice or process-oriented approach, inspired by Cooper and Law's (1995) 'proximal view' and Mol's (2003) praxiography. She contrasts two cases of community formation that both build on an exchange of digital templates dedicated to the sharing of practical examples and expertise among such practitioners. In the first case, the template was prepared by the European Commission and fixed; in the second case, the template was developed by a network of distributed practitioners and open for further adoption. The two cases exhibit different kinds of interaction, circulating different kinds of objects and negotiating different dimensions of collectivity, and they result in different kinds of community. She outlines two 'different levels of sharing and different forms of communality' that emerge in the two settings and concludes that 'knowledge objects such as templates are not simply a means to actively share practices but rather a means for enacting membership'. We are left in limbo, not only by this chapter but also the many other attempts at 'community building' that we observe in the context of proclaimed emerging technosciences that stay ambiguous about their goals (nurturing existing or building new communities as an end in itself or with an overarching ambition) and lack follow up such as reports on later outcomes and related further steps. Continuing to trace through the 'emergence' with longer term analyses is work that remains to be done.

The above-mentioned theoretical ambiguities arise mostly in relation to an unclear character of the social phenomena being addressed. These range from project groups to laboratory teams or organisation; from sub-fields and specialties to disciplines; from collaborative networks to research schools; from proclaimed to manufactured emergences of community. All of these entities are analysed in reference to aspects of community. As, for instance, *Cointe* notes in her chapter: 'projects create communities—these are provisional, more or less loose, and may not outlast the project, but they are communities nonetheless, in that they are united by shared objectives and resources'. In reference to Leonelli and Ankeny (2015) she adds that 'projects also contribute to shaping and reconfiguring existing scientific communities'. The same may apply for the interdisciplinary research collaborations analysed by Ulnicane or for the research laboratory teams analysed by Wylie, whereas proclamation and manufacturing of community feature centrally in the accounts of Blümel and Jarke.

The diversity of phenomena and the ambiguities inherent in these accounts do not help with clarifying the theoretical conception of scientific communities in any categorical sense. At most, they help with clarifying instances of when and how aspects of community (as opposed to aspects characteristic of networks, organisations, or society) become decisive. Going back to Mullins' four stage model it is probably the cluster phase in which this is most perceptible. It follows the stage of communication networks and is later superseded by the institutionalisation of a specialty. According to Mullins, a cluster is 'a collective entity recognised and constituted as such by its members' with 'a shared directory of resources' such as textbooks translated into different languages, new dedicated journals, and symposia. It thereby stretches beyond existing local groups but still possesses a very personal character in 'key meetings' and at 'melting pot sites'. These meetings and sites also serve as a playground for affective, charismatic, or dogmatic dynamics that result in accounts of a distinct habitus within a nascent field, spanning the professional as well as private persona of the protagonists, or the formation of strong social ties so lively depicted in Mullins' account and interview quotations. In a similar vein, Ulnicane depicts 'hot fields' and characterises them as having intense interaction, communication, and rivalry; interdisciplinarity; a strong focus on training; and research groups comprising complimentary expertise. With its charismatic or dogmatic dynamics, the cluster stage or its contemporary technoscientific equivalent of a 'hot field' serve well to establish a common culture as addressed in the comparative reconstructions of scientific fields (Bourdieu 1975), evidential cultures (Collins 1998), epistemic cultures (Knorr-Cetina 1999), or techno-epistemic cultures (Kastenhofer 2013). Both aspects are then combined in the notion of 'epistemic communities' put forward by Meyer and Molyneux-Hodgson (2010).

1.5.4   The Social Engineering of Community

A fourth aspect of addressing scientific community in singular and plural that has already been mentioned above is their link to an instrumental dimension, or, in other words, to a vision of socially engineering science and its outputs. This link is most explicitly drawn by *Jarke* when she refers to Lave and Wenger's 'communities of practice' and the twofold career of this concept in both analytical and interventionist approaches. While the ambiguous causal (and ontological) relationship of practice, interaction, and community within such communities can be addressed analytically by a practice-oriented approach that highlights processes of co-performance and co-production, the interventionist position is held in limbo. The questions of where to intervene, when to intervene, with which approach, and to what ends, hang in the air of technoscientific innovation regimes without clear answers. At the same time, descriptions of 'socially engineering emergence' and their emergent fields pile up. As Ulnicane quotes with regard to bottom-up, self-organised global networks of research, 'these networks cannot be managed; they can only be guided and influenced' (Wagner 2008, p. 105). Attempts at community making or community building can appear as interventions as an end in themselves when the potential to reach a specific goal with a specific intervention stays unclear.

1.6 'Choreographing Identities': From Identity Trouble to Strategic Performance

The chapters in Part I share a primary focus on matters of collectivity and the specificities of community in the sciences. They also, unavoidably, touch upon issues of identity: addressing individual researchers' roles and memberships and identities conferred by social collectives. They demonstrate that whatever the methodological approach—ethnographic or scientometric, with a focus on practices or discourses, on collaborations, networks, or fields—community and identity are intrinsically interwoven. Moreover, the terms 'community' and 'identity' both belong to a kind of vernacular category that research studies cannot totally ignore but that often stays at arm's length to a satisfying theoretical conceptualisation. As such, questions of community and identity have been re-addressed, reformulated, and conceptualised anew by almost every theoretical approach adopted in the analysis of scientific endeavour. This problematic is exacerbated for questions of identity in particular, questions that clearly transgress the disciplinary boundaries of sociology. Identity has been a classical theme for philosophers, psychologists, and cultural scientists, but it is only indirectly, for example with its unavoidable link to community or its inherent political dimension, that it comes within some sociologists' range of vision as a theme of reconstruction or deconstruction. As a result, identity has become a 'suspect category' on epistemic as well as political grounds, while remaining an inevitable theme on empirical grounds.

The chapters of Part II provide us with examples of how identity comes to matter in empirical cases and how it can be conceptualised and inform theoretical discourse. None of the authors would likely consider themselves as identity theorists; rather, they navigate phenomena relating to matters of identity in their empirically grounded work and consequently integrate these into case accounts and case analyses. It is all the more interesting, then, to look at the theoretical approaches to identity that the chapters engage and confront.

1.6.1 Becoming and/or Performing?

The need for scholarly attention to educational trajectories cannot be over stated. Wylie cogently attends to this concern with a focus on the interplay between the identities of 'novice' and 'expert' in a university engineering context. This work throws light on to new ways of thinking about learning as inherently part of identity-making performance.

Sarah Rachel Davies addresses in her chapter the challenging issue of conceptualising identity upfront when she contemplates that she is 'working with a generous definition of identity [. . .] as relating to role, boundary work (or role differentiation), and community.' Still, she does situate her work in a clearly denoted theoretical tradition, building on Goffman's performed self or 'performed character' and his distinction between 'front stage' and 'back stage' performance. She also draws on Hilgartner's (2000) analysis of 'Science on Stage' and his recognition that 'scientists are not 'con men', they simply have 'multiple identities' (ibid, p. 13). Along these lines, Davies' conception of performed identity is processual, contextual, multiple, and multivalent. It necessitates an actor, a character, a theatre, a stage, and an audience. The theatre and stages she focuses on in her participatory ethnography are provided by a science festival that took place over 6 days in Copenhagen in June 2014, zooming in on three particular science communication projects by means of a 'thick description'. Multiplicity, here, refers to performances of individual, group, disciplinary, and organisational identity. Simultaneous performance of, for example, a personal as well as a professional script is possible because both are underdetermined: 'The role of scientist is not highly scripted: one might play it [...] as an addendum to a leisure identity; or [...] as a social character, part of the production of enjoyable community (p. 220).'

But there is also a more reifying layer when Davies speaks of 'identity building' as additional to 'identity performance'. The building of identity as, for example, a university employee is something she experiences herself, back stage during her involvement in a science festival, whereas identity performance is something she observes on the front stages of the science communication theatre (and at several back stages that themselves function as front stages of yet again other kinds of performances). Identity thus is implicitly presented as double sided, just as is explicitly done in *Caitlin Donahue Wylie*'s account (the latter referring to Stevens et al. 2008), but with two different qualities for the two different sides: built (over time as an inner sense of belonging) and performed (ad hoc by 'harried fabricators of impressions', to adopt Goffman's wording). One could argue that with the depiction of identity building processes, Davies highlights a moment 'in which identity-formation becomes more settled and routinized', that maybe contributes to the furthering of 'the narrative infrastructures [Felt 2017] that make [situated identity performances] possible', while the material infrastructure of the festival

> became lived in: things broke and were stuck back together, the marquee took on a distinctive smell, we became familiar with each other and with the tent's soundscape [. . .] The community of the tent, both human and non-human, became deeply familiar (p. 218).

That is, at least, if we assume that the script for 'residing in a tent for six days as a University of Copenhagen employee' is not totally fixed from the beginning.

That both dimensions and relating conceptions of identity—(inner) becoming and (staged) performance—reside closely to each other should not gloss over that a deep theoretical gap runs between the two. In the cautionary words of Lewis P. and Sandra K. Hinchman:

When we spin narratives that form our personal identity, are we creating order out of chaos, i.e. out of a manifold of disordered impressions, sensations, memories, and inner states? Or does the narrative self somehow correspond to, or perhaps develop and articulate, a pre-narrative identity that is already there 'in itself,' antecedent to 'the narrative that constructs it? This obviously Kantian dilemma splits narrativists down the middle (Hinchman and Hinchman 1997, p. xix).

A gap that Wetherell (2010, p. 10) attributes to 'priority disputes between sociology and psychology' (quoted by Cuevas in this volume). It is perhaps owing to Davies' method of participatory observation that both these aspects come to the fore. Her involvement as scientist and university employee, as a member of 'the community of the tent', renders the first aspect conspicuous, although at a second glance the aspect of becoming can also be traced in the accounts of the other two projects in which the author participated as audience only.

1.6.2 Choreographing and Repairing Breaches of Canonical Narratives as Specific Kinds of Identity Work

Andrea Schikowitz's chapter also addresses both becoming and performing, but within a theoretical approach, empirical context, and (longer) time horizon where both aspects merge in multivalent identity work or a practice of 'choreographing'. She builds on Swidler (1986) and Butler (1990) and starts with an historical account of how belonging to disciplinary scientific communities became more fluid and diverse over time to a point where—at least in some realms of science—belonging is characterised by permanent change and necessitates an 'identity beyond community' established by a distinct kind of reactive as much as proactive identity work, performed to 'to establish, adapt, and maintain positions'. In her empirical case, to an even higher degree than in Davies' empirical case, identities and relating scripts are not pre-given, i.e. already established in other contexts). She investigates eleven projects funded within a transdisciplinary sustainability research programme in Austria between 2004 and 2012. Drawing on interviews, focus groups, and ethnographic observation, she highlights the distinct kind of identity work that comes with an 'identity beyond community' and introduces the concept of 'choreographies'. Other than disciplinary identities, these choreographed identities can be taken along, like a suitcase, when travelling from one disciplinary realm to another or when residing in between communities. The account thus evokes a picture of disciplinary residents and transdisciplinary nomads. 'The notion of choreography includes performance, and it directs attention to how different belongings and ideals in different moments might be related through '"dance instead of design" (Law 2003, p. 58)' is to be understood 'as a subtle relation of rules, routines, situated responses, and improvisation'. Four distinct choreographies or choreographed identities emerge from her analysis: an 'explorer', a 'caring broker', a 'moral manager', and a 'polymath', all of which exhibit distinct move patterns, identity work, socialities, and collective orders (see below for the latter).

Carlos Cuevas-Garcia is also concerned with the troubled identities of researchers, specifically those transgressing disciplinary boundaries throughout their careers. He analyses how 27 scientists at a 'large research-oriented British university' and from a variety of (inter)disciplinary fields depict their academic backgrounds in biographical interviews. Like Schikowitz he precedes his empirical analysis with a note on disciplinarity but focuses on the heterogeneous, hybrid, internally fragmented, constantly renegotiated, and to some extent illusionary character of disciplines, rendering them 'real [mostly] in their consequences'. Disciplinary identity is depicted as flexible, rhetoric, contextual, multiple, fluid, and tentative, interdisciplinarity as possibly a common practice rather than a deviant one. The author also invokes the schism between 'personal' and 'social' conceptions of identity—as addressed by Hinchman and Hinchman 1997 and Wetherell 2010, and labelled as becoming and performed above—pondering that all these characterisations might only apply to the latter conception. His own approach is an attempt at abandoning the dichotomy altogether, following the integrative perspective of a discursive psychology. He thus combines an analysis of biographical narratives with an analysis of metaphors capturing inter/disciplinarity. He particularly focuses on canonical narratives and on narratives of breach, trouble, and repair, as conceptualised in Taylor and Littleton's (2006) narrative-discursive analytical approach. With four exemplar interviews, he illustrates major differences, differences that can be explained by particular life trajectories of individual researchers but also by the particularities of the respective core disciplines of the interviewees. Albeit using variations, all interviewees relate to a 'canonical narrative of the single discipline specialist', including an early interest, passion, and skills in a specific discipline; early ambition and clear career goals; training and experience within the same discipline to PhD and beyond; and repair work when breaches occur. Repair, Cuevas-Garcia suggests, allows tensions to be eased, the gap between the fixed and the flexible and fluid to be closed, and an individual position to be developed.

Bettina Bock von Wülfingen's contribution provides a further case of troubled identities. Her analysis of identity work and identity fate at an interdisciplinary German Cluster of Excellence is based on a participatory ethnography and details the external and internal governance regimes determined by national science policy in the steering of the Cluster. She provides an overview of the scholarly discussion of scientific community and the social psychology of belonging. Based on her empirical material, she focuses on short term reactions to the unsettling of disciplinary identities as well as on types of collaborative identity that emerge within the Cluster's interdisciplinary constellation as it establishes its own research community over time. She thereby sketches individual differences that come to matter, such as the 'academic pedigree' or the career stage of the respective scientists, and structural conditions that co-determine the potential for interdisciplinary identity formation. She concludes that the formation of 'collaborative identities' is possible and builds on a more general academic self-conception, beyond academic biographic aspects.

1.6.3 The Affective Dimension

Bock von Wülfingen and Cuevas-Garcia touch on the affective dimension of the unsettling of identity. In the biographical interview sections presented by Cuevas-Garcia, we learn how an affective layer is co-constitutive of the narrated self—even when disciplinary boundaries are not challenged. The emotions featured in the scientists' accounts span a somewhat proscribed spectrum however, from excitement, fascination, enjoyment, being captivated and interested, to surprise and experiences of not feeling stupid anymore, or being able to escape a 'dirty nasty job'. This peculiarity of an omnipresent but highly defined and somewhat meagre affective repertoire is further explored by *Sarah Maria Schönbauer*. She 'decided to explicitly focus on the scientists' emotional engagement with their profession since relating to their community by "associating emotional states with certain activities" (Traweek 1988, p. 76) turned out to be one of the core concerns' of the group leaders and senior postdocs in the biographical interviews she had undertaken as part of an ethnographic field study in 2013 in the United States and Austria. With the specific sample of interviewees, affective narratives covering their own early career stages were a way of stabilising their own meaningful identity, belonging to a respective community (by conforming to its 'emotional culture' and distinct 'feeling style', Parker and Hackett 2012) and relating to a next generation of scientists. After sketching literature that highlights the importance as well as ambivalence of passion and being passionate within science from Max Weber and Robert Merton to contemporary scholars, the author concentrates on the role passion plays in the scientists' accounts of their different career stages, especially when it comes to insecurities, inconsistencies, and discontinuities. She addresses the first phase of choosing a field and curriculum, which is characterised by 'fascination' (comparable to Cuevas-Garcia's account), an ensuing phase of 'committing to science' marked by spatial and temporal concessions, and a third phase of 'imagining a future generation of scientists'. She concludes that 'passionate tales provide some order in and for scientific communities and show that the scientists' emotional relationship with their profession is key to understanding the construction of their identities' (p. 296). Passionate tales 'provide the scientists with continuity by constructing an emotional track record', and passion also seems to provide a drive in the scientists' accounts that allows for overcoming situations of unease. Demonstrating passion thus could also be interpreted by actors in the field as a sign of being able and willing to cope with the adversities of contemporary lives in science and hence translate into a sort of affective capital, just as demonstrating mobility can be interpreted as a proof of commitment to the collective and hence as a value in itself.

A recurring, though not fully elaborated, role in *Schönbauer*'s treatise is assigned to scientists' (assumed) need for security, stability, continuity, coherence, and connection. This resonates with Cuevas-Garcia's reference to some scholars' assumption that 'interdisciplinary researchers have a sense of vulnerability, tension and insecurity'. One might ask whether this—perhaps psychological *a priori*—could be challenged or rephrased in relation to contemporary scholars' acknowledgment of the distinctive role of the unstable, fluid, and fuzzy in the sociology of the sciences.

Davies' account of a science festival adds further hints to the role emotions play in staging science in public. One scientist aims for establishing a specific kind of emotional connection between the lay audience and the beluga whales featured in his exhibition as 'looking straight at you'. Another aims at depicting a science that 'first makes people laugh, and then makes them think' and then 'appreciate the wonders of scientific thinking' to 'understand how fun and interesting and beautiful science is'. Emotions evoked by pictures or performance are thus depicted as a 'natural bridge' between academia and lay audience. In the first case, evoking a certain emotional state of feeling connected to the 'natural world' (and hence caring for it) is often claimed as a central ambition of the science communication project.

1.7 A Preliminary Conclusion

The papers considered here demonstrate that concern and debates regarding collectivity and identity in contemporary sciences are very much alive. Authors tackle a range of contemporary phenomena that intersect with these two main concepts, including: digitalisation; mediatisation; projectification; phenomena of up-sizing (massification and Big Science); hybridisation, post-disciplinarity or 'new disciplinarity' (Marcovich and Shinn 2011); mission orientation; new funding environments; and more. The chapters of the volume thereby renew our attention to both longstanding and novel conceptual matters and, in effect, survey the impact of these shifts in the 'machineries of knowledge production' (Knorr-Cetina 1999), with a view to demonstrating (techno)scientific identity constellations and (techno)-scientific communality as integral parts of such machineries.

It may be more than a mere coincidence that the empirical examples of this volume stem for the most part from synthetic biology, bioinformatics and sustainability sciences. All of these realms of research seem to have more in common than one would assume at first glance (given their quite different disciplinary backgrounds and, arguably, conflicting societal visions). Not only are the field formations highly interdisciplinary, the fields can also be viewed as 'umbrella terms' (just like nanotechnology, cp. Rip and Voß 2013), 'strategic science' (Rip 2002) and as 'new and emerging' (or maybe better: 'ephemeral') science. It would be interesting to reconstruct the technoscientific features they share more broadly in more detail. The analysis of identity and community constellations in these fields could serve as a valuable way in for such an endeavour. We thus hope that this volume can also contribute to a further delineation of technoscience and its distinct qualities.

Technoscientific communities are addressed as emergent and heterogenous, but also as rhetorically fabricated and socially engineered phenomena. Some authors align their analyses of emergence with classical studies that tell us a story about several phases leading to the final maturation and institutionalisation of a field, but hint at a distinct quality of this mature state of contemporary fields (Noël, Blümel);

others underline that contemporary fields are characterised by a kind of 'perpetuated emergence' (Raimbault and Joly). Further terms put forward to signify a qualitative singularity of community in contemporary technosciences are 'project-ed community' (Cointe, addressed also by Ulnicane and Jarke) and 'community of knowledge application' (Degelsegger-Márquez, addressed to some extent also by Jarke). The—traditionally ignored and probably enhanced—agency of early career scientists (Wylie), of 'communities of practice' (Wenger 1998) as sites of (engineered) communality (Jarke) is pointed to in two further texts. Ulnicane allows for discussing 'internal' versus 'external' steering and its effects on scientific community (or, more specifically, collaboration) and Bock von Wülfingen adds an analysis of the effects interdisciplinary clusters of excellence have on community and identity in science.

Contemporary technoscientific identities are characterised as 'choreographed' (Schikowitz), 'troubled' (Cuevas-Garcia, but also Bock von Wülfingen), and 'staged' (Davies)—at first sight leaving meagre or only superficial options for a science worker to identify themselves as 'scientist' or with a distinct scientific field. But on closer look, adopting a choreographed identity seems to just become part of a scientists self-concept, the 'troubled' identity is 'repaired', the process of 'only staging' an identity in public leaves traces on a deeper level of self-identification, leading to for instance to a stronger identification with one's academic institution (Davies). A lack of options for (disciplinary) identification is answered by filling in alternative sources of identity like one's 'academic pedigree' in the short run or an interdisciplinary identity in the longer run (Bock von Wülfingen), a kind of repair work that goes along with a process of regrouping along traits alternative to or at least adding to disciplinary affiliations (also illustrated by Schikowitz and Cuevas-Garcia).

Whereas emotions and affect appear repeatedly in various chapters (even if only with a highly confined spectre, in a highly disciplined manner, and mostly in connection with participatory observation), a further discourse within the contemporary sociology of science literature, namely that on care and caring, is mostly absent from the accounts. *Schikowitz* delineates the figure of the 'caring broker'; Davies' reports on the caring marine biologist and the maintenance work undertaken by the 'community of the tent'; *Raimbault and Joly*'s entrepreneurial scientists engage in the construction of a scientific community while PI's teach, supervise, and mentor novices. The list of empirical instances that would invite a focus on the role of care and caring is long. Still, this dimension remains largely unaddressed. Also, in most chapters it is unclear how the authors came to care about the fields they decided to investigate, in many cases to even participate in. We might deduce that addressing and exhibiting care is just as much a taboo in the sociology of the sciences as it is in the natural science fields studied in this volume. With new discourses emerging on care and responsibility within science studies and beyond, we wonder whether individual actors and communality might again gain more attention.

While many contributions in this volume focus on processes of field emergence rather than 'mature fields' and thereby contribute detailed analyses of a theme less often featured in the sociology of sciences, they do not include clues about the opposite process, the demise, ceasing to be of research fields. Such processes could be interesting especially against a context of fields or communities that seem to be stabilised by external forces, by science funding programmes and other attempts at manufacturing scientific communities. What exactly happens when these forces subside? The same could be asked about identities: when and how does a scientist disengage with a specific identifier and its relating community, when and why does one refuse to take on a specific identity?[8] Are there contexts in science where community membership and identification with one's profession are never an option? How often can one adopt and abandon an identity and community and which traces does such a process leave in a scientific career and community?

Although not one contribution focusses on the more general level of the identity of researchers as scientists, many chapters include analyses that relate to this level when discussing the more specific identities of lab members, project participants or new disciplinarians. It is mostly in the literature on new public management at universities that shifts in identity options and relating tensions between academic and managerial identities are reported and discussed (Henkel 2000; Winter 2009). They are obviously linked to the general trends already enlisted above. Is it then still possible to identify with a traditional academic identity and role? Or have we already ceased to be academics and morphed into research managers a long time (i.e., some major projects and collaborations) ago? Are we part of a republic of (coequal) scholars and the vocation that comes with it, of a scientific community with its distinct ethos and societal status or of a global innovation machinery, based on role differentiation, division of labour and clear output definitions? How do quality/ qualification, ethos and responsibility fare in these contexts? Which stories are told and sold to the public about these (new) identities? And where do we want go from there?

Although the volume is divided into two sections, the entanglements between collectivity and identity are clearly visible, albeit not trivial or tangible in all respects. Much about their interrelation stays implicit and taken for granted. Being a member of a group, a community or an institution, sharing an educational background and socialisation, all provide senses of belonging, go with certain kinds of identification and add to a list of potential identifiers. Identity thus seems to relate individuality and communality in a plurality of ways. Both, (techno)- scientific identity and community also come with an emotional as well as an instrumental dimension, with an aspiration to make sense (who am I? where do I belong to?) as well as strategic desires (how can community be engineered in a way that maximises productivity? How can identity be choreographed so as to provide the best career options?).

---

[8]In July 2019, a workshop organised by Kastenhofer and Vermeulen at the International Society for the History Philosophy and Social Studies of Biology (ISHPSSB) biennial meeting in Oslo addressed such a theme from multiple disciplinary and transdisciplinary angles under the heading 'Should we stay or should we go now?' with a focus on why, when and how scholars stop working on or in an 'emerging field' such as systems biology; a relating paper is in preparation.

The assumption of a co-productive relation between (techno)scientific and social orders, most clearly delineated in Jasanoff (2006), has been addressed at the beginning of this chapter; it is also touched upon in many chapters of this volume. In particular, the impact of science policies and science funding programmes (addressed more broadly in Whitley et al. 2010) and programmatics on (techno)-scientific identity and communality are invoked in various of the case studies. There is, however, less emphasis here on how technoscience co-produces social order beyond the realms of academic research and education. New modes of biosociality, for example, those co-produced by genetic testing and the emerging category of genetic kinship, or, the emerging, datafied understanding of identity related to the discourse on 'identity theft', are just two cases of such co-productions triggered by technoscientific practices and requiring further attention beyond the confines of this volume.[9]

In extra-scientific contexts, an essentialist take on identity and community has of late been severely drawn into question once again with the phenomenon of so-called identitarian movements, necessitating us to re-think what exact status the aforementioned aspirations and desires should be attributed with and how this allows for their instrumentalisation along vested interests and particular kinds of political ideologies. Does it suffice here to allude to the very different context of science as compared to society at large, to a categorical difference between *scientific* community and *scientific* identity as compared to community and identity in society? If so, we would once more need a better understanding of how the concepts of community and identity change when transferred from general sociology to a sociology of science. If not, it will not suffice to just ignore the roles identity and community play and how these are played with in our machineries of knowledge production.

On a final note, this volume is also about the community/communities and identity/identities of a contemporary sociology of science. Addressing longstanding concerns of community and identity also involves a good portion of communitarian activity and identity work in itself. One might assume that such activity and work will need both, space and tolerance for innovation and diversification, but also distinct opportunities and places for stock taking and joint discussions. For both we see this Yearbook series as a valuable resource and hope it will further prosper during the years to come!

---

[9]We thank one anonymous reviewer and the participants of a course on 'identity and community - and how they come to matter in contemporary technoscience' held at the department of science and technology studies at the University of Vienna in 2019 for drawing our attention to this specific limitation of the volume.

References

Abir-Am, P. 1985. Themes, genres and orders of legitimation in the consolidation of new scientific disciplines: Deconstructing the historiography of molecular biology. *History of Science* 23 (1): 73–117.

Arnold, M., and R. Fischer, eds. 2004. *Disziplinierungen. Kulturen der Wissenschaft im Vergleich*. Wien: Turia + Kant.

Balmer, A.S., and K.J. Bulpin. 2013. Left to their own devices: Post-ELSI, ethical equipment and the International Genetically Engineered Machine (iGEM) competition. *BioSocieties* 8 (3): 311–335.

Balmer, A.S., K. Bulpin, and S. Molyneux-Hodgson. 2016. *Synthetic biology: A sociology of changing practices*. Basingstoke: Palgrave Macmillan.

Bastide, F., J.P. Courtial, and M. Callon. 1989. The use of review articles in the analysis of research data. *Scientometrics* 15: 535–562.

Bauman, Z. 2004. *Identity. Conversations with Benedetto Vecchi*. Cambridge: Polity Press.

Beaufaÿs, S. 2003. Wie werden Wissenschaftler gemacht? In *Beobachtungen zur wechselseitigen Konstitution von Geschlecht und Wissenschaft*. Bielefeld: Transcript Verlag.

Becher, T., and P.R. Trowler. 2001. *Academic tribes and territories. Intellectual enquiry and the culture of disciplines*. 2nd ed. Buckingham: SRHE and Open University Press.

Becker, H.S., B. Geer, E.C. Hughes, and A.L. Strauss. 1961. *Boys in white: Student culture in medical school*. Chicago: University of Chicago Press.

Ben-David, J., and R. Collins. 1966. Social factors in the origins of a new science: The case of psychology. *American Sociological Review* 31 (4): 451–465.

Böhme, G., W. van den Daele, and W. Krohn. 1973. Die Finalisierung der Wissenschaft. *Zeitschrift für Soziologie* 2 (2): 128–144.

Bonneuil, C. 2006. Cultures épistémiques et engagement public des chercheurs dans la controverse OGM. *Natures Sciences Sociétés* 14: 257–268.

Böschen, S., K. Kastenhofer, I. Rust, J. Soentgen, and P. Wehling. 2010. Scientific nonknowledge and its political dynamics: The cases of agri-biotechnology and mobile phoning. *Science, Technology, and Human Values* 35 (6): 783–811.

Bourdieu, P. 1975. The specificity of the scientific field and the social conditions of the progress of reason. *Social Science Information* 14 (6): 19–47.

Brown, N. 2003. Hope against hype: Accountability in biopasts, presents and futures. *Science Studies* 16 (2): 3–21.

Bulpin, K., and S. Molyneux-Hodgson. 2013. The disciplining of scientific communities. *Interdisciplinary Science Reviews* 38 (2): 91–105.

Butler, J. 1990. *Gender troubles: Feminism and the subversion of identity*. New York: Routledge.

Calvert, J. 2010. Systems biology, interdisciplinarity and disciplinary identity. In *Collaboration in the new life sciences*, ed. J.N. Parker, N. Vermeulen, and B. Penders, 201–218. Aldershot: Ashgate.

Chari, D.N., R. Howard, and B. Bowe. 2012. Disciplinary identity of nanoscience and nanotechnology research – A study of postgraduate researchers' experiences. *International Journal for Digital Society* 3 (1): 619–616.

Cointe, Béatrice. this volume. The project-ed community. In *Community and identity in contemporary technosciences*, Sociology of the sciences yearbook 31, ed. K. Kastenhofer and S. Molyneux-Hodgson. Cham: Springer.

Collins, H.M. 1981. The place of the 'core-set' in modern science: Social contingency with methodological propriety in science. *History of Science* 19 (1): 6–19.

———. 1998. The meaning of data: Open and closed evidential cultures in the search for gravitational waves. *American Journal of Sociology* 104 (2): 293–337.

Cooper, R., and J. Law. 1995. Organization: Distal and proximal views. *Research in the Sociology of Organizations* 13: 237–274.

Crane, D. 1972. *Invisible colleges: Diffusion of knowledge in scientific communities.* Chicago: University of Chicago Press.

Damm, U., B. Hopfengaertner, D. Niopek, and P. Bayer. 2013. Are artists and engineers inventing the culture of tomorrow? *Futures* 48: 55–64.

Daston, L. 2003. Die wissenschaftliche Persona. Arbeit und Berufung. In *Zwischen Vorderbühne und Hinterbühne. Beiträge zum Wandel der Geschlechterbeziehungen in der Wissenschaft vom 17. Jahrhundert bis zur Gegenwart*, ed. T. Wobbe, 109–136. Transcript: Bielefeld.

Daston, L., and P. Peter Galison. 2007. *Objectivity.* New York: Zone Books.

Daston, L., and H. Sibum. 2003. Introduction: Scientific personae and their histories. *Science in Context* 16 (1–2): 1–8.

De Solla Price, Derek J. 1963/1971. Little science, big science. New York/London: Columbia University Press.

Dixon, J., and N. Kuldell. 2011.Biobuilding: Using banana-scented bacteria to teach synthetic biology. In *Methods in enzymology, Vol 497: Synthetic biology, Methods for part/device characterization and chassis engineering, Pt A*, ed. C. Voigt, pp. 255–271.

Elzinga, A. 1993. Science as the continuation of politics by other means. In *Controversial science. From content to contention*, ed. T. Brante, S. Fuller, and W. Lynch, 127–152. Albany: State University of New York Press.

Felt, U., ed. 2009. *Knowing and living in academic research: Convergences and heterogeneity in research cultures in the European context.* Prague: Institute of Sociology of the Academy of Sciences of the Czech Republic.

———. 2016. Of timescapes and knowledge spaces: Re-timing research and higher education. In *New languages and landscapes of higher education*, ed. P. Scott, J. Gallacher, and G. Parry, 129–148. Oxford: Oxford University Press.

———. 2017. 'Response-able practices' or 'new bureaucracies of virtue': The challenges of making RRI work in academic environments. In *Responsible innovation 3: A European Agenda?* ed. L. Asveld, M. van Dam-Mieras, T. Swierstra, et al., 49–68. Cham: Springer.

Felt, U., J. Igelsböck, A. Schikowitz, and T. Völker. 2013. Growing into what? The (un-)disciplined socialisation of early stage researchers in transdisciplinary research. *Higher Education* 65: 511–524.

Fleck, L. (1983 [1947]). Schauen, sehen, wissen (T. Schnelle, Trans.). In Ludwik Fleck. Erfahrung und Tatsache. Gesammelte Aufsätze, ed. L. Schäfer and T. Schnelle, 147–174. Frankfurt/Main: Suhrkamp.

Fligstein, N., and D. McAdam. 2012. *A theory of fields.* Oxford/New York/Auckland: Oxford University Press.

Frickel, S. 2004. Building an interdiscipline: Collective action framing and the rise of genetic toxicology. *Social Problems* 51 (2): 269–287.

Frickel, S., and N. Gross. 2005. A general theory of scientific/intellectual movements. *American Sociological Review* 70 (2): 204–232.

Frickel, S., and K. Moore, eds. 2005. *The new political sociology of science: Institutions, networks, and power.* Madison: University of Wisconsin Press.

Frow, E., and J. Calvert. 2013. 'Can simple biological systems be built from standardized interchangeable parts?' Negotiating biology and engineering in a synthetic biology competition. *Engineering Studies* 5 (1): 42–58.

Funtowicz, S.O., and J.R. Ravetz. 1992. Three types of risk assessment and the emergence of postnormal science. In *Social theories of risk*, ed. S. Krimsky and D. Golding, 251–273. Westport, CT: Greenwood.

Gibbons, M., C. Limoges, H. Nowotny, S. Schwartzman, P. Scott, and M. Trow. 1994. *The new production of knowledge: The dynamics of science and research in contemporary societies.* London: Sage.

Gieryn, T.F. 1983. Boundary-work and the demarcation of science from non-science: Strains and interests in professional ideologies of scientists. *American Sociological Review* 48 (6): 781–795.

Gläser, J. 2006. Wissenschaftliche Produktionsgemeinschaften. In *Die soziale Ordnung der Forschung*. Frankfurt/New York: Campus.

Gläser, J., G. Laudel, and E. Lettkemann. 2016. Hidden in plain sight: The impact of generic governance on the emergence of research fields. In *The local configuration of new research fields. On regional and national diversity*, Sociology of the sciences yearbook 29, ed. M. Merz and P. Sormani, 25–43. Dordrecht: Springer.

Goodman, C. 2008. Engineering ingenuity at iGEM. *Nature Chemical Biology* 4 (1): 13–13.

Haas, P.M. 1992. Introduction: Epistemic communities and international policy coordination. *International Organization* 46 (1): 1–35.

———. 1994. Do regimes matter? Epistemic communities and Mediterranean pollution control. In *Comparative science and technology policy*, ed. S. Jasanoff, 537–563. Cheltenham: Edward Elgar Publishing.

Hackett, E.J., and J.N. Parker. 2016. Ecology reconfigured: Organizational innovation, group dynamics and scientific change. In *The local configuration of new research fields. On regional and national diversity*, Sociology of the sciences yearbook 29, ed. M. Merz and P. Sormani, 153–171. Dordrecht: Springer.

Hacking, I. 2002. *Historical ontology*. Cambridge/London: Harvard University Press.

Hagstrom, W. 1965. *The scientific community*. New York: Basic Books.

Haraway, D. 1996. *Modest_Witness@Second_Millennium.FemaleMan©_Meets_Oncomouse™*. New York: Routledge.

Harwood, J. 1993. *Styles of scientific thought. The German genetics community, 1900–1933*. Chicago: University of Chicago Press.

Henkel, M. 2000. *Academic identities and policy change in higher education*. London: Jessica Kingsley Publishers.

Hess, D.J. 2005. Antiangiogenesis research and the dynamics of scientific fields: Historical and institutional perspectives in the sociology of science. In *The new political sociology of science: Institutions, networks, and power*, ed. S. Frickel and K. Moore, 122–148. Madison: University of Wisconsin Press.

Hilgartner, S. 2000. *Science on stage: Expert advice as public drama*. Stanford, CA: Stanford University Press.

Hinchman, L.P., and S.K. Hinchman. 1997. Introduction. In *Memory, identity, community: The idea of narrative in the human sciences*, ed. L.P. Hinchman and S.K. Hinchman, xiii–xxxii. Albany: State University of New York Press.

Ibarra, H. 1999. Provisional selves: Experimenting with image and identity in professional adaptation. *Administrative Science Quarterly* 44: 764–791.

Jamison, A. 1997. National styles of science and technology: A comparative model. In *Comparative science and technology policy*, ed. S. Jasanoff, 69–83. Cheltenham: Edward Elgar Publishing.

Jasanoff, S., ed. 2006. *States of knowledge: The co-production of science and social order*. London/New York: Routledge.

Jenkins, R. 2014. *Social identity*. 4th ed. London/New York: Routledge.

Kastenhofer, K. 2013. Two sides of the same coin? The (techno)epistemic cultures of systems and synthetic biology. *Studies in History and Philosophy of Biological and Biomedical Sciences* 44 (2): 130–140.

———. 2016. 'I am a biochemist by training': identity and community in systems biology. *Vienna STS Talk*, 4 May 2016, University of Vienna, Austria. https://sts.univie.ac.at/einzelansicht-news-und-events/news/vienna-sts-talk-karen-kastenhofer/?no_cache=1. Accessed 14 August 2019.

Kelwick, R., L. Bowater, K.H. Yeoman, and R.P. Bowater. 2015. Promoting microbiology education through the iGEM synthetic biology competition. *FEMS Microbiology Letters* 362 (16).

Knorr-Cetina, K. 1982. Scientific communities or transepistemic arenas of research? A critique of quasi-economic models of science. *Social Studies of Science* 12 (1): 101–130.

————. 1999. *Epistemic cultures: How the sciences make knowledge*, 1–25. London/Cambridge: Havard University Press.

Kurath, M. 2010. Negotiating nano: From assessing risks to disciplinary transformations. In *Governing future technologies. Nanotechnology and the rise of an assessment regime*, Sociology of the sciences yearbook 27, ed. M. Kaiser, M. Kurath, S. Maasen, and C. Rehmann-Sutter, 21–36. Dordrecht: Springer.

Kurath, M., and S. Maasen. 2006. Toxicology as a nanoscience? – Disciplinary identities reconsidered. *Particle and Fibre Toxicology* 3: 6.

Kusch, M. 2010. Hacking's historical epistemology: A critique of styles of reasoning. *Studies in History and Philosophy of Science* 41: 158–173.

Lave, J., and E. Wenger. 1991. *Situated learning: Legitimate peripheral participation*. Cambridge: Cambridge University Press.

Law, J. 2003. *Traduction/trahision: Notes on ANT*. Lancester: Lancester University Press.

Lemaine, G., ed. 1976. *Perspectives on the emergence of scientific disciplines*. Chicago: Aldine.

Leonelli, S., and R.A. Ankeny. 2015. Repertoires: How to transform a project into a research community. *Bioscience* 65 (7): 701–708.

Lewis, J., and A. Bartlett. 2013. Inscribing a discipline: Tensions in the field of bioinformatics. *New Genetics and Society* 32 (3): 243–263.

Liebau, E., and L. Huber. 1985. Die Kulturen der Fächer. *Neue Sammlung* 3 (1985): 314–339.

Lorenz-Meyer, D. 2010. Possibilities of enacting and researching epistemic communities. *Sociological Research Online* 15 (2): 1–13.

Marcovich, A., and T. Shinn. 2011. Where is disciplinarity going? Meeting on the borderland. *Social Science Information* 50 (3–4): 582–606.

Mercer, D. 2015. 'iDentity' and governance in synthetic biology: Norms and counter norms in the 'international genetically engineered machine' (iGEM) competition. *Macquarie Law Journal* 15: 83–103.

Merton, R. K. (1993/1942). The sociology of science: Theoretical and empirical investigations. Chicago: University of Chicago Press.

Merz, M., and P. Sormani, eds. 2016a. *The local configuration of new research fields. On regional and national diversity*, Sociology of the sciences yearbook 29. Dordrecht: Springer.

————. 2016b. Configuring new research fields: How policy, place, and organization are made to matter. In *The local configuration of new research fields. On regional and national diversity*, Sociology of the sciences yearbook 29, ed. M. Merz and P. Sormani, 1–22. Dordrecht: Springer.

Meyer, M., and S. Molyneux-Hodgson. 2010. Introduction: The dynamics of epistemic communities. *Sociological Research Online* 15 (2): 14.

Mol, A. 2003. *The body multiple: Ontology in medical practice*. Durham, NC: Duke University Press.

Molyneux-Hodgson, S., and M. Meyer. 2009. Tales of emergence – Synthetic biology as a scientific community in the making. *BioSocieties* 4: 129–145.

Mullins, N.C. 1972. The development of a scientific speciality: The phage group and the origins of molecular biology. *Minerva* 10 (1): 51–82.

Nordmann, A., H. Radder, and G. Schiemann, eds. 2011. *Science transformed?: Debating claims of an epochal break*. Pittsburgh, PA: University of Pittsburgh Press.

O'Malley, M.A., A. Powell, J.F. Davies, and J. Calvert. 2008. Knowledge-making distinctions in synthetic biology. *BioEssays* 30 (1): 57–65.

Osbeck, L.M., and N.J. Nersessian. 2017. Epistemic identities in interdisciplinary science. *Perspectives on Science* 25 (2): 226–260.

Parker, J.N., and E.J. Hackett. 2012. Hot spots and hot moments in scientific collaborations and social movements. *American Sociological Review* 77 (1): 21–44.

Paul, H. 2016. Sources of the self: Scholarly personae as repertoires of scholarly selfhood. *Low Countries Historical Review* 131 (4): 135–154.

Pickstone, J.V. 2001. *Ways of knowing. A new history of science, technology and medicine*. Chicago: University of Chicago Press.

Raimbault, B., J.P. Cointet, and P.-B. Joly. 2016. Mapping the emergence of synthetic biology. *PLoS One* 11 (9): e0161522.

Rip, A. 2002. Regional innovation systems and the advent of strategic science. *Journal of Technology Transfer* 27 (1): 123–131.

———. 2004. Strategic research, post-modern universities and research training. *Higher Education Policy* 17: 153–166.

Rip, A., and J.-P. Voss. 2013. Umbrella terms as mediators in the governance of emerging science and technology. *Science, Technology & Innovation Studies* 9 (2): 39–59.

Serrano, L. 2007. Synthetic biology: Promises and challenges. *Molecular Systems Biology* 3: 158.

Shapin, S. 2008. *The scientific life: A moral history of a late modern vocation.* Chicago, IL: Chicago University Press.

Smolke, C.D. 2009. Building outside of the box: iGEM and the BioBricks Foundation. *Nature Biotechnology* 27 (12): 1099–1102.

Stent, G.S. 1968. That was the molecular biology that was. *Science 160*: 390–395.

Stevens, R., K. O'Connor, L. Garrison, A. Jocuns, and D.M. Amos. 2008. Becoming an engineer: Toward a three-dimensional view of engineering learning. *Journal of Engineering Education* 97 (3): 355–368.

Swidler, A. 1986. Culture in action: Symbols and strategies. *American Sociological Review* 51 (2): 273–286.

Taylor, S., and K. Littleton. 2006. Biographies in talk: A narrative-discursive research approach. *Qualitative Sociology Review* 2 (1): 22–38.

Traweek, S. 1988. *Beamtimes and lifetimes. The world of high energy physicists.* Cambridge/London: Harvard University Press.

Tsui, J., and A.S. Meyer. 2016. Modular projects and 'mean questions': Best practices for advising an international genetically engineered machines team. *FEMS Microbiology Letters* 363 (14).

Vermeulen, N. 2018. The choreography of a new research field: Aggregation, circulation and oscillation. *Environment and Planning A* 50 (8): 1764–1784.

Wagner, C.S. 2008. *The new invisible college. Science for development.* Washington, DC: Brookings Institution Press.

Weber, J. 2010. Interdisziplinierung? Zur Übersetzungspolitik einer neuen Technowissenschaftskultur. In *Interdisziplinierung? Über den Wissenstransfer zwischen den Geistes-, Sozial- und Technowissenschaften*, ed. J. Weber, 83–112. Bielefeld: Transcript.

Weingart, P. 2003. Growth, differentiation, expansion and change of identity – The future of science. In *Social studies of science and technology: Looking back, ahead. Sociology of the sciences yearbook 23*, ed. B. Joerges and H. Nowotny, 183–200. Dordrecht: Kluwer.

———. 2012. The lure of the mass media and its repercussions on science. Theoretical considerations on the medialization of science. In *The sciences' media connection – Public communication and its repercussions, Sociology of the sciences yearbook 28*, ed. S. Rödder, M. Franzen, and P. Weingart, 17–32. Dordrecht: Springer.

Wenger, E. 1998. *Communities of practice. Learning, meaning, and identity.* Cambridge: Cambridge University Press.

Wetherell, M. 2010. The field of identity studies (introduction). In *The Sage handbook of identities*, ed. M. Wetherell and C.T. Mohanty, 3–26. Los Angeles/London/New Delhi: Sage.

Whitley, R. 1984. *The intellectual and social organization of the sciences.* Oxford: Clarendon Press.

Whitley, R., J. Gläser, and L. Engwall, eds. 2010. *Reconfiguring knowledge production: Changing authority relationships in the sciences and their consequences for intellectual innovation.* Oxford: Oxford University Press.

Winter, R. 2009. Academic manager or managed academic? Academic identity schisms in higher education. *Journal of Higher Education Policy and Management* 31 (2): 121–131.

Part I
Synthetic Communities

Chapter 2
Remaining Central and Interdisciplinary: Conditions for Success of a Research Speciality at the University of Strasbourg (1961–2011)

Marianne Noël

2.1 Background and Context

Chemistry is a good example of how modern research is organised in an impressive array of sub-disciplines and hybrid discipline formations, interdisciplinary collaborations, and new experimental systems based on specific methodologies, techniques, and substrates (Meinel in Reinhardt 2001, p. IX). The eclectic mix of objectives, ideas, practices, and ways of thinking that defines chemistry as a science, a profession, or an industry has shaped its academic perception. Bensaude-Vincent and Simon (2008, p. 3) argue that 'chemistry "is an "impure science"; that it mixes science with technological applications, that it eschews high theory, and that it does not hold consistency to be its highest value'. Its hybrid nature, its common interest in the material world, its persistent faith in empirical generalisations instead of rigorous mathematical models (Kovac 2002), and its adhesion to metaphysical constructs (van Brakel 2000) have often put chemistry in a subaltern position in comparison to other sciences, particularly physics. Even as it first established itself as an independent scientific discipline in the eighteenth century, chemistry was regarded as being intellectually inferior to mathematics and physics (Bensaude-Vincent and Simon 2008, p. 3). Stichweh (2001) depicts disciplines as the primary unit of the internal differentiation of science in the late eighteenth and early nineteenth centuries and considers the continuous mutual observation and interaction of disciplines as the most important factor in the dynamics of modern science. In this respect, this chapter addresses the question of how chemists find their place among the many already established positions in the world of research, and, more specifically, what their positioning practices are when they develop new research specialities.

M. Noël (✉)
LISIS, Univ Gustave Eiffel, ESIEE Paris, CNRS, INRAE, Marne-la-Vallée, France
e-mail: noel@ifris.org

© The Author(s) 2021
K. Kastenhofer, S. Molyneux-Hodgson (eds.), *Community and Identity in Contemporary Technosciences*, Sociology of the Sciences Yearbook 31,
https://doi.org/10.1007/978-3-030-61728-8_2

In this chapter, I discuss the development of a research specialty (supramolecular chemistry or SMC) at the University of Strasbourg (France). As it is practised today, SMC is one of the most vigorous and fast-growing fields of chemical endeavours (Steed and Atwood 2009). From the perspective of social studies of science, its history still remains to be written.[1]

This study focuses mainly on the work carried out by and around Nobel Prize laureate Jean-Marie Lehn. Lehn spent the better part of his career in Strasbourg, the university where he had studied. This place is indeed extremely interesting as prestigious awards have been won over a short period of time: Jules Hoffmann (Nobel Prize in Physiology or Medicine in 2011), Martin Karplus (Nobel Prize in Chemistry in 2013) and Jean-Pierre Sauvage (Nobel Prize in Chemistry in 2016) have all developed their research programmes in Strasbourg.

What are the conditions for the success of 'Strasbourg's chemistry'? As Lehn himself stressed, 'Chemistry in Strasbourg, just like the University as a whole, is the product of its history and geography'.[2] In this chapter, I take a socio-historical approach whereby SMC is both a concept and a social object, embedded in an institutional context and shaped by professionals and scientific policies. In so doing, I place the sociology of knowledge in a broader perspective (Bonneuil and Pestre 2015) and examine how *historical* analyses can help improve our understanding of community in *ongoing* disciplinary formations. Within this institutional context (a university setting), the disciplinary framework can't be ignored as disciplines constitute the educational units of organisation (Whitley 1976).

In Strasbourg, the development of SMC has led to the creation of a dedicated institute (a new building) on the university campus, demonstrating a large degree of permanence and legitimacy. Interestingly, this has occurred in a highly centralised state (France), where implementation of an interdisciplinary programme in materials research failed (Bertrand and Bensaude-Vincent 2011). Drawing on historical work devoted to the University of Strasbourg, I will also characterise this singularity.

2.1.1 Revisiting the Emergence of Research Specialties in Chemistry

For reasons similar to those detailed by Raimbault and Joly in their contribution to this book (2021), I consider the notion of discipline as a relevant analytical entry point. In the history of modern chemistry, much attention has been paid to the ways in which chemists demarcated their domain from other domains as well as to their insistence on a chemical epistemology and systems of chemical language and

---

[1]With the exception of Schummer (2006) referring to an unpublished manuscript (footnote 6).

[2]FR Service Archives de l'Université de Strasbourg/Guy Ourisson/GO 150/Centre de Recherche en Chimie de l'Université Louis Pasteur: plaquette de présentation. 'La chimie à Strasbourg', par Jean-Marie Lehn, préface de la plaquette (1992).

imageries (Nye 1993, p. 3). Chemistry stands out as a distinct space in which research follows a logic of cohabitation, thus comparable to the engineering sciences (Bensaude-Vincent and Stengers 2001). Apart from a few notable cases (for instance Gavroglou and Simões (2012) who qualify quantum chemistry as an 'in-between discipline', 'neither physics nor chemistry'), this emphasis on demarcation corresponds with the treatment of disciplines by many sociologists and historians of science who focused on the organisation of nineteenth and early twentieth-century academic scholarship. The underlying 'discipline model' which holds that scientific growth unfolds as a process of differentiation into new specialties and disciplines has recently been challenged in the literature (see Heinze et al. (2013) for a review).

As for epistemic and social patterns of emergence, a fair amount of work on scientific disciplines and their development was carried out in the 1960s and 1970s (for instance, Lemaine 1976), complemented later on by research focusing specifically on chemistry (Nye 1993; Reinhardt 2001).

Two more recent studies are devoted to the emergence of research specialties in France, thereby opting for a concept less rigorous than the established notion of 'discipline'. Voillequin (2008) examines the case of catalysis (1944–2004), which he sees as a paradigmatic case of a techno-science emerging at the crossroads of disciplines. His history of catalysis tells the story of a community in search of legitimacy, aware of the transepistemic importance of its scientific domain (particularly given its ties to industry), constantly wavering between efforts to open up and a tendency towards confinement. In the case of solid-state chemistry, Teissier (2010) provides a model that extends from the formation of a community in the 1930s to its scattering in 2004. Both authors focus on the institutional construction of laboratories and draw on the journey of the 'masters' and the fate of their respective schools of research.

The concept of research schools has prevailed as a cornerstone of reflection on modern chemistry (Servos 1993). It places competition at the heart of the dynamics of scientific activities. The collective in-the-making is presented as a set of competing individuals addressing the subject from incompatible points of view. This latter conception thus offers a rather restricted theoretical frame and runs the danger of wrongfully neglecting contextual factors in the emergence of a specialty, especially the role of pre-existing entities and prevalent social rules.

In this chapter, I consider all three conceptions—that of a discipline, of a research specialty, and of a research school. In the following section, I shortly introduce the classical study of the emergence of molecular biology as a research specialty by Mullins (1972); I refer to Stichweh's (2001) take on the rise of disciplinarity in the formative period of the eighteenth and nineteenth centuries; I then depict how the historical conception of science as organised in disciplinary structures has been adapted to contemporary contexts by Marcovich and Shinn's (2011) notion of a 'new disciplinarity' and how the focus on individual founders of research schools can be replaced by considering a variety of typical trajectories of technoscientific practitioners in the context of a 'new disciplinarity', as put forward by Marcovich and Shinn (2014). Based on this theoretical outline, I present my own empirical case—namely the emergence of supramolecular chemistry in France—and discuss its fit with the presented theoretical approaches.

2.1.2 The Role of Individual Scientists, Coherent Groups, Disciplines and New Disciplinarity

Mullins (1972) not only provided a useful definition and characterisation of specialities as

> an institutional cluster which has developed regular processes for training and recruitment into roles which are institutionally defined as belonging to that specialty. Members are aware of each other's work, although not necessarily deeply involved in communications with one another. They may share a paradigm and a set of judgements about what general work should be done in the field, although the details of those ideas might differ. The specialty, then, has many aspects of a formal organisation, *i.e.*, recruitment procedures, tests of membership, journals, meetings, etc., and the locations which support its work become much more important than they were at earlier stages. (Mullins 1972, p. 74)

He also presented a model of scholarly knowledge development in four stages with distinct, empirically traceable characteristics. The four stages are considered as non-exclusive but sequential patterns that scientific communities exhibit. The importance (and success) of Mullins' model lies in the fact that it combines social and cognitive aspects, including group formation and change. It regards 'coherent groups' as the primary drivers of scientific change (Griffith and Mullins 1972), although neglecting the overall complexity of group dynamics. Building on Mullins' model provides my own analysis with an option to address temporal stratification constituted by various collective mechanisms as well as with the potential to recognise the role of individual researchers and relating groups of researchers. The temporality introduced by Mullins' model is also constitutive of the 'new disciplinarity' put forward much later by Marcovich and Shinn.

To understand 'new disciplinarity' it is first necessary to trace the conception of disciplinarity as it developed in the 1980s and 1990s. For this short overview it should suffice to refer to the historical analysis put forward by Stichweh (1984, 2013). Informed by systems theory, Stichweh traces the emergence of (sub)-disciplinary structures within science, conceived of as a system in constant differentiation during the period his historical study focused on—namely, 1740 to 1890. It is also noteworthy that Stichweh initially focused on physics, which was perceived as a paradigmatic discipline at that time. In the same period, disciplinarity began to be seen as potentially problematic, and debates about disciplinarity, interdisciplinarity, and transdisciplinarity ensued. Around the turn of the last century, a growing number of scholars began to challenge disciplinarity as a valid analytical concept, devising new terms such as 'inter-disciplines' (Weber 2010). As a result, attention seemed to shift away from disciplines back to specialties, research schools, networks, and prominent individuals.

In 2011, Marcovich and Shinn expanded on these discussions by introducing the notion of 'new disciplinarity' as a product of the growing complexity of knowledge and scientific activity. This notion reinstates disciplinary entities as valid points of reference[3] but also highlights the changing prominence of combinatorials,[4] projects,[5] borderlands,[6] displacement,[7] and temporality[8] (Marcovich and Shinn 2011, 2014). The authors first look at the trajectory of Max Delbrück and the phage group as well as the journey of other scientists who, like Delbrück, came from quantum mechanics. They develop the idea that molecular biology was born from a problem-driven combination of multiple projects with a disciplinary grounding, ultimately converging in a concatenation. In this 'new disciplinarity', disciplines remain a practitioner's point of reference, even when he/she 'occupies a "borderland" constituted by research projects that engage alternative specialties' (Marcovich and Shinn 2011, p. 584). The border is understood as a process, a place of convergence, exchange, and creation, not as a simple dividing line (Newman 2006). Marcovich and Shinn (2014) recently further refined their concept. Based on interviews of 47 scientists engaged in nanoscale research in multiple areas of physics and life sciences, they defined four typical trajectories of nanoscale practitioners: (1) beckoning from indoors; (2) extending the repertory; (3) common territory for answering homeland questions; and (4) successive projects, which intertwine the six elements explained above.

2.1.3 *Approach, Methods and Structure of this Chapter*

In the ensuing case study of the emergence of supramolecular chemistry (SMC), I focus on a group of chemists (3 Nobel laureates and their collaborators) that I call a 'paradigm group', resonating with an earlier conception of 'schools of research' and

---

[3]'Disciplines constitute the stable referent of practitioners, providing the intellectual and material resources, language and universe of questions.' (Marcovich and Shinn 2014, p. 172).

[4]'Combinatorials are the association of elements (instruments, materials, concepts and people) whose novel integration often constitutes powerful resources for extending problem solving.' (Marcovich and Shinn 2014, p. 174).

[5]'Projects are to be understood as a crystallisation of questions that are asked in the framework of a discipline but which require resources belonging to other disciplines in order to solve a problem.' (Marcovich and Shinn 2014, p. 175).

[6]The borderland is conceived of as 'an indefinite, fuzzy, narrow swath of terrain contiguous with the well-defined territory of two recognized entities, in our case scientific disciplines.' (Marcovich and Shinn 2014, p. 176).

[7]'Displacement refers to a selective intermittent movement of a scientist into the borderland of his discipline and a subsequent retreat from that terrain.' (Marcovich and Shinn 2011, p. 587).

[8]Temporality relates disciplinary landscapes (with a long-term temporality) to temporal projects (with a short-term temporality) by tracing processes of practitioner displacement. (Marcovich and Shinn 2011, p. 598).

Mullins' notion of 'paradigm groups'.[9] I make use of scientometric tools and trace the co-authorship patterns of a central actor to develop an analytical and sequential model in three periods, building on Mullin's phase model. I also trace the development of SMC as an example of 'new disciplinarity' in action, starting from the trajectory of a prolific and reflexive actor (Jean-Marie Lehn) whose career is marked by a series of shifts or displacements, potentially in line with Marcovich and Shinn's typology. I have particular interest in trajectory (1) (defined as 'traditional' and therefore opposed to a 'new disciplinarity') which for Marcovich and Shinn was originally exemplified by J. Fraser Stoddart, a major actor in SMC, whose research they described as 'remarkably self-referential' as 'he deal[t] endlessly with the same objects and in the same perspective' (Marcovich and Shinn 2014, p. 180). I then discuss whether there are differences in Stoddart and Lehn's respective trajectories and strategies and whether SMC can be positioned within the general framework of a 'new disciplinarity'.

The empirical materials I draw upon for this chapter are the following: for the historical part, I rely on the substantial material underpinning growth of the SMC— review articles,[10] textbooks, publications—including two publications specifically dedicated to the emergence of SMC (Gale 2000; Vicens and Vicens 2011)—press releases, and annual reports. I paid special attention to the Nobel lectures for their biographical richness. I also consulted archives stored at the University of Strasbourg documenting the whole career of Guy Ourisson.[11] For the most recent period, I draw on fieldwork carried out in Strasbourg as part of a wider study (the ANR project PrestEnce), which is supplemented by interviews in Paris and Bordeaux. Twenty-eight semi-structured interviews were collected between April 2010 and November 2013 (25 professors,[12] researchers, or staff belonging to the University and three external interviewees). They were tape-recorded then fully transcribed and

---

[9] 'A paradigm group is the minimal form of a scientific group. Its members have no necessary social connections. [...] The minimal requirement of such an entity is two or more established scientists who have shifted from one viewpoint to another (*Gestalt* shift), and who might or might not be in communication with one another.' (Mullins 1972, 54-55). The paradigm group equates its first step in the proposed four-stage model.

[10] There are more than 3600 review articles indexed in the Web of Science that match the query condition 'Topic = Supramolecular chemistry'.

[11] Guy Ourisson (1926–2006) was a French chemist, a specialist in the chemistry of natural substances. He earned two doctorates, the first at Harvard University in 1952 and the second in Paris in 1954. In 1955, at the age of 29 years, he was appointed Professor at the University of Strasbourg, to which he remained faithful throughout his scientific career. In 1971, he was one of the first founders of the Louis Pasteur University in Strasbourg of which he was also the first president. He had numerous French and international awards and other distinctions. His specialisation was organic chemistry and he worked at the interface of organic chemistry and biology and later also at the interface of organic chemistry and geology (Rohmer 2006). Guy Ourisson has supervised many PhD theses, including that of Jean-Marie Lehn. The fonds Guy Ourisson entered the archives department of the Université of Strasbourg in 2011. References: FR Service Archives de l'Université de Strasbourg/Guy Ourisson.

[12] For various reasons, I didn't interview the main actor (Jean-Marie Lehn).

anonymised and lasted a range of 45 to 90 minutes. Some relevant excerpts of the transcripts are indicated by quotation marks.

The remainder of this chapter is structured as follows: I first present the concepts characteristic of the scientific programme of SMC and their genealogy. Based on Lehn's co-publications profile, I then draw on Mullins' four-step model to propose a three-period chronology. For each of the periods I explore modes of displacement and temporality that are constitutive of Marcovich and Shinn's 'new disciplinarity' and show these modes have a prominent collective dimension. Third, I look at this field's legacy through the organisational study of the Strasbourg chemistry department[13] as it is currently configured. Fourth, I explore the role of architecture and geography. I focus on the ISIS building to illustrate its particular functions: a space that was originally designed to break down disciplinary boundaries and a 'transient' institution where promising young researchers could work before pursuing their career in other research centres. In doing so, I hope to reveal specific patterns of emergence and institutionalisation and show how this results in a new mode or conception of being together.

2.2 SMC, a Speciality 'At the Borders Of'

2.2.1 Concepts Characteristic of the Scientific Programme of SMC

What is SMC? The Larousse dictionary actually provides the following definition: it is a sub-discipline of chemistry which studies weak, non-covalent interactions between molecules. SMC is concerned with the complex entities formed by the association of two or several chemical species linked together by intermolecular forces, whereas molecular chemistry studies the properties of the entities made of atoms linked by covalent forces.

Jean-Marie Lehn was the first to lay its foundations and formalise its concepts in a seminal article published in 1978 (Lehn 1978). This work (especially the synthesis of cryptands performed in his laboratory ten years earlier) earned him the 1987 Nobel Prize for Chemistry, which he shared with Charles J. Pedersen (DuPont) and Donald J. Cram (University of California, Los Angeles) 'for their development and use of molecules with structure-specific interactions of high selectivity'.

SMC is a term coined by Lehn as the '"chemistry beyond the molecule"', bearing to the organised entities of higher complexities that result from the association of two

---

[13]The PrestEnce project endorses the hypothesis that universities and their components (the academic departments) should be considered as potential meso level order and action levels (Paradeise and Thoenig 2013). In France, academic departments in the strict sense of the word do not exist. In this case study, the chemistry department comprises faculty members and research fellows at the CNRS, about 250 permanents researchers grouped in the 'International Center for Frontier Research in Chemistry' created in 2007.

Table 2.1 Number of occurrences of 'supramolecular' in the 1987 Nobel lectures of the three laureates

Laureate	Charles J. Pedersen	Jean-Marie Lehn	Donald J. Cram
Title of the Nobel lecture	Discovery of crown ethers	Supramolecular chemistry – scope and perspectives: Molecules – Supermolecules – molecular devices	The design of molecular hosts, guests and their complexes
Number and types of references	36 (auto-citations, mainly patents)	193 (academic)	54 (academic)
Number of occurrences of 'supramolecular'	0	71	0

or more chemical species held together by intermolecular forces' (Lehn 1987). In an editorial published on the occasion of the 50th anniversary of Pedersen's discovery, Lehn stated that he introduced the term SMC, 'which is now widely accepted and has deeply permeated chemical literature, in order to define, consolidate and generalise the areas of crown ether chemistry, host–guest chemistry and the chemistry of molecular recognition, thus allowing for the emergence of the concepts and perspectives offered' (Lehn 2017). As I will emphasise later, the vocabulary of the SMC stabilised quite late, while a small number of authors like Siegel (1996) or Dance (2003) still questioned its novelty and the usefulness of its concepts in the late 1990s.

Prominent aspects of Lehn's work are its intense concern with chemical semiotics and its wide scope. As show in Table 2.1, this translates into a massive presence of the term 'supramolecular' and a large number of references in Lehn's Nobel lecture compared to the other laureates.

In the late 1960s, a whole range of macro(poly)cyclic composites with spherical cavities were synthesised, and cryptands are now found in chemical product catalogues. After having explored the particular properties of these three-dimensional assemblies in catalysis and molecular recognition, for the transport of ions or molecules, Lehn and colleagues extended their research to the study of 'supramolecular' (a term introduced by Lehn in 1978) entities formed by self-organisation processes using molecular recognition to control and direct the spontaneous formation of complex architectures.

Supramolecular structures are a result of various noncovalent interactions, including electrostatic interactions, hydrogen bonding, van der Waals forces, hydrophobic interaction, coordination, etc., some of which are often cooperating in one supramolecular complex. More importantly, properties of the formed supramolecular complexes are far beyond summation of the individual components. The early 1990s saw the introduction of the notions of adaptation and evolution into chemistry, the extension of self-organisation processes to the selection of the species contributing to it, and the implementation of 'informed' chemical and dynamic diversity. Relying on the dynamic aspects of SMC (named 'by essence' a dynamic chemistry), Lehn proposed the definition of a general concept, covering molecular as well as supra-molecular chemistry—that of Constitutional Dynamic Chemistry (CDC) which introduced a paradigm shift with respect to what he called constitutionally static chemistry (Lehn 2007).

Where does the specialty stand today? Around 20 to 30 per cent of publications in leading chemistry journals such as *Angewandte Chemie, Chemical Communications, Chemistry: A European Journal,* and *The Journal of the American Chemical Society* are concerned with the practical achievements arising from concepts and visions that have been developed in this domain[14] (Diederich 2007). There are currently two facets to the specialty: one is oriented mainly towards synthesis, attempting to design constructions with given molecular bricks, while the other seeks to explain, understand, and partly control existing self-assemblages. The rise of SMC has also been accompanied by a development of new concepts and methods able to monitor the sometimes quite fast dynamics of supramolecular systems. With this, SMC has become fruitful also for other areas in chemistry.

2.2.2   A Three-Period Chronology

As with other specialities (see McCray (2005) on the existence of 'creation stories' that reconstruct the development of a particular idea or invention back to a singularity), SMC comes with narratives about its leaders and 'founding fathers'. In their textbook 'Supramolecular Chemistry', Steed and Atwood (2009) trace its roots to the early days of modern chemistry. In this study, I chose to start in the year of Jean-Marie Lehn's first publications (1961) and cover a period of 50 years.

Jean-Marie Lehn is a prolific author (nearly 900 publications between 1961 and 2011), whose articles have been frequently and progressively cited to this day.[15] Using co-authorship as a proxy for collegiality (Laudel 2002; Larivière et al. 2016), his co-publications profile is reproduced with the 'Demography' module of the CorText Manager tool. The module analyses a .txt file built from a Web of Science request (set of 897 publications released between 1961 and 2011 (Noël 2019), 488 co-authors, number of co-authored publications ≥ 2).[16] Lehn's co-publications profile (Fig. 2.1) identifies a set of co-authors, with whom he has primarily published

---

[14]This represents a considerable number of publications since a journal like the JACS publishes 16,000 articles per year.

[15]At the time of writing, his prospective article entitled 'Perspectives in SMC' published in 1990 in *Angewandte Chemie* has been cited 2767 times.

[16]The 'Demography' module of the CorText Manager tool developed by the IFRIS digital platform (http://www.cortext.net/projects/cortext-manager.html) treats each field of the corpus (authors, journals, terms, etc.) and follows the occurrence of the main elements, which vary in number (20 or 50) depending on the nature of the field. The results are presented as stacked area charts; moving averages are used with time series data to smooth out short-term fluctuations and highlight longer term trends.

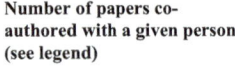

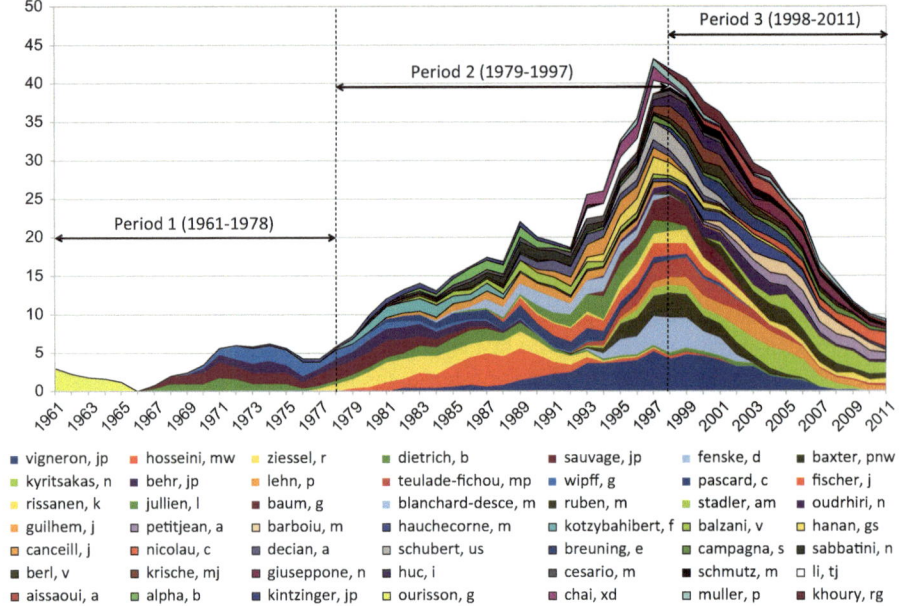

Fig. 2.1 Jean-Marie Lehn's co-publications profile (49 co-authors identified)

(49 names are extracted from a list of 488 co-authors). Their names and profiles are shown in different colours; for instance, on the left hand side of the graph one can see Ourisson's (Lehn's supervisor) contribution in yellow.

This profile features some interesting elements: first, there is an increase in the pace of co-authored publications at the turn of the 1990s (the slope, which seems quite stable since 1966, suddenly increases). This is probably related to the 'Nobel effect' but is also emblematic of the growing importance of teamwork and the increasing division of labour in contemporary chemistry (Cronin et al. 2004). Second, co-authors diversify in the most recent period: they are more numerous and different from those of the previous ones.

Based on slope variations of the profile, I have identified three periods of almost equal duration.

2.2.2.1 Period 1: The Emergence of the Supramolecular Chemistry Paradigm (1961–1978)

The 1960s, when Lehn entered the University, were a time of rapid growth of synthetic organic chemistry. After a thesis on triterpene NMR obtained under Ourisson's supervision, Jean-Marie Lehn left for a year to do a post-doc at Harvard, where he took part in the total synthesis of vitamin B12 (the most complex natural product synthesised to date), finally completed in 1971. Like so many other chemists of this period, Lehn was actively engaged in research related to organic synthesis. He participated in this landmark work in the history of organic chemistry that represented in his eyes 'the modern apogee' (Lehn 1980) since it showed that organic chemists were able to create very complex molecules by forming covalent bonds (Woodward 1973).

Upon his return from Harvard in 1964, Jean-Marie Lehn went to Strasbourg and entered the CNRS to create his own laboratory. He had become a recognised specialist in the use of NMR for understanding the physical properties of organic molecules. Searching for themes different to those he had studied with Ourisson and Woodward, he decided to direct his research towards physical chemistry. NMR studies of the movements of a liquid's molecules ensued. In parallel, he developed an entirely theoretical theme in his group: *ab initio* calculations of structural and conformational properties (Lehn in Kleinpeter and Eastes 2008).

In 1966, Lehn's interest turned to 'how a chemist might contribute to the study of ... highest biological functions' (Lehn 1987, p. 448). In this context, the first synthesis of crown ethers performed by Pedersen, an engineer for DuPont in 1967, supported his idea that it was possible for a cage molecule (which has a cavity) to capture another molecule with a complementary form. This led to the development of the chemistry of cryptands and cryptates in his laboratory.

The first cryptand was synthesised in September 1968 by two researchers doing their PhDs—Jean-Pierre Sauvage and Bernard Dietrich. It had been theoretically designed by Lehn by taking into account the little information available at that time on the complexation and transport of alkali metal ions by natural ionophores. It was a cage molecule, capable of selectively fixing a chemical substance (of the appropriate shape and size, in this case a potassium ion) in its cavity, called a crypt. The cryptand (from the Greek word *kryptos* meaning hidden) was a new molecular object, with a bond between the potassium ion and the crypt which had nothing to do with the covalent bond. Even though the first cryptand was synthesised 'by chance' (as mentioned in the closing lecture on 4 June 2010[17] at the *Collège de France*), this situation was not artificial. The first direct proof of the cryptands' structure was provided by Raymond Weiss, professor at the University of Strasbourg since 1957, using X-ray diffraction techniques.[18] In the following years the power of the concept of 'molecular recognition' and its generality were demonstrated.

Although Lehn made an effort to clarify terms in his 1978 publication, his definition of SMC remained unclear. Mullins (1972) sees this as a condition of emergence, the paradigm being an object to be adjusted and specified under new or stricter conditions.

For his part, Donald J. Cram at UCLA recognised that the work of C.J. Pedersen provided [him/them] 'an entree into a general field' (Cram 1987, p. 419). In his Nobel Lecture (1987, p. 419), he said: 'Although we tried to interest graduate students in synthesizing chiral crown ethers from 1968 on, the efforts were

---

[17]http://www.college-de-france.fr/site/jean-marie-lehn/closing-lecture-2010-06-04.htm. Accessed 24 February 2019.

[18]It was a long-term task because at that time it took from six months to one year to determine the structure of a small molecule.

unsuccessful. In 1970 we insisted that several postdoctoral co-workers enter the field'. By 1974, he published (with his wife, Dr. Jane M. Cram) a first general article entitled 'Host-Guest Chemistry' summarising their thoughts, methods, and results (Cram and Cram 1974). Along their definition, hosts are synthetic counterparts of the receptor sites of biological chemistry and guests the counterparts of substrates, inhibitors, or cofactors. He and his colleagues designed and prepared more than 1000 hosts, each with unique chemical and physical properties (Cram and Cram 1994). Cram underscored the importance of visualisation and the use of Cory-Pauling-Koltun (CPK) molecular models that provide better insights than the usual graphs, tables, and figures.

2.2.2.2 Period 2: The Emergence of a Specialty: From Supramolecular Chemistry to a Science of Supramolecular Systems (1979–1997)

The second period (1979–1997) consisted of two phases: the network phase followed by the cluster phase. In the network phase, certain participants recruited students. In groups of two or three, the scientists began to homogenise their vocabulary, to build fragments of a paradigm: the term 'molecular machine' was fully discussed for the first time in an article by V. Balzani,[19] J. Fraser Stoddart, and collaborators in 1993. They also built new resources which contributed to the dynamic of their relationships: textbooks and curricula, originally written in mother tongue, became key components of the educational agenda.[20] A host of collaborations developed and gave rise to co-publications. In 1987 Vincenzo Balzani gathered contributions (including those of Lehn and Pedersen) collected in a workshop on Capri Island in a textbook entitled 'Supramolecular Photochemistry'. Beyond Strasbourg, Lehn's network extended to colleagues at Collège de France (notably J.-P. Vigneron) where he was appointed in 1979, a position that gave him prestige and time to do research. The radial structure around Lehn expanded to form crosscutting branches between his former students and foreign colleagues—for instance M.W. Hosseini and R. Ziessel co-authored publications with V. Balzani. For their part, J.-P. Sauvage and B. Dietrich entered the CNRS in 1971 and 1974,

---

[19]Vincenzo Balzani is an Italian chemist, now emeritus professor at the University of Bologna. In 1984 he introduced the concepts of supramolecular chemistry in the field of coordination compounds, showing the possibility of controlling the photochemical and photophysical properties *via* an intelligent molecular architecture. Luminescent lanthanide ions complexes were the object of an intense collaboration with Jean-Marie Lehn; in particular Eu3+ and Tb3+ cryptates were considered examples of simple antenna systems. Together with Jean-Marie Lehn he defined supramolecular photochemistry as the incorporation of photoactive units in supramolecular structures, known as receptor/substrate complexes. There have been controversies in Italy considering the absence of Balzani, a pioneer in molecular machines, in the list of Nobel Prize winners in 2016.

[20]*Aspects de la chimie des composés macrocycliques* by B. Dietrich, P. Viout, and J.-M. Lehn, 1991; *Supramolekular Chemie* by F. Vögtle, 1989.

respectively. The Strasbourg group has been able to stabilise, to develop, and recruit new members. Sauvage has developed his own line of research from 1983 onwards.

Mullins stresses the importance of the Cold Spring Harbor summer school as the 'melting pot' site. Part of the group (Atwood and Stoddart) considered the founding conference to have been the one organised in Jachranka, close to Warsaw in Poland. Although the 50 participants were overwhelmingly European, neither Jean-Marie Lehn nor his group was present. Jerry Atwood commented:

> There was a key meeting in Poland in 1980 and that meeting formed the basis for the development of the field of supramolecular chemistry. It was one of those meetings where I knew no one when I went there; all of us came from different areas and in that week I made many of my best friends.[21]

In the cluster phase, decisive changes took place. There was a shift from network to a collective entity recognised and constituted as such by its members. The community produced a shared directory of resources: French and German textbooks were translated into English. This was the time of the creation of journals specifically dedicated to SMC (featuring programmatic editorials), some of them short-lived: the *Journal of Inclusion Phenomena*[22] in 1983, *Supramolecular Chemistry* in 1992. Atwood organised a board of editors that ultimately produced an eleven-volume set (197 chapters, 6672 pages) which covered the gamut of this rapidly expanding area of research (*Comprehensive Supramolecular Chemistry*, Pergamon Oxford, 1996). For his part, Jean-Marie Lehn published *Supramolecular Chemistry: Concepts and Perspectives* in 1995.[23]

Initiatives abounded, as the list of symposia on (and surrounding) SMC attests (Table 2.2). The topics covered include liposomes as well as nanogels or surface science. On that list we also find the first international summer school (a place where students were trained) organised in Strasbourg in 1990. The regular training of students presupposed the existence of job opportunities.

The group began to equip itself with more formal means of communication, thereby becoming a specialty. The resources created became increasingly collective and consubstantial with the group's existence: volume 11 of the manual mentioned above (*Comprehensive Supramolecular Chemistry*) is a cumulative subject index, further aiding the location of specific pieces of information. Between 1981 and 2000, Lehn himself devoted time to extended stays at European universities (as visiting professor in Cambridge, Barcelona, Frankfurt, Karlsruhe, Oxford, etc.) and at

---

[21]ChemComm, An Interview with Professor Jerry Atwood, 2012, https://web.archive.org/web/20121008094334/http://www.rsc.org/Publishing/Journals/cc/News/AtwoodInterview.asp. Accessed February 24, 2019.

[22]Founded by J. Atwood in 1983, it became the *Journal of Inclusion Phenomena and Macrocyclic Chemistry* in 1999.

[23]The book was based on the Baker Lectures at Cornell University in 1978 and the Lezioni Lincee at Accademia Nazionale dei Lincei (Roma) in 1992. Today, it has been cited more than 12,000 times.

Table 2.2 List of symposia on and surrounding SMC (1988–1997), extracted from the Memorandum of Understanding for the implementation of a European Concerted Research Action designated as COST Action D11[a]

Date	Conference	Location
September 1988	13th International Symposium on Macrocyclic Chemistry	Hamburg, Germany
September 1989	Meeting on Organised Molecular Systems	Parma, Italy
September 1990	First International Summer School on Supramolecular Chemistry	Strasbourg, France
July 1991	Supramolecular Chemistry, Towards Self-Organisation	Le Bischenberg, Obernai, France
1991	EUCHEM Conference on Supramolecular Reactivity and Catalysis	Padua, Italy
May 1992	Micelles and Liposomes Research in Switzerland and Europe	Zürich, Switzerland
June 1994	Supramolecular Structures and Self-replication	Maratea, Italy
August 1994	Research Conference on Supramolecular Chemistry	Mainz, Germany
September 1995	Liposomes: State of the Art	Freiburg, Germany
March 1996	Nanogels and Sol-gel Processing: New Approaches Towards better materials	Zürich, Switzerland
June 1996	11th International Symposium on Surfactant in Solution	Jerusalem, Israel
September 1996	Convegno Nazionale GICI, scienza e tecnologia dei sistemi organizzati	Bari, Italy
July 1997	9th International Conference on Surface and Colloid Science	Sofia, Bulgaria

[a]COST Action D11 Supramolecular chemistry (1999–2003) https://www.cost.eu/actions/D11/#tabs|Name:overview.Accessed|https://www.cost.eu/actions/D11/#tabs|Name:overview.Accessed 24 February 2019

Harvard. His co-authorship pattern is much more diverse (particularly in the cluster phase as shown on Fig. 2.1) than it was before 1978.

Around 1996 the specialty stabilised. With the support of chemical learned societies, Jean-Marie Lehn created *Chemistry: A European Journal*, which became the main forum for SMC advances. *Chemistry* also stands out due to its 'concepts' section hosted since September 1996. One of the main aims of *Chemistry* was to highlight and support the outstanding research produced by groups across Europe.

As part of the European programme COST-chemistry, facilitated by a chemist from Toulouse, the D11 action[24] was launched in 1998, with funding that boosted the Strasbourg teams.

---

[24]Launched in 1971, COST is a pan-European intergovernmental initiative aimed at strengthening scientific and technical research in Europe by supporting cooperation between European researchers. COST is a networking instrument (it doesn't fund research itself). In 1990, a chemist from Toulouse, Gilbert Balavoine, was entrusted with chairing the COST-chemistry programme. COST Chemistry Actions were launched, numbered D1, D2, ... which try to cover as best as possible the scientific field of chemistry. D11 action gathered 60 research groups from some

2.2.2.3   Period 3: Institutionalisation Processes (1998–2011)— Materialising Concepts

The third period was the post-Nobel Prize and institutionalisation period. Scientists moved from a project with an initial research question to another. Jean-Marie Lehn worked on many topics that shared one core research question: 'what does SMC mean?'. He endeavoured to materialise the concept of SMC in its successive senses and devoted part of his efforts to ensuring the long-term survival of the bodies created: the Supramolecular Science and Engineering Institute and the thematic network of advanced research (RTRA).[25] He was involved in many and diverse collaborations and travelled extensively.

In December 2011, ISIS counted nine laboratories and four industrial branches. This original structure is constituted of senior laboratories, headed by recognised and internationally renowned scientists, and junior laboratories, where researchers beginning their career develop independent research as part of a project that is not scheduled to last more than six years. ISIS stands out in the French research ecosystem: the recruitment of foreign professors on the international market concurs with the strong local roots of this organisation.

In 1987, Jean-Marie Lehn defined SMC as the 'chemistry beyond molecules'. It is a polysemic object, both a concept and a specialty, even a discipline, with the main characteristic of being situated 'at the borders of'. That is in fact the name (in French) of the RTRA (the Centre International de Recherche aux Frontières de la Chimie which has been translated as International Centre for Frontier Research in Chemistry).

2.2.3   An Original Conceptual and Organisational 'Heritage'

Which social processes encouraged the emergence of this specialty in Strasbourg? Based on my fieldwork and historical documents, I list a few of them, looking for patterns of a general scope.

---

20 European countries and was endowed around 40 million euros over 5 years. Balavoine's correspondence with Ourisson attests to the links maintained between the two researchers about this programme. (FR Service des archives de l'Université de Strasbourg; fonds Guy Ourisson; GO454; Programmes de coopération scientifique).

[25]The RTRA CIRFC (*Centre International de Recherche aux Frontières de la Chimie*) was created in 2006. The associated scientific research foundation FRC (*Fondation pour la Recherche en Chimie*) was established in 2007 with the University of Strasbourg, the CNRS, BASF-France, and Bruker-Biospin as funders and with the dotation of the Ministry of Higher Education and Research. The administration of the RTRA and of the foundation is located in the ISIS building, where the laboratory of its director is also localised.

2.2.3.1   A Polymorphic, Though Coherent and Organised Community, Attentive to Its Position

From both inside and outside, chemistry is perceived as a coherent and organised entity: the multiple stories researchers shared collectively (from the awarding of the Nobel Prize to the national competition for RTRA) have nurtured the disciplinary framework. As Lehn mentioned in his Nobel Lecture, chemists in Strasbourg, as a whole, contributed to the Prize:

> I wish to thank very warmly my collaborators at the Université Louis Pasteur in Strasbourg and at the College de France in Paris whose skill, dedication and enthusiasm, allowed the work described here to be realized. Starting with B. Dietrich et J.-P. Sauvage, they are too numerous to be named here, but they all have contributed to the common goal. (Lehn 1987, p. 484).

The community as a whole derived benefits from these successes. Today, the chemistry department (*Faculté de Chimie*) offers a wide range of specialties. As expressed by a high-level person with policy-making responsibilities at CNRS, 'it's a mosaic'.[26] But for an executive head of the university,

> compared to other disciplines, chemistry is easy to position: there are industrial partners, it has a good image within society … a strong symbolic value with Jean-Marie Lehn. This introduces strong subjective elements, other than the quality of the person, the quality of the science which is done. It's simple with respect to the outside world. In communication efforts, when I say 'Jean-Marie Lehn Nobel Prize', it's very easy![27]

This symbolic dimension has been undoubtedly strengthened by awarding the Nobel Prize to other chemists from (or affiliated with) Strasbourg whose work is more or less connected to that of Lehn and SMC.[28] The quotation above shows that, despite the epistemic diversity of the department (a mosaic), the disciplinary framework created and creates a sense of unity. A sense of allegiance to chemistry, a characteristic of the discipline emphasised by Lenoir (1997, pp. 48–49), was coupled earlier in Lehn's discourse with attention to its centrality: 'Chemistry plays a *central role* both in the natural sciences and in knowledge, and in its economic importance and omnipresence in our daily lives' (Lehn 1980, emphasis by the author). This positioning practice of chemists, referring to chemistry as the central science, persists until today (Bertozzi 2015). It is not specific to Strasbourg but has undoubtedly been taken up in the emergence of SMC.

---

[26]Interview 2, 25/05/2012.

[27]Interview 3, 22/11/2011.

[28]I refer here to Jules Hoffmann (2011), Martin Karplus (2013), and Jean-Pierre Sauvage (2016). In his Nobel lecture, Sauvage explicitly thanked '[his] mentor and friend Jean-Marie Lehn, the teachers who had a strong influence on [his] scientific interests, Guy Ourisson and Raymond Weiss'.

2.2.3.2   A Strong Local Rootedness with Great International Openness

The University's international openness can be rooted in its history (Craig 1984; Olivier-Utard 2010; Crawford and Olff-Nathan 2005). Following the annexation of Alsace-Moselle by Prussia, authorities encouraged the recruitment of foreign teachers from 1871 onwards. This was also the case after World War I, when Alsace came back under the auspices of the French State. A professor explained:

> For more than 150 years, the University appeared to be sometimes French, sometimes German, sometimes French.... In its relation to Germany, France considered the University as a 'showcase' and the other way round. There are always been a recruitment of quality people. Since these early periods, people have been carefully selected to import researchers coming from abroad who settled here in Strasbourg.[29]

The aim was to build an exceptional setting, which remained extraneous to the local people and required a number of years to fully appropriate. As many testimonies collected during the fieldwork attested, there is still a large incongruity between foreign and locally-trained professors (the latter constituting a large part of interviewees in our sample). For instance, the need for teaching in French at the undergraduate level creates a clear differentiation in terms of access to master's programmes, which has been described as problematic by interviewees.

In addition to these features came the development of institutional structures (first the ISIS, then the RTRA) supported by stakeholders, especially in the Alsace Region.[30] The Strasbourg members of the 'paradigm group' (almost all are members of the French *Académie des Sciences*) embedded knowledge within a political dimension, which gave it the power of both a social and a symbolic link (Jacob 2007). The Alsatian local rootedness translates into the capacity to act in and at the service of the region as well as at the local, national, and international level. The career of Guy Ourisson, the 'ultimate godfather of Strasbourg chemistry',[31] is emblematic thereof.

2.2.4   The ISIS Building as a Mediator Between Epistemic Practices and Politics

In the late 1990s Lehn created the Supramolecular Science and Engineering Institute (*Institut de Science et d'Ingénierie Supramoléculaire*, ISIS), which was hosted in a new building from 2002 onwards. This institute was specifically designed for the

---

[29]Interview 4, 27/04/2010.

[30]Created in 1991, the *Pôle universitaire européen de Strasbourg* (European university centre in Strasbourg) symbolised the rapprochement of the three existing universities by associating them with the three local authorities (Alsace Region, Bas-Rhin department, Communauté Urbaine de Strasbourg).

[31]Interview 3, 22/11/2011.

Fig. 2.2 The ISIS building at the University of Strasbourg

further development of his research, based on a vision he himself had defined: that of a scientific project incubator, where 'the most brilliant young researchers in the supramolecular sciences will be able to express themselves freely, before pursuing their career in other research centres'.[32] This new institute signified a major achievement in his career and was thought of as an organisational innovation in the French research scene. It was supported by the Louis Pasteur University in partnership with the CNRS and funded by the local government.

The entrance hall of the five-story building (Fig. 2.2) bears a plate with the engraving 'ISIS is the project of Jean-Marie Lehn, Nobel Prize for Chemistry, supported by the Louis Pasteur University in partnership with the CNRS'.

Such personalisation is quite rare in the world of French research. In a report to the Alsace Region in 1990,[33] Guy Ourisson noted that

> chemistry research in Alsace presents very particular characteristics, which need to be properly identified in order to understand that the level of this research is not a result of

---

[32]Declaration by Claudie Haigneré, French Minister for Research and New Technologies, inauguration of the ISIS, 9 December 2002, http://discours.vie-publique.fr/notices/033000507.html. Accessed 1 March 2019.

[33]FR Service Archives de l'Université de Strasbourg /Guy Ourisson/GO 480/Livre Blanc de la recherche et de la technologie en Alsace, p. 30 (1990).

Jean-Marie Lehn receiving the Nobel Prize, but the reason why this Nobel Prize was possible, without him having to leave the country!

The ISIS building was designed by the French architect Claude Vasconi (1940–2009). Vasconi was born in the same city as Lehn and was of the same generation. He graduated from the Ecole Nationale Supérieure des Arts et de l'Industrie in Strasbourg. ISIS' architecture was jointly designed with two purposes: first, priority was given to the horizontal circulation in a volume organised around a central atrium in order to meet a strong demand for decompartmentalising disciplines. As for a whole generation of science architecture in the 2000s (Yaneva 2010), the atrium became an important interactive space (a connecting mechanism, ibid., p. 143). The building invited the creation of new types of associations among researchers from different (sub-)disciplines or specialties and among public research and industry. A senior professor summarised it in this form:

> A specialist in a discipline, a related area [to SMC], a platform. Four platforms and a fifth for common machines, this was [Jean-Marie Lehn's] concept.[34]

Second, ISIS was also conceived of as a 'transient' structure, where independent young researchers could work before they gained a permanent position in the academic system. The concept was that of a space where 'mobile professors' work/compete with each other. This particular setting refers to a phase in the researchers' careers rather than to the substance of their work (Normark 2015).

Based on Strasbourg's experience, public research organisations (e.g. CNRS, INSERM) and universities encouraged the creation of such 'research hotels' (*hôtels à projets*) in France. In Bordeaux, the European Institute of Chemistry and Biology (Institut Européen de Chimie et de Biologie, IECB) was built with a specific reference to Lehn's project. The local authorities provided support, as had happened in Alsace, and the IECB opened in 2003. The 'research hotels' model has since experienced ups and downs: the issue of incongruity between 'local' and foreign researchers that benefit from the provided resources to different degrees was also raised by an interviewee in Bordeaux.

2.3   Concluding Remarks

This chapter proposes a historical chronology in three periods, which corresponds to the way Jean-Marie Lehn describes his work. Starting from fieldwork and the prominence of an academic scholar in *a specific place,* I tried to de-centre the historical narrative from the heroic 'he' to the collaborative 'they' by looking at patterns of collaboration and co-authorship of *an individual* (Nye 2014). This

---

[34]Interview 5, 02/03/11.

enabled me to understand the *collective* development of this specialty.[35] The chapter highlights a process of collective realisation through emphasis on a 'great man' and enrolment mechanisms (Callon and Law 1982). These achievements also came with the constitution of a mythology, which is vital to legitimise a disciplinary community (Nye 1993). I was able to measure the weight of the figurehead and the significance of the Nobel Prize as a symbolic resource: 80 per cent of interviewees spontaneously cited his name (or rather his first name, Jean-Marie) in the interviews. The analysis reveals how the local Alsatian culture also shaped the development of SMC. The influence of local conditions (and the special place of Strasbourg) is tangible in this historical reconstruction of emergence of SMC in France. The way scientists collectively used the institutional context at local, national, and international levels was crucial in the dynamics of emergence and institutionalisation.

The concepts and language of SMC can be to a large extent attributed to the scientific creativity of Lehn (Bowman-James et al. 2012, p. 1). His great innovation was to argue that the synthesis of molecules was primarily guided by new functions and essentially by the information contained in the molecules. Away from a growth model based on demarcation and constant differentiation, SMC found its place thanks to its conceptual developments and encouragement towards openness rather than confinement. This study borrows a lot from Mullins' work. The phasing resulting from the combination of scientometrics and the Mullins' model made the temporal processes visible, which were far from static: on the individual level, I've insisted on Lehn's multiple displacements, whether conceptual, methodological, or physical (such as periods abroad or marked by implementation of a new building). As far as the Strasbourg core was concerned, projects were multiple, as attested by the list of publications and the countless number of supramolecular systems that were created: cryptands, helicates, multicompartmental nanocylinders, 'grid-type' entities, ordered polymetallic arrays, etc. There was a strong material dimension to this emergence: the synthesis of new compounds, the contribution of models, and instrumental techniques like NMR and X-Ray diffraction, which were developed and made available in Strasbourg,[36] led to a multitude of combinatorials. Temporalities were varied, from punctual to long-term disciplinary anchor. Using an approach privileging cognition (with a co-authorship model that emphasises the structure of communication), I argue that Lehn's career exemplifies the 'new disciplinarity', a form of disciplinary regime that is characterised by specific modes of temporality and displacement. Given the abundance of his activities, Lehn is difficult to situate in the typology discussed in Sect. 2.1.2 (I propose a mix of trajectories (2),

---

[35]There are limitations to this approach, which may not be extended to every case study: it is not always possible to identify a single person or group that originates a field; a Nobel Prize (that awards a maximum of three people) is not always awarded to fields that gain great prominence.

[36]The development of shared facilities has always been a priority of budgetary policies, as shown in the minutes of meetings (1966–1980) at the Institut de Chimie de Strasbourg. FR Service des archives de l'Université de Strasbourg/Guy Ourssion/GO167/Gestion des Relations au sein de l'Institut de Chimie (1966–1980). Today the RTRA continues to competitively allocate important resources to shared facilities.

(3), and (4)). The elements describing Stoddart's career path in the reference paper are too succinct to allow for a comparative analysis, suggesting a more detailed analysis (examining both epistemic and social emergence) is needed to refine the four patterns developed by Marcovich and Shinn in their 'new disciplinarity' model.

In this work, I have tried to make the circulation of concepts (but also of terms, methods, and visions) visible. These concepts were not fixed; *they travel between disciplines, between individual scholars, between historical periods, and between geographically dispersed academic communities. Between disciplines, their meaning, reach, and operational value differ*' (Bal 2002, p. 24). The emphasis on circulating entities makes it possible to reveal heterogeneous actor-networks in which contemporary technosciences are deployed, where concepts (a 'chemistry beyond molecules', the introduction of the notions of adaptation and evolution, etc.) or visions (that of mobile professors) can even materialise in buildings and institutions. The case study shows that disciplinary structures are very much alive, as evidenced by the allegiance of scientists to the discipline, which is coupled with the attention paid to its centrality. In light of the increasing complexity of scientific knowledge and activities, this investigation of SMC offers a perspective on how to live together as part of the 'new disciplinarity'.

Acknowledgments I wish to thank Pierre-Benoît Joly, Catherine Paradeise, Soraya Boudia, Morgan Meyer, and David Demortain, as well as the members of PrestEnce project, for their insightful comments and support. Earlier versions of this chapter have been discussed at conferences and in workshops and benefited from valuable comments and key inputs (in particular from Davide Donina, Susan Molyneux-Hodgson, Niki Vermeulen, and Karen Kastenhofer). I am also deeply grateful to the anonymous reviewers and to the interviewees.

This chapter is an output of the PrestEnce project funded by the Agence Nationale de la Recherche (ANR-09-SOC-011). This work has also received travelling support from the Société Chimique de France, Club d'Histoire de la Chimie, and from the IFRIS.

References

Bal, M. 2002. *Travelling concepts in the humanities: A rough guide.* Toronto: University of Toronto Press.

Bensaude-Vincent, B., and J. Simon. 2008. *Chemistry: The impure science.* London/Singapore: Hackensack/Imperial College Press. Distributed by World Scientific Pub.

Bensaude-Vincent, B., and I. Stengers. 2001. *Histoire de la chimie.* Paris: La Découverte.

Bertozzi, C.R. 2015. The centrality of chemistry. *ACS Central Science* 1 (1): 1–2.

Bertrand, E., and B. Bensaude-Vincent. 2011. Materials research in France: A short-lived national initiative (1982–1994). *Minerva* 49 (2): 191–214.

Bonneuil, C., and D. Pestre, eds. 2015. *Histoire des sciences et des savoirs. 3. Le siècle des technosciences (depuis 1914).* Paris: Éditions du Seuil.

Bowman-James, K., A. Bianchi, and E. García-España, eds. 2012. *Anion coordination chemistry.* Wiley-VCH-Verl: Weinheim.

Callon, M., and J. Law. 1982. On interests and their transformation: Enrolment and counterenrolment. *Social Studies of Science* 12 (4): 615–625.

Craig, J.E. 1984. *Scholarship and nation building: The universities of Strasbourg and Alsatian society, 1870–1939.* Chicago: University of Chicago Press.

Cram, D.J. 1987. *Nobel Lecture. Nobel Media AB 2019*. https://www.nobelprize.org/uploads/2018/06/cram-lecture.pdf. Accessed 24 Feb 2019.

Cram, D.J., and J.M. Cram. 1974. Host-guest chemistry: Complexes between organic compounds simulate the substrate selectivity of enzymes. *Science* 183 (4127): 803–809.

Cram, Donald J., and J.M. Cram. 1994. *Container molecules and their guests*. Cambridge: Royal Soc. of Chemistry.

Crawford, E.T., and J. Olff-Nathan, eds. 2005. *La science sous influence: l'université de Strasbourg enjeu des conflits franco-allemands, 1872–1945*. Strasbourg: Nuée Bleue.

Cronin, B., D. Shaw, and K.L. Barre. 2004. Visible, less visible, and invisible work: Patterns of collaboration in 20th century chemistry. *Journal of the American Society for Information Science and Technology* 55 (2): 160–168.

Dance, I. 2003. What is supramolecular? *New Journal of Chemistry* 27 (1): 1–2.

Diederich, F. 2007. 40 years of supramolecular chemistry. *Angewandte Chemie International Edition* 46 (1–2): 68–69.

Gale, P.A. 2000. Supramolecular chemistry: From complexes to complexity. *Philosophical Transactions of the Royal Society of London. Series A: Mathematical, Physical and Engineering Sciences* 358 (1766): 431–453.

Gavroglou, K., and A. Simões. 2012. *Neither physics nor chemistry: A history of quantum chemistry*. Cambridge, MA: MIT Press.

Griffith, B.C., and N.C. Mullins. 1972. Coherent social groups in scientific change. *Science* 177 (4053): 959–964.

Heinze, T., R. Heidler, R.H. Heiberger, and J. Riebling. 2013. New patterns of scientific growth: How research expanded after the invention of scanning tunneling microscopy and the discovery of Buckminsterfullerenes: New patterns of scientific growth. *Journal of the American Society for Information Science and Technology* 64 (4): 829–843.

Jacob, C., ed. 2007. *Lieux de savoir*. Paris: Albin Michel.

Kleinpeter, E., and R.-E. Eastes. 2008. *Comment je suis devenu chimiste*. Paris: Le Cavalier bleu éd.

Kovac, J. 2002. Theoretical and practical reasoning in chemistry. *Foundations of Chemistry* 4 (2): 163. https://link.springer.com/article/10.1023%2FA%3A1016035726186. Accessed 03 May 2019.

Larivière, V., N. Desrochers, B. Macaluso, P. Mongeon, A. Paul-Hus, and C.R. Sugimoto. 2016. Contributorship and division of labor in knowledge production. *Social Studies of Science* 46 (3): 417–435.

Laudel, G. 2002. What do we measure by co-authorships? *Research Evaluation* 11 (1): 3–15.

Lehn, J.-M. 1978. Cryptates: Inclusion complexes of macropolycyclic receptor molecules. *Pure and Applied Chemistry* 50 (9–10): 871–892.

———. 1980. *Leçon inaugurale faite le vendredi 7 mars 1980, Collège de France, Chaire de chimie des interactions moléculaires*. Paris: Collège de France.

———. 1987. *Nobel Lecture. Nobel Media AB 2019*. https://www.nobelprize.org/uploads/2018/06/lehn-lecture.pdf. Accessed 24 Feb 2019.

———. 1995. *Supramolecular chemistry: Concepts and perspectives: A personal account built upon the George Fisher Baker lectures in chemistry at Cornell University [and] Lezioni Lincee, Accademia nazionale dei Lincei, Roma*. Weinheim/New York: VCH.

———. 2007. From supramolecular chemistry towards constitutional dynamic chemistry and adaptive chemistry. *Chemical Society Reviews* 36 (2): 151–160.

———. 2017. Supramolecular chemistry: Where from? Where to? *Chemical Society Reviews* 46 (9): 2378–2379.

Lemaine, G., ed. 1976. *Perspectives on the emergence of scientific disciplines*. The Hague/Aldine: Chicago/Mouton.

Lenoir, T. 1997. *Instituting science: The cultural production of scientific disciplines*. Stanford University Press.

Marcovich, A., and T. Shinn. 2011. Where is disciplinarity going? Meeting on the borderland. *Social Science Information* 50 (3–4): 582–606.

―――. 2014. *Toward a new dimension: Exploring the nanoscale.* Oxford: Oxford University Press.

McCray, W.P. 2005. Will small be beautiful? Making policies for our nanotech future. *History and Technology* 21 (2): 177–203.

Mullins, N.C. 1972. The development of a scientific specialty: The phage group and the origins of molecular biology. *Minerva* 10 (1): 51–82.

Newman, D. 2006. The lines that continue to separate us: Borders in our 'borderless' world. *Progress in Human Geography* 30 (2): 143–161.

Noël, M. 2019. *Set of 897 publications released between 1961 and 2011 (Web of Knowledge request, Author=Lehn, JM), Mendeley Data, V1.*

Normark, D. 2015. Flexibility or inexactitude? The 'Lab 60' at Karolinska Institutet: From medical disciplines towards the modern biomedical complex. *Ambix* 62 (2): 167–188.

Nye, M.J. 1993. *From chemical philosophy to theoretical chemistry: Dynamics of matter and dynamics of disciplines, 1800–1950.* Berkeley: University of California Press.

―――. 2014. Mine, thine, and ours: Collaboration and co-authorship in the material culture of the mid-twentieth century chemical laboratory. *Ambix* 61 (3): 211–235.

Olivier-Utard, F. 2010. L'université de Strasbourg de 1919 à 1939: s'ouvrir à l'international mais ignorer l'Allemagne. *Les Cahiers de Framespa* 6: 401.

Paradeise, C., and J.-C. Thoenig. 2013. Academic institutions in search of quality: Local orders and global standards. *Organization Studies* 34 (2): 189–218.

Raimbault, B., and P.-B. Joly. 2021. *The emergence of technoscientific fields and the new political sociology of science* (In this book).

Reinhardt, C., ed. 2001. *Chemical sciences in the 20th century: Bridging boundaries.* Weinheim/New York: Wiley-VCH.

Rohmer, M. 2006. Guy Ourisson (1926–2006). *Angewandte Chemie International Edition* 45 (48): 8088–8088.

Schummer, J. 2006. Gestalt switch in molecular image perception: The aesthetic origin of molecular nanotechnology in supramolecular chemistry. *Foundations of Chemistry* 8 (1): 53–72.

Servos, J.W. 1993. Research schools and their histories. *Osiris* 8: 2–15.

Siegel, J.S. 1996. Molecular collectives: Supramolecular chemistry. *Science* 271 (5251): 949–949.

Steed, J.W., and J.L. Atwood. 2009. *Supramolecular chemistry.* 2nd ed. Chichester: Wiley.

Stichweh, R. 1984. *Zur Entstehung des modernen Systems wissenschaftlicher Disziplinen: Physik in Deutschland, 1740–1890 (1. Aufl.).* Frankfurt am Main: Suhrkamp.

―――. 2001. Scientific disciplines, history of. In *International encyclopedia of the social & behavioral sciences*, 13727–13731. Oxford: Elsevier.

―――. 2013. *Wissenschaft, Universität, Professionen: soziologische Analysen (Neuaufl.).* Bielefeld: Transcript.

Teissier, P. 2010. Solid-state chemistry in France: Structures and dynamics of a scientific community since World War II. *Historical Studies in the Natural Sciences* 40 (2): 225–258.

van Brakel, J. 2000. *Philosophy of chemistry: Between the manifest and the scientific image.* Leuven: Leuven University Press.

Vicens, J., and Q. Vicens. 2011. Emergences of supramolecular chemistry: From supramolecular chemistry to supramolecular science. *Journal of Inclusion Phenomena and Macrocyclic Chemistry* 71 (3–4): 251–274.

Voillequin, B. 2008. *Contribution à l'histoire de la catalyse en France (1944–2004). Dynamiques disciplinaires et régimes de production de savoir.* Paris: Université Paris Nanterre.

Weber, J., ed. 2010. *Interdisziplinierung? zum Wissenstransfer zwischen den Geistes-, Sozial- und Technowissenschaften.* Bielefeld: Transcript.

Whitley, R. 1976. Umbrella and polytheistic scientific disciplines and their elites. *Social Studies of Science* 6 (3–4): 471–497.

Woodward, R.B. 1973. The total synthesis of vitamin B12. *Pure and Applied Chemistry* 33 (1): 145–178.

Yaneva, A. 2010. Is the atrium more important than the lab? Designer buildings for new cultures of creativity. In *Geographies of science*, ed. P. Meusburger, D. Livingstone, and H. Jöns, vol. 3, 139–150. Dordrecht: Springer Netherlands.

Chapter 3
What Synthetic Biology Aims At: Review Articles as Sites for Constructing and Narrating an Emerging Field

Clemens Blümel

3.1 Introduction

Analyses of scientific communities and collectives are central to Science and Technology Studies (STS) and sociological studies of science. In this article, I aim to elaborate on how the analysis of review articles can contribute to the analysis of synthetic biology as an epistemic community. Particularly in synthetic biology, community-building seems to benefit from growing societal awareness (Cserer and Seiringer 2009) as well as from new forms of organisational and scientific infrastructure. Community-building in this field, I suggest, can not only be explored by studying material spaces or events, such as conferences, workshops, or laboratories, but it may also be traced by exploring specific textual practices or representations within the scholarly discourse, in online commentaries, reports, or position papers.

Although much research on the topical landscape of societal and scholarly discourses in synthetic biology has been undertaken in recent years, these analyses have scarcely taken the specific contexts of publication—that is, the different scholarly or non-scholarly genres—into account. In this chapter, I will focus on the structure and content of review articles in synthetic biology. Characteristic framings of synthetic biology such as the notions of 'engineering biology' or 'synthesis and biology' (Kastenhofer 2013; Molyneux-Hodgson and Meyer 2009) have been articulated in this genre of scholarly literature (Benner and Sismour 2005; Endy 2005; Purnick and Weiss 2009). Other research on review articles (Myers 1991) has shown that they can contain presentations or characterisations of a research entity, which have an effect on the perception of a novel scientific

C. Blümel (✉)
Research area 'Research Systems and Science Dynamics', German Centre for Higher Education Research and Science Studies (DZHW), Berlin, Germany
e-mail: bluemel@dzhw.eu

© The Author(s) 2021
K. Kastenhofer, S. Molyneux-Hodgson (eds.), *Community and Identity in Contemporary Technosciences*, Sociology of the Sciences Yearbook 31,
https://doi.org/10.1007/978-3-030-61728-8_3

collective. Moreover, the mere existence and the relative share of review articles may be understood as an indicator of the state of a scientific field (Bastide et al. 1989). Thus, an analytical focus on review articles may be elucidating in the reconstruction of synthetic biology and of research collectives in general.

Based on a theoretical framework combining approaches from linguistics with sociological text analysis, I hence explore the relative structure and position of review articles in the publication landscape, the different types of review articles one finds in the field, how these review articles deal with research in the field, and how these reviews present the field as a whole. The chapter draws on bibliometric analysis and content analysis of scholarly documents. It is structured as follows: in the subsequent section, I will discuss different approaches to the analysis of review articles and develop a theoretical framework on how the existence or non-existence of this genre, its different types, and specific textual strategies can inform the study of scientific fields. After presenting my empirical approach in section three, I present major findings in section four. I conclude with a discussion of how findings of the study may contribute to the analysis of synthetic biology as a research field and to wider debates in STS.

3.2 Review Articles, Genre Analysis and the Study of Scientific Fields

The review article can be considered as a neglected genre in the analysis of scholarly discourse (Azar and Hashim 2014, p. 76; Bastide et al. 1989, p. 535). Contrary to the classic research article, which has often been framed as the 'master narrative of our times' (Swales 2003), the scientific review article has not received similar attention among scholars in both the sociology of science and in linguistics. With a few exceptions (Bastide et al. 1989; Myers 1991), one finds almost no contributions of how the analysis of reviews can be used to study specific scientific fields. Most research on review articles has taken the perspective of genre analysis in science (Swales 1990), which focuses on the ways specific textual formats in science correspond to recurring situations by studying textual structure, argumentative patterns, and the usage of specific vocabulary in specific forms of academic discourse (Berkenkotter and Huckin 1993).

What is known from the literature so far is that there is no such thing as *the* scientific review (Azar and Hashim 2014; Virgo 1971; Woodward 1974). Rather, one finds different forms and types of review articles, which have been mainly established in the twentieth century. In 1961, John Adams (1961) distinguished between disciplinary review articles, which emerged from the annual report of the nineteenth century, and categorical review articles, which aim at selectively discussing specific solutions to specific problems. It is widely held that the latter depended much more on the personal styles of their authors (Azar and Hashim 2014, p. 76). Ten years later, Judy Virgo (1971, p. 279) complained about the rising number of these articles and the fact that one could hardly control this vastly increasing genre of scholarly writing. Linguistic research into the different practices of writing established that the typology of review articles in science can be further differentiated: Woodward (1974) has found that there are interpretive, critical, speculative, or even popular review articles. According to Azar and Hashim (2014), these findings show that review articles can appear in various forms, and their writing can comprise very different academic practices such as evaluating, problematising, or reporting (ibid., p. 77).

While scholars of linguistics have established different types and characteristics of review articles, little is known about the specific uses and functions of these types of text in the various disciplines and fields. Virgo (1971), for instance, argued that review articles in medicine are of a specific type and characteristic (meta reviews or systematic reviews) because they are read by a specific audience (ibid., p. 279). Few studies, however, have systematically taken the opposite approach to explore how reviews might inform the analysis of research collectives. One of these exemptions is the work of Bastide et al. (1989) who argue that the prevalence of reviews may be understood as an indicator of the state of a research field. They have proposed that the non-existence of reviews would indicate a 'low degree of maturity and organization', while the 'existence of reviews manifests a certain degree of development' (Bastide et al. 1989, p. 556). In addition, they also claimed to find specific types of reviews in novel fields and disciplines (ibid.).

If these findings are applied to what is known in science studies about the field of synthetic biology, one may expect a rather low number of reviews since the field is mostly understood as still emerging (Raimbault et al. 2016) and rather heterogeneous with various different epistemic orientations (Kastenhofer 2013). In order to establish a relationship between the type or content of reviews and the state of the field, I propose to more closely examine the specific ways of how review articles present, order, or deal with research in the field. Ordering activities are central to the textual practices of scientific review articles as they are supposed to report on trends within the epistemic trajectories of specific fields (Virgo 1971). As established above, reviews provide different forms of ordering when presenting, selecting, and highlighting research by establishing abstract categories, providing temporal relationships, or carving out fields of application.

One possible strategy to contribute to these discussions is to focus on self-characterisations of the field. According to Luhmann, self-characterisations are the means by which systems describe differences between themselves and their environment (Luhmann 1990). The concept of self-characterisation has recently been applied to explore the degree specific social entities reflect on themselves (Kaldewey 2013). It can be argued that the analysis of self-characterisations also contributes to the analysis of the states of scientific fields because different patterns of self-characterisation provide information about degrees of maturity and types of organisation in scientific collectives.

Increasingly, one finds research on self-characterisations which is related to identity work in science—that is, the ways how scientific fields establish and maintain views on themselves, their subject, or their mode of research. One of the means by which scientific collectives are constructed as valuable and meaningful entities is through categories or meta-categories of science (Shapin 2001), which often appear in secondary scholarly writing such as reviews (Myers 1991, 2003). These categories allow for characterising and comparing scientific fields by attributing to them specific qualities, which are then constructed as relevant and which most often signify wider use, such as interdisciplinarity, innovativeness, or novelty. The construction of categories as a means of valuation is particularly apparent in what Bastide et al. (1989) have called programmatic review articles which contain hierarchised lists of research questions. Along this line of reasoning, I will focus on how the field is presented to the reader in such a way that it appears as a credible, legitimate, relevant, or valuable entity and process. This can be attempted by relating specific forms of textual organisation to patterns of self-characterisation.

But there is another sort of persuasive textual repertoire—namely, narrative forms of ordering. It can be argued that such textual repertoires may be particularly apparent in historical or narrative review articles. Narrative forms of ordering in scholarly writing provide meaning and value by identifying and relating inaugural events, thereby constructing larger collectives. Lepenies and Weingart (1983) established that narrative presentations of science can contribute to a field's identity by providing a dominant frame or narration to which readers can attach. By focusing on different histories on molecular biology, Abir-Am (1985) found that the field had been constructed as 'rocket science' by presenting its history as a history of heroes. From these different bodies of research—genre studies (Bazerman 1988; Myers 1991, 2003) on the one hand as well as studies of disciplinary histories on the other (Abir-Am 1985)—it can be argued that reviews may contain different ways of ordering, presenting, and legitimating a scientific field, contributing to an understanding of the establishment and reproduction of scientific collectives. To empirically account for the role of review articles in the construction of scientific fields, one needs to more closely examine not only the position of reviews in relation to other types of text but also the modes of textual presentation in scientific fields.

Based on these considerations, I further specify the questions to be addressed in this chapter. First, what is the place of review articles within the publication landscape of synthetic biology in quantitative terms (e.g. what is the share of review articles in the field; what can be said about the citation figures of review articles within synthetic biology)? Second, what different forms and types of review articles do we find in synthetic biology publications? Third, what textual strategies are employed to present and value the field in these articles, and do these strategies change over time? And fourth, how do differences in presentation relate to the legitimation or valuation of the field? In the conclusion my empirical data will be used to derive implications for a better understanding of synthetic biology as a research field.

3.3 Methods

The above-mentioned research questions require a careful integration of different perspectives and methods. More specifically, I combine content analysis of scholarly texts with quantitative analyses of publications.

In a first step, I analysed how review articles are represented in the field of synthetic biology in order to gain insights into the publication structure of the field and the role of review articles therein. To account for the field as a set of publications, I conducted a bibliometric analysis, paying particular attention to specific document types by relying on a topical search query strategy, using Web of Science (WoS) as a database. This search strategy was based on keywords combined with Boolean operators and additional components such as 'years'. I thereby followed the topical search approach of Tunger (2009), drawing on and further modifying the search strategy of Pei et al. (2012).[1] Articles were collected for the years 2002–2012 in order to allow for the analysis of citation effects. I further refined the search by focusing on science and technology research domains. The full search algorithm reads as follows:

> 'Synthetic Biology' OR 'artificial cell' OR 'minimal genome' OR 'artificial system AND biology*' OR 'artificial ecosystem' OR 'XNA' OR 'Computational design NOT Engineering' AND 'Artificial Life'.

The query was iteratively complemented (date of the search: 10 December 2015). Finally, the resulting corpus was cleaned, resulting in 3406 synthetic biology articles.[2] Relying on the information on documents types provided by Web of Science, this corpus allowed for comparisons between the different types of publications in the field, relating to the question of how frequent review papers are compared to other publication types in the field.

In a second step, I analysed a smaller set of review papers which had been generated from the initial corpus using citation counts as selection criterion in order to explore whether these had specific textual strategies of ordering knowledge and whether these strategies coincide with other textual structures, thus allowing for the identification of specific types of reviews. Relating to the work of Swales and Naijar (1987), I particularly concentrated on introductions and conclusions (Swales and Naijar 1987; Swales 1990).[3] The material underwent a qualitative content analysis, utilising MAXQDA as a software package to support coding activities. The code system was developed from a mixture of inductive and deductive coding strategies (Kuckartz 2014). Theoretical categories were constructed by integrating concepts of genre analysis and studies of scientific texts, such as categories for order

---

[1] Further such studies include Oldham et al. (2012) and Raimbault et al. (2016). According to my re-analysis of the output generated by the search term of Pei et al. (2012), I found that the use of 'artificial system' as a key word can lead to systematic overestimation of search results and false positives.

[2] This figure was less than the outcome of the study conducted by Pei et al. (2012).

[3] Similar arguments for this selection have been engaged by Bastide et al. (1989).

and structure of the text (Bastide et al. 1989) and for argumentation patterns (Bazerman 1988). The goal was to explore how different types of review articles are related to different ways of presenting and valuing the field of synthetic biology. Different analytical strategies were applied for identifying types of reviews in the field: a) textual markers of purposes in the text were analysed in order to account for explicitly mentioned textual goals, and b) specific modes of organising and ordering research in the field were explored which would lead more indirectly towards typifying review articles. Citation counts of the articles were considered in order to compare the different types of reviews regarding their visibility in the scholarly landscape. In a third step, different forms of textual organisation were related to specific field descriptions in order to account for how the field is constructed as a valuable entity. Referring to the concept of self-characterisation, I aimed at identifying and exploring particular passages which allowed for establishing specific relationships between textual organisation and ways of legitimating synthetic biology. Lastly, these different patterns were brought into a temporal order so as to account for the dynamics of synthetic biology in the period studied.

3.4 Results

3.4.1 Reviews in the Field of Synthetic Biology

The analysis of scientific literature revealed that the term 'synthetic biology' has been used as in titles, abstracts, or full texts in 3406 publications in 520 different journals.

Table 3.1 shows the percentage of different publication types in the above-mentioned database. The most common document type is the classical research article with almost 1800 hits for this database and period (2002–2012). More than 16% of the articles in Synthetic Biology, however, are categorised as review articles in the WoS database, thus rendering this type of article the second most frequent document type in the corpus.[4] Such a high share seems puzzling in a field that has

Table 3.1 Document types in synthetic biology

Publication type	Absolute figures	Relative (in per cent)
Review articles	564	16.6
Research article	1798	52.8
Data set	308	9.0
Editorial material	185	5.4
Other	551	16.2
Total	3406	100

Source: Web of Science/Web of Knowledge (accessed May 2016)

---

[4]The share of review articles in synthetic biology is higher than in the total Web of Science (WoS) database output in the respective period (2002–2012), which is 1.9%. Of course, the comparison

Table 3.2 Top ten review articles of synthetic biology by times citation (TC)

Year	Author	Title	Source	TC
2005	Endy	Foundations for engineering biology	Nature	558
2009	Li et al.	Drug discovery and natural products: End of an era or an endless frontier?	Science	416
2005	Benner	Synthetic biology	Nat Rev Genet	381
2006	Andrianantoandro et al.	Synthetic biology: New engineering rules for an emerging discipline	Mol Syst Biol	296
2009	Purnick/Weiss	The second wave of synthetic biology: From modules to systems	Nat Rev Mol Cell Biol	271
2008	Kell	Synthetic biology for synthetic chemistry	ACS Chem. Biol.	219
2010	Khalil et al.	Synthetic biology: Applications come of age	Nat Rev Genet	214
2006	Bhattacharya et al.	Domains, motifs, and scaffolds	Annu. Rev. Biochem	212
2008	Veening et al.	Bistability, epigenetics, and bet-hedging in bacteria	Annu. Rev. Microbiol.	199
2005	Sprinczak	Reconstruction of genetic circuits	Nature	188

Source: Web of Science (accessed 18 March 2016)

been repeatedly qualified as emerging when compared to the assumption put forward by Bastide et al. (1989).[5]

In order to gain more information about what these figures might mean—that is, what is actually established in these types of text—I conducted a content analysis of a set of review articles. As mentioned in the methods section of this article, review articles for content analysis were selected by citation count. Based on the WoS citation data, 80 of the most highly cited review articles from the original database (N = 3406) were selected, of which 77 were subsequently analysed more closely. Table 3.2 shows the distribution of these selected publications over time, with a visible bias for earlier papers resulting from the selection criterion (i.e. more recent papers naturally accrue fewer citations in a shorter time span). The highest number of review articles within the sample dates from the year 2009 (Fig. 3.1).

Twenty-five of the most highly cited review articles were also among the 50 most highly cited articles in the total synthetic biology database,[6] which is in line with previous findings that review articles have a higher probability of being cited than other document types (Knottnerus and Kottnerus 2009). The high percentage of

---

with the total output of WoS publications is problematic because the WoS database covers many fields and disciplines which differ greatly in terms of maturity and dynamic. But the share of review articles was also lower in fields of similar size and state of emergence, such as organic solar fuel cells, where the share of reviews comprised 3.2% (WoS base) in the period under study.

[5]The assumption was that the frequency of review articles mirrors the maturity of a field; see Sect. 3.2.

[6]For full citation details, see the supplementary material of this article.

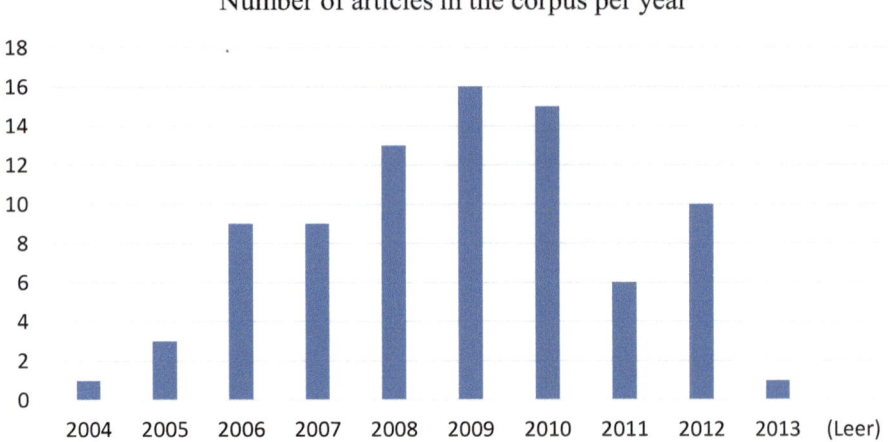

Fig. 3.1 Number of review articles in the sample corpus per year (2002–2012). (Source: Web of Science (accessed March 2016))

heavily cited review articles allows for framing the identified review as relevant for the analysis of the field, as they have been referred to extensively by the scholarly community.

In order to explore whether, and if so, how, these structural figures relate to the intellectual and social state of the field, I analysed the content and the textual strategies within the articles in more detail. If the existence and prevalence of specific types of review articles changes in the period studied, then one would assume corresponding changes in the maturity or organisation of the research field as proposed in the theoretical section of this chapter.

3.4.2   Results of the Content Analysis

3.4.2.1   Types of Reviews in the Corpus

Different types of reviews were identified by mapping codings from different categories—first, the (most often explicitly mentioned) purpose of the article, and second, the mode of textual presentation of research results (the latter with a set of relevant subcategories described in the next section). The analysis of purpose in documents is widely used in genre analysis because academic documents are primarily understood to be highly structured and purpose-oriented types of texts (Swales 1990). Hence, stated purposes were in most cases (64 out of 77) *explicitly* mentioned. Specific phrases directly related the purpose of the article to the document type, such as 'in this review we aim at providing an overview' (Purnick and Weiss 2009) or 'the purpose of this review is a discussion of another side of genomics' (Stähler et al. 2006) or 'in this review, we describe recent accomplishments'. In order to account for the diversity of the purposes provided within the set of articles, I analysed the verbs which were used to specify the focus of an article. Rather neutral phrases ('describe', 'providing [an] overview', and to a lesser extent, 'discuss') were common, coinciding with what Virgo (1971) has established as the functions of reviews—that is, to select and to provide oversight. In some cases, however, I also found explicit claims to establish goals and future directions of the research field, particularly in the early years (2005–2008):

> In this review, we describe what has been done thus far in synthetic biology and how the field is actively being moved forward, as well as survey various prospective views that articulate different visions of synthetic biology's future. (Drubin et al. 2007)

Another source for the identification of different types of reviews was their modes of textual organisation. These modes were coded inductively following two main strategies: firstly, I analysed how the texts established author-reader relationships, using specific repertoires of connecting the different parts of the text—that is, specific repertoires which aim at broadening the readership by referring to merely public knowledge concepts and science policy semantics. And second, I explored the ways in which the reviews dealt with the different contributions to the research field. I found that these different strategies often lead to codings of similar textual passages.

Thus four different modes emerged: *evaluation*—passages that aimed at highlighting and valuing specific contributions in the research field; *classification*—passages that related research to existing or novel epistemic categories; *exemplification*—passages that exemplify specific research to make a broader claim about the field; and *narration*—passages providing a meaningful sequence of events. I then continued to produce case specific summaries of codings in order to decide whether a specific mode of textual organisation dominates a text as a whole (Kuckartz 2014, p. 70). Based on these findings, I explored the relation between the different codes and categories—that is, modes of textual organisation, structure, and purpose of the text. I came up with three different types of reviews.

One set of review articles was dominated by exemplification. Endy (2005), for instance, does not provide a full discussion of research contributions but rather proposes a new frame of what engineering in biology means by providing examples of research projects which illuminate his vision of biology. Exemplifying articles often provided a principle or an abstract criterion (Andrianantoandro et al. 2006; Khalil and Collins 2010) or even a set of overarching principles (Benner and Sismour 2005; Endy 2005; Pleiss 2006) in order to organise research. This type of review can be found particularly in the years between 2005 and 2008. The respective articles share further features, such as a specific use of expectations[7] in their introductions, relating the above-mentioned abstract criteria with a vision of how

---

[7]It has already been established elsewhere (Kastenhofer 2013; Blümel 2016) that synthetic biology is strongly characterised by using expectations. In this article, I established in particular that reviews are important sites for articulating visions, which may have affected their visibility in the publication landscape.

the field may influence or impact on other fields or societal realms as in the following example:

> Synthetic biology will revolutionize how we conceptualize and approach the engineering of biological systems. The vision and application of this emerging field will influence many other scientific and engineering disciplines, as well as affect various aspects of daily life and society. (Andrianantoandro et al. 2006)

I labelled this type 'prospective review'. This type included the most highly cited articles in the corpus (Endy 2005; Benner and Sismour 2005; Andrianantoandro et al. 2006; Isaacs et al. 2006),[8] potentially indicative of raising the most awareness among (different) scientific communities. They can claim *novelty* for describing a recently emerging entity, e.g., a new collective.[9] Given that many of the prospective review articles in this corpus cite each other, the prevalence of these articles may also indicate a kind of collective willingness to establish 'futures' of synthetic biology. What is more, the majority of review articles in the corpus can be referred to this type (N = 35). Different to what Bastide et al. (1989) have established, the prevalence of such reviews would then indicate the emergence of scientific fields instead of maturity of a scientific field.

By contrast, another set of review articles was dominated by narrative sequences. Following Arnold (2012), a narrative structure is understood as an interrelated pattern of events that allows for identifying a subject (synthetic biology) and a specific object (e.g. the construction of novel biosystems) in such a way that the various moments which hindered or enabled the solution of problems (lack or existence of resources, technologies, collaborations) are meaningfully presented. In such reviews (Khalil and Collins 2010; Purnick and Weiss 2009), several events were closely related as being part of similar technological advances, predecessors, or consequences of innovations in the field. Thus, by classifying specific review articles in the sample corpus as narrative, I found that major contributions in the research field were presented not only in a temporal order but also embedded in a larger (hi)story. In various review articles (Drubin et al. 2007; Mukherji and van Oudenaarden 2009), distinct epistemic events, such as the construction of genetic devices (Elowitz and Leibler 2000), were connected to construct a specific presentation of synthetic biology. Some of these articles even provided their own history of the field (Cheng and Lu 2012; Khalil and Collins 2010; Luisi et al. 2006; Mukherji and van Oudenaarden 2009; Stephanopoulos 2012).These review articles appear to be more frequent in the period from 2007 onwards. And, on average, they receive

---

[8]The above-mentioned article by Andrianantoandro et al. 2006 has been cited almost 300 times (according to the WoS database). Similar citation counts can be found for an article by Purnick and Weiss, which has been cited 271 times (data obtained from WoS).

[9]More research into the reception of scientific literature would be needed in order to establish a relationship between the use of visionary language and citation patterns.

fewer citations than prospective review articles,[10] which might indicate that these articles attract less attention among scientific communities.

A third set of review articles was dominated by evaluative textual passages. (see also Sect. 3.2 of this article). Often, these articles highlighted benefits of synthetic biology for specific applications or application areas, such as biofuels, materials, clinical applications, or chemical compounds. As the citation figures show (Table 3.2), only one article of this type (Veening et al. 2008) features among the ten most highly cited articles in this sample. It is more difficult to attribute these review articles to a specific period, but they became more frequent in the years after 2010.

Based on these different purposes and ways of organising and structuring the text, I identified three different types of review articles in the sample corpus: prospective or programmatic reviews, categorical reviews, and narrative or historical reviews. The distribution of citation patterns suggests that particularly prospective and to a lesser extent narrative review articles became objects of intense discussion in the scholarly community. The high number of prospective reviews in the early period suggests that a specific type of review articles may play an important role in the emergence of the field.

3.4.2.2 'What Synthetic Biology Aims At': Strategies of Self-Characterisation

After having established these different types of review articles and their characteristics, I now come to the third question of how these texts differ in presenting the field.

Relying on the theoretical positions established in Sect. 3.2, I argue that in order to study the state of a research field by applying rhetoric and argumentative studies, one may need to focus on reconstructing the different ways the field is presented. On a conceptual level, these passages can be understood as self-characterisations of the field. Based on the findings of Pnina Abir-Am (1985), it can be argued that presentations of a field are seldom neutral but allow for valuing research in a respective area. To explore presentations of the field, I relied on a specific but frequent (N = 64) type of textual passage, which I have termed 'field description', which highlights the boundaries and objects of research. These passages often contained a definition of the field and highlighted a specific goal towards which 'synthetic biology would head' or 'aim at' (Drubin et al. 2007; Pleiss 2006). These field descriptions were highly connected with other codes (referring to modes of textual organization) in that they contain a high number of rhetoric strategies, like metaphors, analogies, or comparisons. Following the theoretical considerations presented in Sect. 3.2, these passages were suitable candidates for exploring what has been termed the field's

---

[10]The effect that review articles published earlier are more likely to receive higher citations has been taken into account in this analysis.

self-characterisation. The function of the field descriptions in the organisation of the review is particularly apparent in the following example:

> The evolving discipline of synthetic biology overlaps with efforts in Metabolic Engineering and provides a complementary framework for the de novo design of new biosynthetic pathways. Synthetic Biology aims to frame the engineering of biology in a manner that is analogous to other engineering disciplines and relies on core principles of design and characterization to facilitate and the rapid and reproducible deployment of biological machines. (Prather and Martin 2008)

In this example, the field of synthetic biology was related to and demarcated from other fields (such as metabolic engineering). The presentation of the field came with a set of analogies, core principles, and possible application fields which allowed for qualifying and valuing research in this area.

A closer examination of other field descriptions from different periods reveals that field descriptions differed in how they are textually arranged and how they present synthetic biology as a collective entity. The set of prospective reviews (covering mainly the period from 2005 to 2008) started with stating the potential of synthetic biology (Andrianantoandro et al. 2006; Heinemann and Panke 2006), which was often connected to the articulation of a goal which the field should accomplish in the future (Benner 2003; McDaniel and Weiss 2005). Furthermore, the field was characterised by applying meta-categories such as 'highly interdisciplinary' (Endy 2005), 'emerging' (Isaacs et al. 2006), 'novel' (Sprinzak and Elowitz 2005), 'innovative', or 'application oriented' (Kaznessis 2007).

Yet, in the narrative or historical type of reviews, which were mainly published after 2008, field descriptions had a slightly different structure in that they highlight epistemic categories as attributes of the field, such as xeno-biology (Schmidt 2010) or advanced genetic circuitry (as in, for instance, Nandagopal and Elowitz (2011)). There were also changes in the tense describing the field from future tense (N = 13)—'what the field ought to be' to present tense—'what synthetic biology is' (see, for instance, as an example). Moreover, the review articles in the period from 2008 onwards often provided their own history of synthetic biology (Khalil and Collins 2010; Mukherji and van Oudenaarden 2009), supported by a greater use of pictures and graphical material (such as timelines). This shows that different types of reviews also accompany differences in presenting the field, reflecting the differences which have been established for these texts as a whole.

3.4.2.3   From Authoritative to Narrative Forms of Community-Building: Changes in Legitimating the Field

Finally, I address the questions of how these different ways of presenting research result in different practices of legitimating the field, and what this could mean for the use of reviews in the analysis of scientific fields and communities. It has been established in the theoretical section of this article that the articulation and construction of self-characterisation can play an important role in establishing collective identities of scientific fields. The ways by which novel or emerging scientific collectives are characterised can have performative effects on the perception of the community. I argue that accounts of self-characterisation also relate with attempts of legitimating the field as a credible or relevant community. Yet, the frequency of self-characterisations and the repertoires which are used to legitimate a field in these contexts may vary and may indicate different levels of maturity. Therefore, I have elaborated on the different repertoires of legitimation in these contexts.

Prospective and narrative review articles differed in that regard. Prospective reviews mainly employ authoritative legitimating strategies deriving the field's credibility from endeavours established earlier or by referring to success stories of other areas. One of the ways by which legitimacy is sought can be termed *niche building*—that is, situating the field of synthetic biology in the larger context of scientific discoveries (McDaniel and Weiss 2005, p. 4769).

Rhetorically, the field of synthetic biology was also presented as one solution to a problem that had been established earlier (e.g. Andrianantoandro et al. 2006, p. 1). Thus, the field was placed within a larger history of scientific progress. In this example, synthetic biology was presented as a consequence of various sequences of innovations which ultimately led to the emergence of the field. The mode of locating synthetic biology within various credible endeavours also attributed significance to its current state. Synthetic biology was hence presented as an extension of genetic engineering on the systems level. It was these (external) movements and discoveries from which the legitimacy and credibility of synthetic biology (as an extension thereof) is derived.

Another rhetorical means by which the field was legitimated is the use of comparisons and analogies to other scientific fields and disciplines. The use of historical comparisons for legitimating and establishing scientific collectives can also be found in other fields (Abir-Am 1985; Lepenies and Weingart 1983) and has already been studied for synthetic biology (Bensaude Vincent 2013). Generally, the use of comparisons with established fields has various effects: it establishes the author-reader relationship by referring to taken-for-granted knowledge, at least among large parts of the readership (Berger and Luckmann 1969/2013). Moreover, providing examples in scientific writing is often used to broaden the readership of articles beyond the narrowly defined research community.

It can be argued that the textual use of comparisons is to ensure that a broad readership can follow the argument if comparisons and analogies refer to taken-for-granted repertoires of social knowledge which are widely shared in the social world (Berger and Luckmann 1969/2013). Another use of comparisons may also be as a means of presenting a given but perhaps contested issue as more realistic or less problematic (Lepenies and Weingart 1983).[11] In these cases, the comparison functions as a transfer of credibility which allows for framing the novel issue as a plausible way of doing or knowing things. Connecting established fields to (potentially) promising implementations in synthetic biology is not a single phenomenon

---

[11]For a general argument about comparisons in the context of synthetic biology, see also Torgersen and Schmidt (2013).

but a recurring issue in the 2005–2008 reviews (Heinemann and Panke 2006). The comparison to an established practice then appears to be less relevant for characterising the field than for emphasising what synthetic biologists will soon be capable of. This mode of relating an established scientific practice to a potentially relevant but yet not existing implementation of synthetic biology concepts was frequently found in the sample.

In the second period after 2008, the rhetorical means by which the field was legitimated (except those that have been classified as categorical) can be termed *narrative normalisation*. What differed to the strategies of valuation and legitimation was that they did not rely—or did so to a lesser extent—on external resources, such as analogies or comparisons to other fields. Instead, these articles and the self-characterisations of synthetic biology often contained their own history of the field with more specific problems and challenges. They thereby established a specific agency and frame of the field by providing initial and foundational events. Purnick and Weiss (2009), for instance, presented a foundational disciplinary history in which simple genetic circuitries (oscillators and switches) had a specific role. The authors stated that their mode of textual presentation predominantly exemplified specific phases of the field:

> We begin this review by examining the first wave in synthetic biology, a phase that has focused on creating and perfecting genetic devices and small modules. We do not provide a comprehensive discussion of synthetic biology projects, but rather a description of several informative examples. (Purnick and Weiss 2009)

Different to other prospective reviews in the sample, research contributions were not presented to demonstrate the state of the community; rather, each of the presented examples had a specific role in narrating the field. This fits well with what Arnold (2012, p. 21) has called the narrative structure of texts, which is the way a subject (e.g. synthetic biology) is established to master a specific goal. Purnick and Weiss listed problems the field had already mastered and specified new ones which needed to be addressed if the field were to reach its overarching goals (Purnick and Weiss 2009, p. 411). Hence, Purnick and Weiss combined a foundational history of the field with an articulation of future challenges and current necessities.

Khalil and Collins (2010) performed a similar form of valuation and legitimation by providing another narration of the field. In their introductory section (ibid., p. 367), the description of the field was related to a history about the first genetic devices of synthetic biology. Such framing by first discoveries played a less dominant role than in the field characterisations of the earlier prospective review articles (McDaniel and Weiss 2005; Benner 2003; Sismour and Benner 2005; Andrianantoandro et al. 2006; Heinemann and Panke 2006; Pleiss 2006). Khalil and Collins additionally framed the field by referring to founding events. In this narrative of self-characterisation, these events appeared as a coherent story in which a specific narrative of engineering appeared as a common denominator of all descriptive elements. Different to earlier review articles (Pleiss 2006, p. 736; Heinemann and Panke 2006; Benner and Sismour 2005) which had accounted for different disciplinary orientations and 'flavors' of synthetic biology, the overarching narrative of engineering was not contested any more. Such narrative streamlining of the field's development became most apparent in the latest article of the database, which described forward engineering as the only goal and identity of synthetic biology. Cameron et al. (2014) did so by providing 'milestones' of the field which fit this framing (ibid., p. 382) and devising phases of the field accordingly. Specific importance was attributed to the 'foundational' years (ibid.) 2000–2003. In this way, a dedicated narrative of synthetic biology as an engineering field in the making was deployed, establishing and stabilising the field as a quasi-discipline by providing it with its own history. This parallels Abir-Am's (1985) analysis of the way several scholars and spokesmen contributed to the consolidation of molecular biology by providing it with a specific history (ibid., p. 75). Other than this study, the narrative review articles discussed in this chapter make less use of foundational fathers, heroes, or spokesmen (e.g. Delbrück or Lura) but rather draw on 'inaugural devices' (Khalil and Collins 2010), such as repressilators, oscillators, and other technical tools that first exemplified ideas of what has later been termed synthetic biology (Blümel 2016).

Thus, differences in presenting synthetic biology were linked to different strategies and uses of legitimation and valuation. While the earlier prospective review articles relied on authoritative strategies of legitimating the field by focusing on broader trajectories of scientific discovery or on analogies to other fields, the more recent narrative reviews relied on quasi-disciplinary histories, depicting a story-line fraught with challenges, actors, and events. For the analysis of synthetic biology as a research field, this suggests that the field has undergone changes in the way it is perceived, presented, and discussed by its proponents. Thus, the field has undergone processes of institutionalisation and stabilisation. The finally dominant narrative focus on 'heroic objects' can be perceived as a story which contributed to the establishment of field specific sources of self-perception, legitimation, and credibility.

3.5   Conclusion

In this article, I explored how a discourse analysis focusing on review articles can contribute to the understanding of synthetic biology as an evolving research field. Based on explorations of previous research on this genre, I proposed that the analysis of review articles is a suitable way to explore self-characterisations of synthetic biology, allowing for reconstructing ways of attributing the field's credibility, legitimacy, and value. I articulated four different research questions and subsequently addressed them in reference to my empirical findings.

In order to account for the first question pertaining to the role of review articles in the publication landscape of synthetic biology, I conducted a bibliometric analysis, referring to the distribution of different publication types in the field. The share of review articles proved to be relatively high compared to average figures in the WoS database and to other comparable fields. Different to what (Bastide et al. 1989) proposed, review articles were highly prevalent in all stages of the emerging field, albeit with changes to their argumentative scheme (see below).

In order to elaborate in more detail the ways of ordering, presenting, and legitimating synthetic biology in review articles, I applied a qualitative content analysis which was informed by the aforementioned bibliometric analysis. Addressing the second research question, I established different types of reviews along different articulated purposes as well as differences in ways of ordering and textual organisation: a prospective type, which provided visions and prospects of the field; a categorical type, focusing on specific applications; and, third, a narrative type of review articles. Since the prevalence of these different types in the selected sample of the 77 most-cited review articles corresponds to specific phases of the field, it can be argued that the character of review articles has changed within the period at hand (2002–2012).

In order to account for the third question, I analysed the textual strategies employed to present the field as a whole. I found that textual strategies of presenting the field differed in accordance with the aforementioned types of reviews. Prospective reviews dealt with the depicted contributions to the field by means of exemplifying or illuminating abstract criteria, while narrative reviews ordered the reported contributions to the field along founding events which mirrored the progress of the field. I argued that these different strategies coincide with different ways of valuing or legitimating the field. Whereas the early 'prospective' review articles often provide comparisons and analogies to established disciplines in order to legitimate and justify the field, narrative review articles strongly relied on founding tales of the field, establishing and stabilising specific contributions as milestones within a specific history of the field.

Based on these findings concerning the ways of ordering, presenting, and legitimating the field in review articles, implications for the study of synthetic biology and other emerging scientific fields can be established. First, from analysing the content of the review articles, it can be concluded that the self-characterisation of the field is a recurring issue in synthetic biology; many textual passages address what the field aims at and what it stands for, indicative of a contested or tentative identity. The different structures and strategies of the three dominant types of review articles, however, point towards the prevalence of different developmental stages of the field. Early 'prospective' review articles aim at establishing synthetic biology as a valuable and legitimate field by referring to external resources. By contrast, the later narrative review articles from the years 2009 onwards indicate that the field may have already experienced and successfully managed conflicts over its intellectual orientation. Existing literature on synthetic biology has already shown that proponents of the field make use of futuristic visions (Kastenhofer 2013) in order to legitimise the field or to raise awareness among different audiences (Cserer and Seiringer 2009; Pei, Gaisser, and Schmidt 2012). Findings from this article contribute to this literature but show that different strategies of acquiring legitimacy have been used in the emergence of the field at different stages.

These findings also have implications for research on specific rhetorical devices and their role in establishing the identity of synthetic biology. Recently, Nerlich and

McLeod (2016) pointed out that the field has made great use of such devices to promote itself. Other research highlighted how presentations of synthetic biology benefit from the use of comparisons and analogies to other fields (Bensaude-Vincent 2013). The findings presented in this chapter contribute to this line of research by arguing for a stronger consideration of different types and contexts of publications.

Thirdly, the findings presented here also relate to a larger body of research on narratives and narrations in science. Recently, Mary Morgan (2017) established that narratives and narrations are a widespread and pervasive practice across different scholarly domains. Narrative ordering thus is a way to 'answer how and why things happen' and plays an important role in many scientific contexts. Narrative forms of ordering also play a particular role in legitimating and valuing novel scientific fields. For instance, Abir-Am (1985) and Lepenies and Weingart (1983) provided insights into how the narrative presentation of scientific fields contributed to their establishment. Yet, while Abir-Am (1985) elaborated on how the narrative construction of founding heroes contributed to creating collective identities in molecular biology, the founding tales of synthetic biology culminate in presenting inaugural devices as 'heroic objects'. I argue that the construction of these founding stories can be understood as a form of *narrative community-building*, contributing to the fields' collective understanding of identity and configurations of agency. Thus, analyses of collective narrations in review articles can provide a different perspective on the reconstruction of emerging scientific fields such as synthetic biology.

References

Abir-Am, P. 1985. Themes, genres and orders of legitimation in the consolidation of new scientific disciplines: Deconstructing the historiography of molecular biology. *History of Science* 23 (1): 73–117.

Adams, S. 1961. The review literature of medicine. *Bibliography of Medical Reviews*: 6.

Andrianantoandro, E., S. Basu, D.K. Karig, and R. Weiss. 2006. Synthetic biology: New engineering rules for an emerging discipline. *Molecular Systems Biology* 16: 1–14.

Arnold, M. 2012. Erzählen. Die ethisch-politische Funktion narrativer Diskurse. In *Erzählungen im Öffentlichen: Über die Wirkung narrativer Diskurse*, ed. M. Arnold, G. Dressel, and W. Viehöver, 17–64. Springer.

Azar, A.S., and A. Hashim. 2014. Towards an analysis of review article in applied linguistics: Its classes, purposes and characteristics. *English Language Teaching* 7 (10).

Bastide, F., J.P. Courtial, and M. Callon. 1989. The use of review articles in the analysis of a research area. *Scientometrics* 15: 535–562.

Bazerman, C. 1988. *Shaping written knowledge: The genre and activity of the experimental article in science*. Madison/Wisconsin/London: The University of Wisconsin Press.

Benner, S.A. 2003. Synthetic biology: Act natural. *Nature* 421: 118.

Benner, S.A., and A.M. Sismour. 2005. Synthetic biology. *Nature Reviews Genetics* 6: 533–543.

Bensaude Vincent, B. 2013. Discipline-building in synthetic biology. *Studies in History and Philosophy of Science Part C: Studies in History and Philosophy of Biological and Biomedical Sciences* 44 (2): 122–129.

Berger, P., and T. Luckmann. 2013. *Die gesellschaftliche Konstruktion der Wirklichkeit*. Frankfurt am Main: Fischer. (Original work published 1969.).

Berkenkotter, C., and T.N. Huckin. 1993. Rethinking genre from a sociocognitive perspective. *Written Communication* 10 (4): 475–509.

Blümel, C. 2016. Enrolling the toggle switch: Visionary claims and the capability of modeling objects in the disciplinary formation of synthetic biology. *NanoEthics* 10 (3): 269–287.

Cameron, E.D., C.J. Bashor, and J.J. Collins. 2014. A brief history of synthetic biology. *Nature Reviews Microbiology*: 381–390.

Cheng, A.A., and T.K. Lu. 2012. Synthetic biology: An emerging engineering discipline. *Annual Review of Biomedical Engineering* 14: 155–178.

Cserer, A., and A. Seiringer. 2009. Pictures of synthetic biology: A reflective discussion of the representation of synthetic biology in the German media and by SB experts. *Systems and Synthetic Biology* 3 (1–4): 27–35.

Drubin, D.A., J.C. Way, and P.A. Silver. 2007. Designing biological systems. *Genes & Development* 21: 242–254.

Elowitz, M., and S.A. Leibler. 2000. A synthetic oscillatory network of transcriptional regulators. *Nature* 403: 335–338.

Endy, D. 2005. Foundations for engineering biology. *Nature* 438: 449–453.

Heinemann, M., and S. Panke. 2006. Synthetic biology—Putting engineering into biology. *Bioinformatics* 22 (22): 2790–2799.

Isaacs, F.J., D.J. Dwyer, and J.J. Collins. 2006. RNA synthetic biology. *Nature Biotechnology* 24 (5): 545–554.

Kaldewey, D. 2013. *Wahrheit und Nützlichkeit*. Bielefeld: transcript.

Kastenhofer, K. 2013. Synthetic biology as understanding, control, construction, and creation? Techno-epistemic and socio-political implications of different stances in talking and doing technoscience. *Futures* 48: 13–22.

Kaznessis, Y.N. 2007. Models for synthetic biology. *BMC Systems Biology* 1: 47.

Khalil, A.S., and J.J. Collins. 2010. Synthetic biology: Applications come of age. *Nature Reviews. Genetics* 11 (5): 367–379.

Knottnerus, J.A., and B.J. Kottnerus. 2009. Let's make the studies within systematic reviews count. *The Lancet* 373 (9675): 1605.

Kuckartz, U. 2014. *Qualitative Inhaltsanalyse, Methoden, Praxis, Computerunterstützung*, 2. Auflage. Weinheim/Basel: Juventa.

Lepenies, W., and P. Weingart. 1983. Introduction. In *The functions and uses of disciplinary histories*, ed. L. Graham, W. Lepenies, and P. Weingart, IX–XX. Dordrecht: D. Reidel Publishing Company.

Luhmann, N. 1990. *Die Wissenschaft der Gesellschaft*. Frankfurt am Main: Suhrkamp.

Luisi, P.L., C. Chiarabelli, and P. Stano. 2006. From never born proteins to minimal living cells: Two projects in synthetic biology. *Origins of Life and Evolution of Biospheres* 36: 605–616.

McDaniel, R., and R. Weiss. 2005. Advances in synthetic biology: On the path from prototypes to applications. *Current Opinion in Biotechnology* 17: 476–483.

Molyneux-Hodgson, S., and M. Meyer. 2009. Tales of emergence—Synthetic biology as a scientific community in the making. *BioSocieties* 4 (2–3): 129–145.

Morgan, M.S. 2017. Narrative ordering and explanation. *Studies in History and Philosophy of Science* 62: 86–97.

Mukherji, S., and A. van Oudenaarden. 2009. Synthetic biology: Understanding biological design from synthetic circuits. *Nature Reviews Genetics* 10: 859–871.

Myers, G. 1991. Stories and styles in two molecular biology review articles. In *Textual dynamics and the professions: Historical and contemporary studies of writing in Professional communities*, ed. C. Bazerman and Paradis, 45. Madison: The University of Wisconsin Press.

———. 2003. Discourse studies of scientific popularization: Questioning the boundaries. *Discourse Studies* 5 (2): 265–279.

Nandagopal, N., and M.B. Elowitz. 2011. Synthetic biology: Integrated gene circuits. *Science (New York, N.Y.)* 333 (6047): 1244–1248.

Nerlich, B., and C. McLeod. 2016. The dilemma of raising awareness 'responsibly': The need to discuss controversial research with the public raises a conundrum for scientists: When is the right time to start public debates? *EMBO Reports* 17 (4): 481–485.

Oldham, P., S. Hall, G. Burton, and J.A. Gilbert. 2012. Synthetic biology: Mapping the scientific landscape. *PLoS One* 7 (4): e34368.

Pei, L., S. Gaisser, and M. Schmidt. 2012. Synthetic biology in the view of European public funding organisations. *Public Understanding of Science* 21 (2): 149–162.

Pleiss, J. 2006. The promise of synthetic biology. *Applied Microbiology and Biotechnology* 73: 735–739.

Prather, K.L.J., and C.H. Martin. 2008. De novo biosynthetic pathways: Rational design of microbial chemical factories. *Current Opinion in Biotechnology* 19 (5): 468–474.

Purnick, P., and R. Weiss. 2009. The second wave of synthetic biology: From modules to systems. *Nature Reviews Molecular Cell Biology* 10: 410–422.

Raimbault, B., J.-P. Cointet, and P.-B. Joly. 2016. Mapping the emergence of synthetic biology. *PLoS One* 11 (9): e0161522.

Schmidt, M. 2010. Xenobiology: A new form of life as the ultimate biosafety tool. *BioEssays* 32: 322–331.

Shapin, S. 2001. How to be antiscientific? In *The one culture? A conversation about science*, ed. J.K. Labinger and H.M. Collins, 99–115. Chicago: University of Chicago Press.

Sismour, M., and S. Benner. 2005. Synthetic biology. *Nature Reviews Genetics* 6: 533–543.

Sprinzak, D., and M. Elowitz. 2005. Reconstruction of genetic circuits. *Nature* 438.

Stähler, P., M. Beier, X. Gao, and J.D. Hoheisel. 2006. Another side of genomics: Synthetic biology as a means for the exploitation of whole-genome sequence information. *Journal of Biotechnology* 124: 206–212.

Stephanopoulos, G. 2012. Synthetic biology and metabolic engineering. *ACS Synthetic Biology* 1 (11): 514–525.

Swales, J. 1990. *Genre analysis: English in academic and research settings*. Cambridge: Cambridge University Press.

———. 2003. *'That master narrative of our time': The research article revisited.* The 14th European symposium on language for special purposes: Communication, culture, knowledge, University of Surrey, Guildford, UK.

Swales, J., and H. Naijar. 1987. The writing of research article introductions. *Written Communication* 4 (2): 175–191.

Torgersen, H., and M. Schmidt. 2013. Frames and comparators: How might a debate on synthetic biology evolve? *Futures* 48: 44–54.

Tunger, D. 2009. *Bibliometrische Verfahren und Methoden als Beitrag zu Trendbeobachtung und -erkennung in den Naturwissenschaften.* Schriften des Forschungszentrums Jülich Band 19: Jülich.

Veening, J.-W., W.K. Smits, and O.P. Kuipers. 2008. Bistability, epigenetics, and bet-hedging in bacteria. *Annual Review of Microbiology* 62: 193–210.

Virgo, J. 1971. The review article: Its characteristics and problems. *The Library Quarterly* 41 (4): 275–291.

Woodward, A. 1974. Review literature: Characteristics, sources and output in 1972. *ASLIB Proceedings* 26: 367–376.

Chapter 4
The Emergence of Technoscientific Fields and the New Political Sociology of Science

Benjamin Raimbault and Pierre-Benoît Joly

4.1 Introduction

Although use of the term synthetic biology in the scientific literature dates back to the early twentieth century, contemporary synthetic biology started to bloom around the turn of the new millennium and has been presented as novel—perhaps even revolutionary—and 'cool'. Like most emerging fields, synthetic biology has been defined in numerous different ways by members of a self-selected community in the making. The European Commission (EC) opinion on synthetic biology (2014) identifies 35 published definitions and proposes the following: 'SynBio is the application of science, technology and engineering to facilitate and accelerate the design, manufacture and/or modification of genetic materials in living organisms (p5)'. Synthetic biologists suggest that compared to modern biotechnology (e.g. genetic engineering, genomics, high throughput biology, etc.), the epistemic novelty of synthetic biology (Synbio) lies in the systematic use of engineering approaches to intentionally design artificial organisms.

Among the numerous analyses of Synbio, many focus on the role of actors active in the field's construction. The American civil engineer Drew Endy, who originally summarised Synbio as aimed 'to make biology easier to engineer', is often invoked as the heroic entrepreneur who played a key role in the emergence of this new field (Campos 2013). Social scientists have investigated the emergence of Synbio using mostly ethnographic approaches or in collaborations with Synbio scientists. The field of enquiry then becomes the (American) lab and/or boundary spaces such as the *International Genetically Engineered Machine (iGEM)* competition, a very popular

B. Raimbault (✉) · P.-B. Joly
Laboratoire Interdisciplinaire Sciences Innovations Sociétés (LISIS), CNRS, INRAE, Université Gustave Eiffel, Marne-la-Vallée, France
e-mail: raimbault.benjamin6@gmail.com; joly@inra-ifris.org

© The Author(s) 2021
K. Kastenhofer, S. Molyneux-Hodgson (eds.), *Community and Identity in Contemporary Technosciences*, Sociology of the Sciences Yearbook 31,
https://doi.org/10.1007/978-3-030-61728-8_4

student contest in Synbio, or governance institutions. As enlightening and rich as those works are, few—with the notable exceptions of Bensaude-Vincent (2013) and Molyneux-Hodgson and Meyer (2009)—offer a more generic perspective of the emergent scientific knowledge at the scale of the discipline or specialty.

In line with the new political sociology of science (NPSS) and the call for renewed attention to institutional factors and asymmetries of power (Frickel and Moore 2006), this chapter proposes to analyse the emergence of Synbio as a *techno*scientific *field* (TSF). The phrase TSF summarises the two theoretical influences we draw on in this work. First, the term techno-scientific refers to the constructivist approach in science and technology studies (STS) which analyses scientific production as a coproduction process—social and epistemic, cognitive and material, cultural and natural, human and non-human. STS covers a heterogeneous set of theories, but we focus mainly on actor-network theory (ANT). Second, our reference to *field* is grounded in the sociological theory developed by Pierre Bourdieu (1975). We draw also on recent developments by Neil Fligstein and Doug McAdam who proposed a theory of strategic action fields (SAF). They define an SAF as a 'constructed mesolevel social order in which actors interact with one another on the basis of shared understandings about the purposes of the field, relationships to others in the field (including who has power and why), and the rules governing legitimate action in the field' (Fligstein and McAdam 2012, p. 9). Thus, field theory allows analysis of emergence from the perspective not only of the production of knowledge and artefacts but also of the emergence of a social order that involves a common understanding of what is at stake, the power relations and hierarchies, and the rules that govern interactions.

Our objective is both methodological and theoretical. The emergence of scientific disciplines has long been a major issue for sociology and the philosophy of science.[1] However, interest in it faded after the 1980s for two reasons. On the one hand, scholars for whom the discipline constitutes a key form of the organisation of scientific production generally consider disciplines to be stable organisational and cognitive entities. They are not much interested in studying their emergence. On the other hand, STS scholars generally do not consider disciplines to be relevant entities: 'Anti-disciplinary partisans' (the expression coined by Marcovitch and Shinn (2011)) see a radical discontinuity between disciplines from the present and those from the past and, at times, consider that we are living in a post-disciplinary world (Gibbons et al. 1994; Nowotny et al. 2001).

We suggest that it is time to re-engage with the emergence of scientific fields in a new way. First, a number of fields have emerged since the 1970s as technoscientific fields as they are characterised by a strong coupling of basic knowledge and technological tools. Biotechnology and nanotechnology are emblematic examples. However, scholars interested in technoscience (Latour, Hottois, etc.) generally side with the anti-disciplinary camp and hence do not focus on the emergence of new

---

[1]See, inter alia, Ben-David and Collins 1966; Kuhn 1962; Lemaine et al. 2012; Mullins 1972.

fields[2]). We suggest that the emergence of technoscientific fields is conditioned by both internal (e.g. scientific credibility) and external (e.g. societal relevance) dynamics. In contrast to scholars such as Marcovich and Shinn (2011) who take the rise of a 'new disciplinarity' seriously but focus exclusively on internal dynamics, we consider that this hybrid nature must be integrated upfront in the analytical framework. Second, the availability of data and new algorithms allows the design and implementation of innovative approaches that match quantitative and qualitative methods. In addition, the new political sociology of science perspective leads to serious consideration of the role of pre-existing entities (institutions, values, norms, infrastructures) in the constitution of scientific fields.

We begin by describing our approach to the emergence of a technoscientific field and then outline our original digital inquiry methodology. The final section of the chapter presents the results of our analysis of Synbio to illustrate the fecundity of new scientometric methods applied within an analytical frame which draws on both STS and field theory.

4.2 Theoretical Framework—The Emergence of Technoscientific Fields Revisited

In their review published in the 2016 edition of the *Handbook of Science and Technology Studies*, Ed Hackett and his colleagues point to the different reasons which might explain the emergence of a new scientific field: coherent networks that arise around potentially generative questions or phenomena; new research instrumentation which allows the knowledge field to be extended; original research ideas that capture the attention of influential groups; the promise of new uses for new scientific knowledge; and, of course, the cumulative effect of these factors which often condition the institutionalisation of a new field (Hackett et al. 2016, p. 740).

The first of these reasons has been explored in depth in the literature. Thomas Kuhn's book *The Structure of Scientific Revolution* points to the ways that social processes and events shape the internal content of intellectual and scientific inquiry (Kuhn 1962). Nicholas Mullins adopts a similar perspective in his paper on the origins of molecular biology. Mullins presents a model of emergence based on socio-cognitive dynamics (Mullins 1972). The process of emergence begins with identification by a group of researchers of a 'central intellectual problem' and ends with the completion of the 'specialty' as 'an institutionalised cluster which has developed regular processes for training and recruitment into roles which are institutionally defined as belonging to that specialty. Members are aware of each other's work, although not necessarily deeply involved in communications with one another' (Mullins 1972, p. 74). The role of a small coherent network of scientists has also been studied by several different scholars, including Diana Crane who

---

[2]With Bensaude-Vincent (2013) again being the exception.

systematically analyses the communication of scientific knowledge in two scientific fields (maths, rural sociology). She identifies the crucial role of a small set of scientists—the 'invisible college'—who monopolise scientific recognition and play a key role in establishing cognitive and social norms (Crane 1972). In contrast, Harry Collins (1974) developed a critique that would become influential in STS. His main argument was that the interactions within scientific communities are not a matter of exchanges of codified information but rather exchanges of tacit informa- tion and practices that are not captured well using scientometric approaches. How- ever, like Crane, Collins considers that a subset of scientists play a key role. He proposed the concept of 'core-set'—defined as scientists 'who are actively involved in experimentation or observation, or making contributions to the theory of the phenomenon, or of the experiment, such that they have an effect on the outcome of the controversy' (Collins 1981, p. 8). More recently, Frickel and Gross (2005) highlighted the key role played by actors with high scientific capital together with younger scholars whose intellectual genealogies connect them to high-status net- works. Their argument is close to Fligstein and McAdam (2012) who suggest that field changes (including the emergence process) are triggered by an alliance between some incumbents and some challengers.

Drawing on these contributions, we also suggest that the dynamics of the co-evolution of a coherent network of key actors and a set of central research questions is the main driver of the emergence process. However, our approach focusses on two key elements. First, we assume that the emergence of a technoscientific field is a multi-scalar dynamic in which the micro, meso, and macro levels are intertwined, while simultaneously conserving their specificity. This is a major difference to approaches based on ANT.[3] Second, we suggest that interactions between the internal dynamics of the emergent field and its broader environment (including the non-scientific environment) constitute a key component of the emergence process. This is a major difference with a Bourdieusian analysis of scientific field which praises the autonomy of the field as guaranteeing the purity of science.

4.2.1 Emergence as a Multi-Scalar Process

Whereas networks are characterised by their fluidity, multi-membership, and lack of clear frontiers, action fields are bounded, and the actors are strongly interdependent and share a common understanding of the stakes and rules of the field. Local interactions and network dynamics (intensification of communication, extension of interactions) are key components but are both constrained and constitutive of the

---

[3] See the recent paper by Bruno Latour and colleagues (2013) who develop digital approaches to analyse technoscientific dynamics. In line with ANT, they use numerical profile databases to test the Tardian hypothesis of a monadic world that dissolves the micro and macro scales.

meso-level order.[4] This focus on the meso-level order properties that characterise SAF is instrumental.

In Bourdieusian field theory, change is the result of struggles between groups of actors whose positions (e.g. dominators vs dominated) depend on the distribution of capital. The capital of the scientific field is constituted mainly by the recognition of peers-competitors (Bourdieu 1975). Also, Fligstein and McAdam (2012) point to the role of internal struggles and suggest more specifically that field emergence generally requires an alliance among the challengers—who have specific social skills related to the settlement of a new field—and the incumbents. They insist also on the influence of the broader environment. Fields do not exist in a vacuum, and relations with other fields play a key role in their dynamics. Following Fligstein and McAdam, we suggest that two types of relations have a major influence on scientific fields. On the one hand, fields have a Russian doll-like structure: fields are embedded within other fields, with any number of smaller fields nested inside the larger ones. A specific illustration might be: science/biology / plant biology /plant genetics /corn genetics (etc.). Smaller fields have partial autonomy; the repertoire of rules in the fields in which they are embedded in is both a constraint and a result of the combined actions of the actors involved. The emergence of a new field is accompanied by a change to the rules, shared meanings, and collective identity. On the other hand, scientific fields have some relations to non-scientific fields. For instance, in line with Fligstein and McAdam (2012), Elisabeth Popp Berman (2014) shows that the marketisation of US academic science is related to the combined role of a changing environment and bottom-up internal strategies. We expand on this below.

Fligstein and McAdam (2012, p. 89) characterise emerging fields as arenas where agreement over the basic conditions of the SAF have yet to emerge.[5] They identify four elements that shape the process of field formation: mobilisation (in our case, identification of an original central intellectual problem which steers the actors to engage in new directions); social skills that favour the stabilisation of a new order; state intervention (for instance, funding of new specific research programmes); and the creation of new organisations (in our case scientific associations, journals, annual conferences, etc.)

Considering the specialised literature on the emergence of a scientific field as well as Fligstein and McAdam (2012), we assume that a small collective of scientists plays a core visioning role. We call this collective the *core-group* in order to insist on (1) the central (core) position of scientists belonging to it and (2) the interactions and alliances among the members of this group. Social skills of members of this group are related first to the identification of the new research questions which constitute

---

[4]This representation of multi-scalar dynamics draws on Anthony Giddens's (1984) structuration theory and his notion of duality of the structural.

[5]The four basic conditions are: (1) a shared understanding of what is going on in the field, i.e. what is at stake; (2) a set of relatively fixed actors in the field whose roles and comparative status/power are consensually defined by others in the SAF; (3) a set of shared understandings about the nature of the 'rules' that will govern interaction in the field; (4) a broad interpretive frame that individual and collective strategic actors introduce to make sense of what others within the SAF are doing.

existential cement. Social skills are needed also for creating the new organisations that support the existence of the field.

Members of the emerging field come from neighbouring fields, and the process of emergence is conditioned by their ability to recruit new members while reinforcing the boundaries of the field. Here, following Thomas Gieryn (1995), we consider emergence as a boundary-work process based on stabilisation of a new set of rules which are both epistemic and social. Epistemic rules bring together the establishment of an epistemic culture, notably a set of research practices and standards of proofs and a common research agenda (Knorr-Cetina 1999), whereas social rules refer to common values and norms that define, for example, relations with industry, relations with other disciplines, intellectual property rights, etc. We concur with Marcovich and Shinn's (2011) observation that in the case of nanoscience, the scientists involved remain strongly grounded in their referent disciplines. The process of emergence of a scientific field proceeds through progressive specialisation of a community of scientists. The process involves participation of a fuzzy and evolving population. As Frickel (2004) mentions, the boundaries of an emerging field are porous. In the early phases of the emergence process, actors may be experimenting with and testing new research ideas and new social rules while remaining strongly attached to their referent disciplines. In these early phases, outcomes are uncertain and depend on the ability of the emergent field to demonstrate its credibility and legitimacy.

How do the members of this core-group identify one another? How does the external analyst identify the group? We assume scientific publications—because of their role in the attribution of priority rights and in the allocation of scientific capital—may facilitate the external analyst's identification of the core-group. A given scientist will signal his or her interest in the Synbio field through the use of specific terms in the most visible parts of the paper and by quoting members of the core-group. This process of literary inscription (Latour and Woolgar 1986) provides traces which the analyst can use to analyse the emergence process.

In contrast to the literature on the emergence of a scientific field which insists on a shared understanding of the stakes related to the new field, we argue that cognitive heterogeneity is constitutive of the process of emergence of a technoscientific field.[6] Our argument is based on two characteristics of technoscientific fields (Joly 2013). First, the engagement of actors in a new emerging field is motivated by strong expectations and the promise of success, which means that it has to build on a breakthrough—a radical novelty—while also being scientific credible. Second, in the process of emergence of a technoscientific field it is necessary to enrol diverse audiences beyond scientists, such as venture capitalists, big companies, regulators, policy makers, and the wider public. Heterogeneity is crucial for meeting the diverse and somewhat contradictory constraints.

---

[6]This does not mean that heterogeneity is associated only with emergence. As one reviewer pointed out, heterogeneity can also signal fragmentation or dissolution. However, the theoretical point here is that heterogeneity plays an active role in the process of emergence for the reasons discussed.

4.2.2   Strategic Control of External Relations

In the case of the emergence of scientific fields, the process is fostered by perceived scientific opportunities related to a shift in the research frontier. Hackett et al. (2016) suggest that other complementary reasons may explain the emergence of a new field, and their combination is generally necessary for the institutionalisation of a new field. Here we want to focus on a factor that is essential to a technoscientific field: the need to address societal problems. Following this, we assume that the work performed at the boundary of the field is instrumental to the emergence of a new field. A new field does not emerge only as a result of interactions within the scientific field. As already mentioned, control of external resources (including state intervention) is necessary for this process to occur. This recalls Latour and Woolgar's (1986) proposition in their *Laboratory Life* of an extended credibility cycle which takes into account the transformation of external resources into scientific capital. Following this intuition, research on the science/industry links shows that, depending on their position in the scientific field, laboratories may exploit the links with industry (Joly and Mangematin 1996). It has been shown also that these links induce a higher propensity for explorative research, whereas peer regulation reinforces normal science (Evans 2010). More generally, these observations are related to the notion of the strength of weak ties: control of external links constitutes a strategic resource (Granovetter 1973).

In this respect, the idea of an extended credibility cycle is important. Therefore, within the perspective of emergence, it is necessary to analyse how the strategies of boundary spanners secure access to critical resources.

Three propositions to describe and analyse the emergence of a TSF can be made. They are related to the three arguments as outlined: (1) the role of heterogeneity. Our first proposition claims that heterogeneity (of disciplines, research questions, visions, and social norms) plays a key role in the emergence phase of a new technoscientific field.; (2) the way actors' positions (or hierarchies) can influence the process. This leads us to a second proposition where alliances between different types of scientific entrepreneurs (incumbents and challengers) are instrumental to the process of emergence; and finally (3) the strategic interactions between the studied field and other fields (autonomy) connected to the last proposition highlighting the importance of strategies for the control of external resources in the process of emergence.

4.3 Use of Advanced Scientometrics and Qualitative Methods

The divide between quantitative (mainly scientometric) and qualitative (mainly ethnographic) approaches is deeply entrenched in STS (Wyatt et al. 2016).[7] However, this divide is being challenged by the availability of data and a new generation of algorithms that lead to old theoretical questions being revisited.

This chapter contributes to bridging the gap between quantitative and qualitative methods based on an articulation between a scientometric analysis and a five-year investigation of the Synbio community conducted by one of the authors (BR). Our original methodology is presented in a previous publication in a more technical way (Raimbault et al. 2016). Here we focus on the main principles of the study. The qualitative investigation is based on a multi-sited ethnography involving France and the USA (Marcus 1995). The 10-month participant observation at three Synbio centres[8] was completed by attendance at the relevant scientific conferences and by information from over 130 interviews with scientists, technicians, administrative support staff, policy makers, entrepreneurs, and industry project managers.

Based on claims that emergence is a multi-scalar process, and strategic external relations are crucial, we develop three key methodological considerations.

4.3.1 Delineation

Since boundary work in an emerging field is work in progress, delineation of the corpus has received considerable attention. In the case of Synbio, we built an initial core corpus comprising articles extracted from the Thomson Reuters Web of Science scientific publications database between 2000 and 2012. The articles matched the simple query condition: 'Topic = synthetic biology'. Since we assume that authors working in the domain may not use the expression 'synthetic biology' systematically, we decided to perform a lexical expansion (Mogoutov and Kahane 2007) which allowed us to identify 11 supplementary terms specific to the field. Thus, our corpus was 4605 scientific articles related to Synbio (Raimbault et al. 2016). As we argue elsewhere, the descriptive characteristics of this corpus are similar to those outlined in Oldham et al.'s (2012) quantitative analysis.[9] This strategy is relevant to

---

[7]In the 1980s, bridging the qualitative and quantitative approaches was a core objective of Michel Callon, who developed scientometric analysis based on ANT to map the dynamics of science (Callon et al. 1986).

[8]These are the Center for Integrative Synthetic Biology (Massachussetts Institute of Technology, USA), the Institute of Systems and Synthetic Biology (Genopole, France), and Toulouse White Biotechnology (TWB, France).

[9]Shapira et al. (2017) observe that our search strategy is far more restrictive than that exploited by Hu and Rousseau (2015).

the premises of field theory according to which the Synbio space is not a construction of the analyst but must correspond to the shared understanding of the actors involved.

4.3.2   *Heterogeneity*

We use original and rigorous approaches based on co-citation and lexical networks to identify a set of epistemic clusters and their relative positions. Citation analysis is a traditional scientometric technique which makes it possible to visualise clusters of related citations as the constituent sub-domains of a given research field (Fig. 4.1). The rationale underlying co-citation analysis is that the structure of the co-citation network indexes the epistemic foundations of the research community which

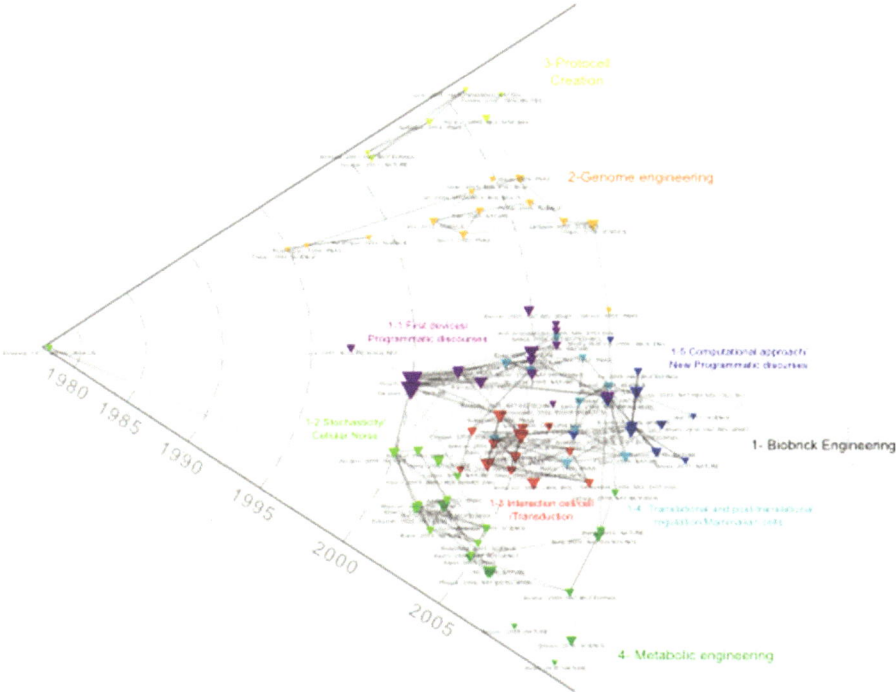

Fig. 4.1 References co-citation map. The 100 most cited references are considered to build a co-citation network. Node sizes scale with the number of citations received by references. Node colour depends on their cluster assignment. Edge widths are proportional to the strength of the link between references. Cluster labels (bold and upper-case) were manually added and correspond to a subfield in Synbio. During this labelling process, we gathered five very tightly connected clusters under the same subfield. We identify four subfields projected on a historical axis, allowing us to follow the dynamic and the weight of each subfield: 1.X DNA-based device construction, 2 Protocell creation, 3 Genome engineering, 4 Metabolic engineering

consists of groups of seminal works shared by the various sub-communities and not by the *prestige* or the *intentionality* of the 'most obvious' authors, ignoring knowledge-production in practice.[10] Clusters are detected by using an automatic community detection algorithm which determines the multi-level structure of the citation network.

4.3.3 Distribution of Scientific Capital

We selected three different indicators to determine the position of each author in the corpus on the assumption that association with those indicators was a good approximation for scientific capital: (1) author centrality (an indicator commonly used in network analysis to measure the extent to which a node is likely to be a passage point for other nodes in a network); (2) cumulative impact of the author's publications in the corpus; and (3) author productivity (number of articles in the corpus). These three indicators are represented on the graph in Fig. 4.3: centrality is represented on the horizontal axis, impact on the vertical axis, and productivity is represented by the size of the circle.

Finally, we completed the formal identification of the core-group, including a set of complementary data related to the building of the socio-economic environment constituting the field. In the STS literature, scientists' attempts to bridge natural and social worlds are central, whereas Bourdieusian field theory is focused more on field purity. In contrast (and according to the frame proposed by Fligstein and McAdam), we are interested in the boundary-spanning dimension of the core-group. Thus, we examine core-group members' involvement in three different sets of activities based on the biographical information of the principal investigator (PI). For each dimension we identify important activities in which the PI either participates or not. The sum of this binary information results in three different scores for each author. The resulting table provides the overall positioning of the members of the core set according to the three indicators (Table 4.1):

(1) their interactions with companies that use new knowledge produced in Synbio and transforms it into new products and processes;
(2) their contribution to the establishment of organisations that contribute to stabilisation of the emergent field;
(3) their contribution to initiatives aimed at building Synbio governance—for example, regulation as well as national and international research programmes. In 2011, Zhang et al. (2011) listed 39 reports produced between 2004 and 2011 dedicated to Synbio governance. This list increases the number of such reports to 58 based on the literature review and our field investigation.

---

[10]In 1974, Collins challenged Crane's questionnaire based methodology, writing that 'there is serious doubt that the most important contributors to the ideas of a scientist are necessarily the most obvious contributors' (Collins 1974, p. 169).

Table 4.1 Repartition of core-group members according their contribution to the Synbio subfields (BE = Biological Engineering, ME = Metabolic Engineering, GE = Genome Engineering and PC = Protocell Creation) identified previously and their involvement in non-academic activities based on three indicators: Business Development (Bus.), Core-Organisation (Inst.), and Governance (Gov.)

Principal Investigator	Country	Subfield	Institutions	Bus.	Org.	Gov.
C. A. Voigt*	USA	BE	MIT	3	4	1
P. Silver*	USA	BE	Harvard	2	4	2
D. Endy*	USA	BE	Stanford	3	3	6
J. Keasling*	USA	BE, ME	Berkeley	3	3	2
G. Church*	USA	BE, ME, PC	Wyss institute	3	3	2
J. Collins*	USA	BE	MIT/Wyss institute	3	2	1
R. Weiss*	USA	BE	MIT	2	2	2
A. Arkin*	USA	BE	Berkeley	2	2	0
W. Lim*	USA	BE	UCSF	1	2	1
M. Fussenegger	Switzerland	BE	ETH Zurich	3	2	0
F. Arnold	USA	BE	Caltech	3	2	0
C. Smolke	USA	BE	Stanford	3	2	1
J. Hasty	USA	BE	UCSD	2	1	0
T. Lu	USA	BE	MIT	3	1	0
X. Wang	USA	BE	Arizona State	1	1	0
D. Mcmillen	Canadian	BE	Toronto	1	1	0
W. Weber	Switzerland	BE	BIOSS (Freiburg)	3	1	0
L. You	USA	BE	Duke	0	1	0
G. Stephanopoulos	USA	BE	MIT	3	1	0
T. Segall-Shapiro	USA	BE, GE	MIT	1	0	0
J. Liao	USA	BE, GE, ME	UCLA	3	0	0
M. Elowitz	USA	BE	Caltech	2	0	1
S. Benner	USA	BE	FAME	3	0	0
V. De Lorenzo	Spain	BE	CNB Madrid	3	0	2
Y. Zhang	USA	BE, ME	Virginia Tech	2	0	0
C. Hutchison	USA	GE	JCVI	2	0	0
C. Venter'	USA	GE	JCVI	3	0	1
H. Zhao	USA	GE, ME	Illinois University	2	0	1
C. Moya	Spain	PC, GE	University of Valencia	0	0	0

The presence of a * indicates the participation of the author in the Synthetic Biology Engineering Research Center (SynBERC)

4.4   Mapping the Emergence of Synbio as a Technoscientific Field

4.4.1   Heterogeneity

P1–Heterogeneity (of disciplines, research questions, visions, and social norms) plays a key role in the emergence phase of a new technoscientific field

As discussed above, we argue that heterogeneity is not only present in but is constitutive of the emergence process. First, subfields are complementary in terms of the formulation of promises and establishment of credibility. Second, heterogeneity is important for addressing the needs of diverse audiences.

Our empirical analysis shows that Synbio does not correspond to a set of research oriented toward a common question. The related research can be better described as a set of distinct research programmes gathered under one umbrella (Rip and Voß 2013). As already mentioned, engineering biology is the central reference bounding the Synbio field and the Synbio community. However, the meanings attached to engineering are diverse, and biologists and social scientists working in Synbio acknowledge this heterogeneity. Frequently, they distinguish among three main approaches: DNA-based construction (another naming convention for a 'biobrick engineering approach'), genome-driven cell engineering, and protocell creation (O'Malley et al. 2008).

Co-citation analysis allows us to identify four subfields (or research programmes) which are plotted in Fig. 4.1. Three have been identified in previous work (O'Malley et al. 2008)[11]—biobrick engineering, genome engineering, and protocell creation. The fourth subfield is 'metabolic engineering' which corresponds to a scientific approach established during the late 1990s.

Biobrick engineering is the most central subfield. It uses an explicit electronic metaphor, introduces a set of new concepts such as gene circuits, genetic switches, biobricks, chassis, etc., and refers to modelling. The biobrick approach is the most visionary approach, meaning that its vision for Synbio (programmatic discourses and direction for the field) affects the interpretation of biology (e.g. modularity) or scientific practices (e.g. standardisation). The ambition is to provide a new biology ontology and push for new disruptive practices.

The second subfield, protocell creation, has been diminishing since 2006. Its scope includes basic, even classic, scientific interrogations, such as the origins of life, rather than applications.

Genome engineering tackles two main issues: (1) synthesis of large nucleic acid sequences; and (2) research on a minimal genome. Interestingly, and although Venter is considered a pioneer of Synbio, genome engineering does not connect directly to biobrick engineering.

---

[11]The construction of this typology is actually very different. O'Malley et al.'s classification is based on group intentions and routine practices, whereas ours is based on an automatic process.

Finally, metabolic engineering stands out in the sense that this cluster refers to an existing and already stabilised technoscientific field. Work on this approach is characterised by a strong focus on applications (e.g. production of artemisinin) and optimisation of expression of metabolic pathways.

Rather than seeing this heterogeneity as a weakness in the structuration of Synbio, epistemic diversity is a resource contributing to the stabilisation of the field. Epistemic disruption is promoted by the most promising approach (biobrick engineering), whereas the cohesion between approaches is related to an already established field (metabolic engineering). The biobrick approach provides vanguard visions of the field (Hilgartner 2015). This approach is well-known for its ability to promise and to stimulate utopian discourse, which is efficient for recruiting practitioners and attracting particular policy makers and funders. On the other hand, metabolic engineering (and to a lesser extent genome engineering) provides credibility because of its demonstrated ability to deliver technological applications. Promises allow resources to be mobilised and used to explore new research avenues. At the same time, they put pressure on the scientists and the institutions involved to engage in the design of practical applications.

4.4.2 Hierarchy

P2–Alliances between different types of scientific entrepreneurs (incumbents and challengers) are instrumental to the process of emergence

Our scientometric dataset provides critical information on the evolving population of scientists active in Synbio. From a quantitative perspective, publications and newcomers show high rates of growth (Fig. 4.2). In 2014 (the last year for which the dataset can be considered complete) there were 3044 unique authors, 2133 of whom were newcomers (never previously published in the field). Hence, although decreasing, the proportion of scientists that had published only once in Synbio was above 60% in 2014. We can assume that an important proportion of those authors who have published once in Synbio would not consider themselves synthetic biologists; they more likely feel a sense of belonging to some other field.

However, over time, the internal connectivity of the field increases sharply. In Fig. 4.2, connectivity is measured by the size of the 'largest component'—defined as the largest subset of researchers connected through a direct path in the collaboration network (green curve in Fig. 4.2). This increase can be interpreted as signalling an intensive process of community-building.

The concomitance of population growth and its connectivity is a sign of maturation. However, it tells us little about the part of the community that considers itself attached to Synbio and which is active in building the field. To identify this subgroup, we applied three measures systematically: network centrality, scientific recognition (impact measured by the number of citations received), and productivity (number of publications in Synbio). We identified a group of 30 individuals with the

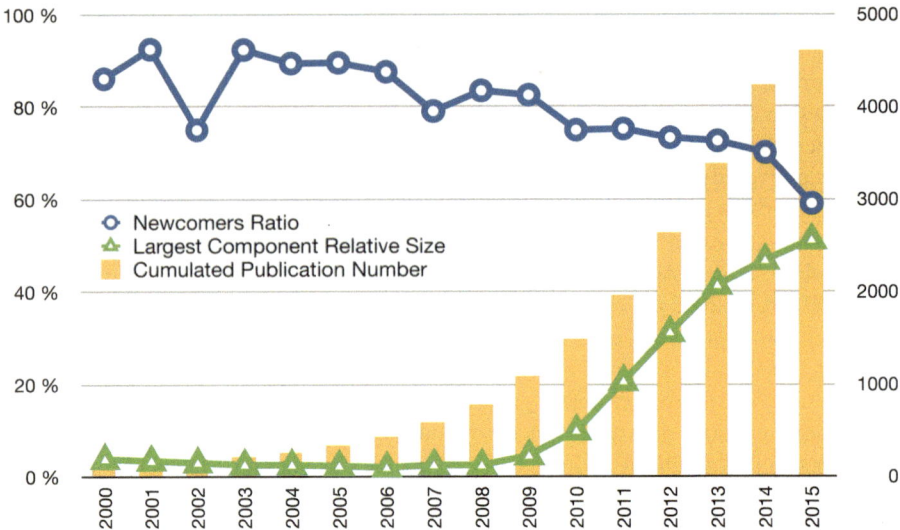

Fig. 4.2 Global population statistics for the synbio community over time. Cumulated number of publications (bar chart), ratio of newcomers, and largest component relative size. While the number of publications follows a typical exponential growth, the newcomers ratio—while very high—is decreasing with time, and the largest component relative size has been growing significantly since 2010, indicating progressive structuration of the Synbio community

highest global score obtained as the product of author centrality and impact,[12] allowing us to bridge across field and network effects. We call this collective the core-group.

Figure 4.3 shows how stratified the population is: more than 10,000 scientists (over 11,321 scientists who have published at least once in Synbio) record both very low impact and centrality. They are concentrated in the bottom left corner of Fig. 4.3. The 30 scientists with the highest scores are plotted on a pie diagram showing the distribution of their articles among the four different approaches (green for biobricks engineering, yellow for genome engineering, purple for metabolic engineering, and pink for protocell creation). Four descriptive comments are needed. First, we observe an over-representation of US-based scientists: among the 30 members of the core-group, 25 work in US institutions. Second, members of the core-group all work in prestigious institutions. Their academic career paths reflect the involvement of these institutions in Synbio. Third, the epistemic landscape is strongly defined by members of this core-group, who are co-authors of 77 of the 100 most cited articles in the corpus. Fourth, members of the core-group are highly concentrated in the central approach of biobrick engineering, which is coherent with our analysis.

---

[12]This product is equivalent also to computation of the product of centrality, average impact per paper, and number of publications.

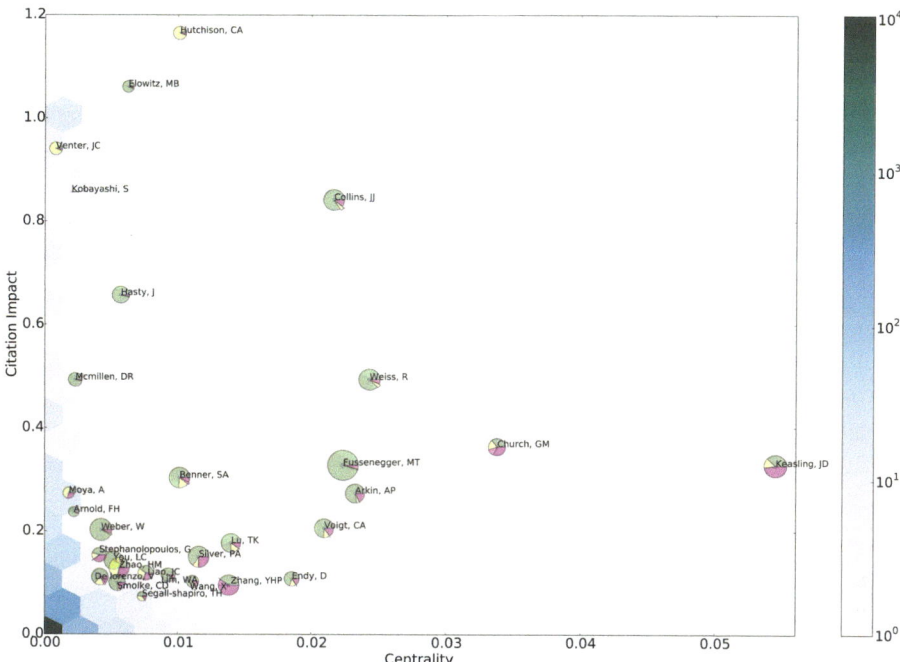

Fig. 4.3 Position analysis of the main actors in the synbio community according to centrality and impact

The composition of capital among the core-group is also heterogeneous. Looking at the network, we see that some scientists may not be central but are highly influential (Venter, Elowitz); some are very central and highly influential (Keasling, Church); others are not very influential but are very central (Endy); etc. This heterogeneity in the composition of capital is important since the core-group is composed of scientists with different skills and positions, both well-established scientists and challengers.

It is important to note that the social skills of core-group members do not consist only of their capacity to formulate attractive research questions. They are also linked closely to community-building. Our indicator 'core-organisations' includes the involvement of core-group members in the main stabilising organisations of Synbio already identified in social science analysis. They are: the iGEM competition (judge, advisor, or laboratory registered by the contest), which is considered 'the most important community-building activity in the biobricks field' (Calvert 2012); the ERC programme SynBERC (PI and affiliated PI); and some international conferences such as SBX.0 (organiser and member of the steering committee) and SEED— Synthetic Biology: Engineering, Evolution, Design (organiser and member of the steering committee).

PIs in the biobrick approach are the most active: 21 out of 25 PIs linked to this approach are involved in at least one of these organisations (Table 4.1). The most active PIs in the three organisations (score of 3 or 4) belong to the biobrick approach; adherents to other approaches are comparatively less systematically involved in these kinds of organisations. All those subscribing to the genome engineering approach, for example, have a score of less than or equal to 1. It is interesting that although J. C. Venter is often considered to be one of the founders of the Synbio domain, according to our indicators he is not involved in the construction of the Synbio community (score = 0). Four of six PIs linked to the metabolic engineering approach have scores lower than or equal to 1. This is a reflection not of lack of involvement in the scientific organisations of Synbio but merely of the fact that in this approach representatives are already invested in specific metabolic engineering organisations. For example, Stephanopoulos, a pioneer in the field, edits *Metabolic Engineering* and was one of the founders of the international conference cycle 'Metabolic Engineering', whose fourth conference was co-organised by Liao. Finally, the centrality of Jay Keasling can be explained by his strong participation in Synbio organisations (he is director of SynBERC) and metabolic engineering (he was a member of the editorial board of *Metabolic Engineering* and has participated in many metabolic engineering conferences, etc.).

Although the scientists in the core-group have different profiles, different visions, and different positions in the field, they regularly combine their efforts to promote the field by setting up new projects or devices which support community-building. An emblematic example is the creation of the Synthetic Biology Engineering Research Center (SynBERC), established at Berkeley (Rabinow and Bennett 2012).

Hence, the core-group has a central position and plays a very active role in community-building. Importantly, the alliance strategies of incumbents and outsiders are decisive in this process. Two comments are due here. First, although defined by very formal criteria, this core-group includes the informal networks identified in the literature as at the heart of development of the domain. Second, and contrary to the Synbio literature, our analysis puts Drew Endy's role into perspective. The combination of strategies related to different positions appears to be a crucial element in the emergence process.

4.4.3 Autonomy

P3–The importance of the strategies for the control of external resources in the process of emergence

Analysis of the interactions between Synbio and non-scientific areas draws on the core-group identified based on the criteria of centrality and scientific recognition discussed above. Our qualitative analysis shows that these 'star scientists' also play important roles as boundary spanners. We consider the contributions of the members of the core-group to relations with industry and Synbio governance.

First, members of the core-group are heavily involved in activities aimed at realising the commercial potential of Synbio. Since the 1980s, the hybrid nature of scientists who are both researchers and entrepreneurs (Shapin 2009) has been emphasised. According to Kleinman and Vallas (2001), this is part of a broader phenomenon of asymmetric convergence related to alignment between the codes and practices of academia and industry. Thus, asymmetry refers to the uneven nature of the circulations between the two social worlds to the benefit of industry. In relation to its effect, particularly on the world of biology, asymmetric convergence has been studied extensively but seldom considered a source of credibility for the establishment of an emerging field. Mackenzie (2013) points out that one of the strategies of synthetic biologists to establish the credibility of the domain consists of building intermediate technical achievements. However, Mackenzie does not consider the commercial character of these technical achievements, whereas we assume it to be central as per the most recent report from the American Academy of Sciences concerning Synbio entitled *Industrialization of Biology*.[13] Our indicator 'business development' draws on three kinds of information: participation in a start-up scientific advisory board (SAB), creation of a start-up, and ownership of patents as the inventor. Almost all the PIs (27 out of 29) are involved in the economic sector, and two-thirds of them score high for 'business development'. Almost all PIs (27 out of 29) have applied for a patent as an inventor. This is an exceptionally high score despite the patenting of academic research results becoming routine in biotechnology (Sauermann and Stephan 2012). Finally, 20 members of the core-group are members of at least one SAB, and 17 have created a start-up. This involvement is extremely high for members of the core set who were identified by their centrality in scientific production.

Second, members of the core-group are very active in the setting up of organisations to promote the development of Synbio and build its legitimacy. We consider four criteria: (1) being a PI in the Synthetic Biology Engineering Research Center (SynBERC),[14] considered the seminal organisation of the field (Rabinow and Bennett 2012; Schyfter and Calvert 2015); (2) being a member of a lab entered in the iGEM contest, which annually gathers thousands of students in Boston (Calvert 2012; Keller 2009)[15]; (3) organisation of one of the International Meetings on Synthetic Biology organised by the Biobrick Foundation (SBX.O) (Campos 2012); and (4) organisation of the annual SEED conference, which has become one of the main events for the international Synbio community following its creation in 2014.

Our 'governance' indicator captures the participation of members of the core-group in the preparation of reports dedicated to Synbio governance and advocacy for state intervention. The literature highlights the early work related to legitimising Synbio in intensive production of grey literature (Zhang et al. 2011). This work is

---

[13]*Industrialization of biology: A roadmap to accelerate the advanced manufacturing of chemicals.* (2015). Retrieved from https://www.nap.edu/catalog/19001/industrialization-of-biology-a-roadmap-to-accelerate-the-advanced-manufacturing

[14]Between 2015 and 2016, 11 interviews were conducted with the main PIs in SynBERC.

[15]*Synthetic Biology Labs Registered.* (n.d.). Retrieved from http://igem.org/Lab_List

conducted in multiple spaces: regulatory (final opinion on synthetic biology I, II, III[16]), ethical (Presidential Commission for Bioethical Issues[17]), scientific (conferences, iGEM competition, academies of sciences), and industry (report on the regulation of synthetic genes[18]). It is common to include the synthetic biologists among the authors of these reports. For instance, to create this indicator we listed 58 reports which reflect the importance of governance issues in Synbio. About half of the core-group (13 PIs) has coordinated at least one report dedicated to one aspect of Synbio governance. This involvement remains relatively targeted since only one of the 13 PIs concerned has been involved in more than two reports.[19] Governance issues constitute a collective and institutionalised element of discussion on Synbio development and are maturing alongside the scientific and economic environment. The most representative example of the heterogeneous agenda alignment is undoubtedly Drew Endy (author of 6 reports). The profile of activities of J. C. Venter should also be mentioned. Although he was personally involved in only one—albeit decisive—report (by the Presidential Commission for the Study of Bioethical Issues), his J. Craig Venter Institute is very active in governance issues. Indeed, staff in this centre who are trained in law and political science have prepared reports on the regulation of Synbio.

4.5   Conclusion

This chapter conceptualises the emergence of a TSF as a multiscalar and progressive establishment of a new set of epistemic and social rules. In order to account for this phenomenon, we draw on STS and field theories to propose an original methodological and conceptual framework which is attentive to the interactions between structure and agency in the analysis of change.

Our conceptual contribution relies on the design of an approach focused on three core propositions for characterising the emergence process of a TSF. We use innovative scientometric tools to create specific variables that allow us to investigate both network and field structural dynamics. Hence, this chapter contributes to the

---

[16] *Final Opinion on Synthetic Biology.* (2014). https://ec.europa.eu/health/scientific_committees/consultations/public_consultations/scenihr_consultation_21_en

[17] *The Ethics of Synthetic Biology: Guiding Principles for Emerging Technologies.* (2011). Retrieved from https://president.upenn.edu/meet-president/ethics-synthetic-biology-guiding-principles-emerging-technologies

[18] *Synthetic biology and the US biotechnology regulatory system: Challenges and options.* (2014). Retrieved from https://www.jcvi.org/sites/default/files/assets/projects/synthetic-biology-and-the-us-regulatory-system/full-report.pdf

[19] Note that we used a very discriminating criterion for building our 'governance' indicator because we considered only strong implications in the various reports (e.g. organisation, editing, coordination)

current stream of research that aims to bridge qualitative and quantitative approaches.

We use the emergence of synthetic biology to illustrate this approach and test our methodology. This novel approach could obviously be applied to other emerging fields. Beyond validation aspects, a comparative exercise of diverse emergent TSF would allow classical STS issues to be revisited in relation to what is called 'the new production of knowledge' (Gibbons et al. 1994; Nowotny et al. 2001).

First, a comparative study of different emergent TSF would make possible a more precise investigation of science-industry relations and knowledge production in the context of applications. While the rise of commodification of science is now commonly identified (Mirowski et al. 2008), we still lack studies on the way that it affects research content (Gläser and Laudel 2016) and how industry is a crucial site for the establishment of credibility. The use of scientometrics would facilitate comparison, while strict ethnographic studies are often closely attached to a specific site (Kleinman 2003). This could help to identify various types of commodification and to study how far this is related to the differentiation of contemporary sciences.

Second, the emergence of TSF is also a political process. Our study showed how the involvement of core scientists in realising the commercial potential of the field cannot be dissociated from involvement in governance issues. Here, the collective strategy has been highlighted at a micro level. Comparative studies between emergent TSF engaging political science literature would allow a macro-level strategy to be reached. In particular, it would be interesting to consider the constitution of scientific communities as a site where national science policies and globalised circulation of knowledge articulate or confront each other (Jasanoff 2011).

Finally, comparison does not have to be restrained to emergent TSF. Our framework makes it possible to study scientific production dynamics in line with David Hess' proposition for 'a more deeply historised sociology of scientific knowledge'. (Hess 2006). Synbio is not the first field claiming to engineer biology, and a comparison with the structuration of metabolic engineering—a field already settled in the 1990s—shows how fields are intertwined and combined each other (Raimbault 2018). Heterogeneity is thus also a temporal heterogeneity, and novel fields could be considered as the reformulation of existing practices, concepts, or organisations.

Acknowledgments The authors wish to thank the Laboratoire Interdisciplinaire Science Innovation Société at Université Paris-Est for the research assistance and more especially David Demortain, Marianne Noël, Jean-Philippe Cointet and Catherine Paradeise for discussing previous versions of the chapter. This work also benefited greatly from comments by the two reviewers and the editors.

References

Bensaude-Vincent, B. 2013. Discipline-building in synthetic biology. *Studies in History and Philosophy of Science Part C: Studies in History and Philosophy of Biological and Biomedical Sciences* 44 (2): 122–129.

Berman, E.P. 2014. Field theories and the move toward the market in US academic science. In *Fields of knowledge: Science, politics and publics in the neoliberal age*, ed. S. Frickel and D.J. Hess, 193–221. Bingley: Emerald Group Publishing Limited.

Bourdieu, P. 1975. La spécificité du champ scientifique et les conditions sociales du progrès de la raison. *Sociologie et sociétés* 7 (1): 91–118.

Callon, M., J. Law, and A. Rip. 1986. Qualitative scientometrics. In *Mapping the dynamics of science and technology*, ed. M. Callon et al., 103–123. London: Palgrave Macmillan.

Calvert, J. 2012. Ownership and sharing in synthetic biology: A 'diverse ecology' of the open and the proprietary? *BioSocieties* 7 (2): 169–187.

Campos, L. 2012. The BioBrick™ road. *BioSocieties* 7 (2): 115–139.

———. 2013. Outsiders and in-laws: Drew Endy and the case of synthetic biology. In *Outsider scientists: Routes to innovation in biology*, ed. O. Harman and M.R. Dietrich, 331–348. University of Chicago Press.

Collins, H.M. 1974. The TEA set: Tacit knowledge and scientific networks. *Social Studies of Science* 4 (2): 165–185.

———. 1981. The place of the 'core-set' in modern science: Social contingency with methodological propriety in science. *History of Science* 19 (1): 6–19.

Crane, D. 1972. *Invisible colleges: Diffusion of knowledge in scientific communities*. Chicago: University of Chicago Press.

European Commission. The Scientific Committees on consumer safety, on emerging and newly identified health risks, on health and environmental risks (2014) Final opinion on Synthetic Biology.

Evans, J.A. 2010. Industry collaboration, scientific sharing, and the dissemination of knowledge. *Social Studies of Science* 40 (5): 757–791.

Fligstein, N., and D. McAdam. 2012. *A theory of fields*. Oxford/New York/Auckland: Oxford University Press.

Frickel, S. 2004. Building an interdiscipline: Collective action framing and the rise of genetic toxicology. *Social Problems* 51 (2): 269–287.

Frickel, S., and N. Gross. 2005. A general theory of scientific/intellectual movements. *American Sociological Review* 70 (2): 204–232.

Frickel, S., and K. Moore. 2006. *The new political sociology of science: Institutions, networks, and power*. University of Wisconsin Press.

Gibbons, M., C. Limoges, H. Nowotny, S. Schwartzman, P. Scott, and M. Trow. 1994. *The new production of knowledge: The dynamics of science and research in contemporary societies*. SAGE.

Giddens, A. 1984. *The constitution of society: Outline of the theory of structuration*. University of California Press.

Gieryn, T.F. 1995. Boundaries of science. In *Science and the quest for reality*, ed. A.I. Tauber, 293–332. London: Palgrave Macmillan.

Gläser, J., and G. Laudel. 2016. Governing science: How science policy shapes research content. *European Journal of Sociology/Archives Européennes de sociologie* 57 (1): 117–168.

Granovetter, M.S. 1973. The strength of weak ties. *American Journal of Sociology* 78 (6): 1360–1380.

Hackett, E.J., J.N. Parker, N. Vermeulen, and B. Penders. 2016. The social and epistemic organization of scientific work. In *The handbook of science and technology studies*, ed. U. Felt, R. Fouché, C.A. Miller, and L. Smith-Doerr, 4th ed., 733–765. Cambridge, MA: MIT Press.

Hess, D.J. 2006. Antiangiogenesis research and the dynamics of scientific fields: Historical and institutional perspectives in the sociology of science. In *The new political sociology of science:*

Institutions, networks, and power, ed. S. Frickel and K. Moore, 122–148. University of Wisconsin Press.

Hilgartner, S. 2015. Vanguards, visions and the synthetic biology revolution. In *Science and democracy: Making knowledge and making power in the biosciences and beyond*, ed. S. Hilgartner, C. Miller, and R. Hagendijk, 33–56. Routledge.

Hu, X., and R. Rousseau. 2015. From a word to a world: The current situation in the interdisciplinary field of synthetic biology. *PeerJ* 3: e728.

Jasanoff, S. 2011. *Designs on nature: Science and democracy in Europe and the United States.* Princeton University Press.

Joly, P.-B. 2013. On the economics of techno-scientific promises. In *Débordements: Mélanges Offerts à Michel Callon*, ed. M. Akrich, Y. Barthe, F. Muniesa, and P. Mustar, 203–222. Presses des Mines via OpenEdition.

Joly, P.-B., and V. Mangematin. 1996. Profile of public laboratories, industrial partnerships and organization of R & D: The dynamics of industrial relationships in a large research organization. *Research Policy* 25 (6): 901–922.

Keller, E.F. 2009. What does synthetic biology have to do with biology? *BioSocieties* 4 (2–3): 291–302.

Kleinman, D.L. 2003. *Impure cultures: University biology and the world of commerce.* University of Wisconsin Press.

Kleinman, D.L., and S.P. Vallas. 2001. Science, capitalism, and the rise of the 'knowledge worker': The changing structure of knowledge production in the United States. *Theory and Society* 30 (4): 451–492.

Knorr-Cetina, K. 1999. *Epistemic cultures: How the sciences make knowledge.* Harvard University Press.

Kuhn, T.S. 1962. *The structure of scientific revolutions.* Chicago: University of Chicago Press.

Latour, B., and S. Woolgar. 1986. *Laboratory life: The construction of scientific facts.* Princeton University Press.

Lemaine, G., R. Macleod, M. Mulkay, and P. Weingart. 2012. *Perspectives on the emergence of scientific disciplines.* Walter de Gruyter.

Mackenzie, A. 2013. Realizing the promise of biotechnology: Infrastructural-icons in synthetic biology. *Futures* 48: 5–12.

Marcovich, A., and T. Shinn. 2011. Where is disciplinarity going? Meeting on the borderland. *Social Science Information* 50 (3–4): 582–606.

Marcus, G. 1995. Ethnography in/of the world system: The emergence of multi-sited ethnography. *Annual Review of Anthropology* 24: 95–117.

Mirowski, P., E.-M. Sent, E.J. Hackett, O. Amsterdamska, M. Lynch, and J. Wajcman. 2008. The commercialization of science and the response of STS. In *Handbook of science and technology studies*, ed. E.J. Hackett, O. Amsterdamska, and M. Lynch, 635–689. Cambridge, MA: MIT Press.

Mogoutov, A., and B. Kahane. 2007. Data search strategy for science and technology emergence: A scalable and evolutionary query for nanotechnology tracking. *Research Policy* 36 (6): 893–903.

Molyneux-Hodgson, S., and M. Meyer. 2009. Tales of emergence—Synthetic biology as a scientific community in the making. *BioSocieties* 4 (2–3): 129–145.

Mullins, N.C. 1972. The development of a scientific specialty: The Phage Group and the origins of molecular biology. *Minerva* 10 (1): 51–82.

Nowotny, H., P. Scott, and M. Gibbons. 2001. *Re-thinking science: Knowledge and the public in an age of uncertainty.* London: Polity Press.

O'Malley, M.A., A. Powell, J.F. Davies, and J. Calvert. 2008. Knowledge-making distinctions in synthetic biology. *BioEssays* 30 (1): 57–65.

Oldham, P., S. Hall, and G. Burton. 2012. Synthetic biology: Mapping the scientific landscape. *PLoS One* 7 (4): e34368.

Rabinow, P., and G. Bennett. 2012. *Designing human practices: An experiment with synthetic biology.* University of Chicago Press.

Raimbault B. 2018. *A l'ombre des biotechnologies: Reformuler la production de savoirs par la bio-ingénierie en France et aux Etats-Unis.* PhD dissertation, Université Paris-Est Marne la Vallée.

Raimbault, B., J.-P. Cointet, and P.-B. Joly. 2016. Mapping the emergence of synthetic biology. *PLoS One* 11 (9): e0161522.

Rip, A., and J.-P. Voß. 2013. *Umbrella terms as mediators in the governance of emerging science and technology.*

Sauermann, H., and P. Stephan. 2012. Conflicting logics? A multidimensional view of industrial and academic science. *Organization Science* 24 (3): 889–909.

Schyfter, P., and J. Calvert. 2015. Intentions, expectations and institutions: Engineering the future of synthetic biology in the USA and the UK. *Science as Culture* 24 (4): 359–383.

Shapin, S. 2009. *The scientific life: A moral history of a late modern vocation.* University of Chicago Press.

Shapira, P., S. Kwon, and J. Youtie. 2017. Tracking the emergence of synthetic biology. *Scientometrics* 112 (3): 1439–1469.

Wyatt, S., S. Milojević, H.W. Park, and L. Leydesdorff. 2016. The intellectual and practical contributions of scientometrics to STS. In *The handbook of science and technology studies*, ed. U. Felt, R. Fouché, C.A. Miller, and L. Smith-Doerr, 4th ed., 87–113. Cambridge, MA: MIT Press.

Zhang, J., C. Marris, and N. Rose. 2011. *The transnational governance of synthetic biology: Scientific uncertainty, cross-borderness and the 'art' of governance*, Report no. 4. London: City, University of London. Retrieved from http://openaccess.city.ac.uk/16098/1/Transnational %20Governance%20SynBio%202011.pdf.

Chapter 5
Self-Organisation and Steering in International Research Collaborations

Inga Ulnicane

> *Self-organizing networks that span the globe are the most notable feature of science today. These networks constitute an invisible college of researchers who collaborate not because they are told to but because they want to, who work together not because they share a laboratory or even a discipline but because they can offer each other complementary insight, knowledge or skills. (Wagner 2008, p. 2)*

5.1 Introduction

An important characteristic of the scientific community is international collaboration among its members. Some of the key works on the scientific community, such as Diane Crane's 1972 book *Invisible Colleges* and Warren O. Hagstrom's 1965 work *The Scientific Community,* demonstrate the major role that collaboration plays in advancing and diffusing knowledge in scientific communities. While collaboration, including international collaboration, has a long tradition in the scientific community since its early professionalisation in the seventeenth and eighteenth centuries (Beaver and Rosen 1978; Crawford et al. 1993), today it is intensifying, as demonstrated by an increasing share of international co-authorships from 10.14% in 1990 to 24.55% in 2011 (Wagner et al. 2015). Many factors, both internal and external to science, account for this growth, including increasing specialisation, growing costs of scientific instruments, the development of information and communication technologies, the need to address cross-border problems, and policy support for internationalisation at national and regional levels (e.g. the European Research Area initiative; see Chou and Ulnicane 2015). International collaborations lead to higher impact research (Adams 2013; Wagner and Jonkers 2017; Wagner et al. 2018).

I. Ulnicane (✉)
Centre for Computing and Social Responsibility, De Montfort University, Leicester, UK
e-mail: inga.ulnicane@dmu.ac.uk

© The Author(s) 2021
K. Kastenhofer, S. Molyneux-Hodgson (eds.), *Community and Identity in Contemporary Technosciences*, Sociology of the Sciences Yearbook 31, https://doi.org/10.1007/978-3-030-61728-8_5

As the growth of international collaboration today is driven by scientific and political reasons, a better understanding of the tension between self-organisation and steering discussed in studies on international research collaboration (Engels and Ruschenburg 2008; Wagner 2008; Wagner and Leydesdorff 2005) becomes increasingly relevant. While these studies have indicated the positive role of self-organisation and the more challenging one of steering for launching productive collaborations, so far little is known about how self-organisation and steering work and interact in international collaborations.

In order to deepen the understanding of processes of self-organisation and steering, this article undertakes an explorative study to empirically identify relevant features of self-organisation and steering, their role, and interaction. To do so, two contrasting in-depth case studies of international research collaboration in nanosciences are compared. In the first case study of long-term informal international research collaboration, it can be expected that self-organisation dominates; while in the second case study of collaboration within a European project, it can be expected that steering plays the main role. While steering of research collaborations can come from diverse sources, including university governance, this contribution focuses on steering from grant-giving agencies.

Concerns about external interference in research collaboration are well known in the literature on the scientific community. In his 1965 study of the scientific community in the United States, Hagstrom analysed rapid changes in 'the traditional organisation of basic research in universities' where it was assumed that 'the scientist is essentially free to choose problems and techniques as he pleases, that he is free to accept or refuse collaboration with others' (Hagstrom 1965, p. 140). He suggested that 'a more complex form of organisation may be supplanting free collaboration' (ibid.) due to a number of changes, including dependence on grant-giving agencies, which can inhibit the flexibility of research and discourage risk-taking and long-term projects.

Since 1965, when Hagstrom's concern about dependence on grant-giving agencies was published, the share of project funding in overall science funding has increased and today accounts for a quarter to more than half of total public research funding (Steen 2012). This growing role of external funding particularly affects international collaboration because the majority of project funding is distributed nationally. Gradually increasing but still relatively limited funds for international collaborative projects are allocated, for example, within the EU Framework Programme (FP) launched in the 1980s (Ulnicane 2015). Traditionally, a lot of international collaboration has been 'informal', i.e. outside dedicated international projects, and has only shown up in international co-authorships (Georghiou 1998). Thus, the intensification of international collaboration in times when science is increasingly based on external (predominantly national) funding and is simultaneously expected to address cross-border societal challenges (Ulnicane 2016) requires better understanding of the role that self-organisation and steering plays in international collaborations.

By exploring self-organisation and steering, this article aims to contribute to studies on managing scientific collaboration (Bozeman and Youtie 2017) and theory development of research collaborations (Shrum et al. 2007).

This paper proceeds as follows. First, key claims in the existing literature on self-organisation and steering in international research collaboration are presented. Second, the methodology and data sources are explained. Third, processes of self-organisation and steering in two in-depth case studies are examined. Finally, insights on self-organisation and steering are discussed.

5.2 Self-Organisation and Steering in International Research Collaborations

Several studies on (international) research collaboration have indicated that the self-organisation of scientists plays a major role in creating productive collaborations, while the role of policy steering is limited or even counter-productive (Engels and Ruschenburg 2008; Melin 2000; Wagner 2008; Wagner and Leydesdorff 2005). In his study on collaboration, Melin (2000, p. 39) finds a high degree of self-organisation where collaborations take place 'without other initiators than the researchers themselves, and also the forms of the ventures are the products of researchers' own organisation'. Wagner and Leydesdorff (2005, p. 1616), in their research on international scientific networks, reach a similar conclusion on the major role played by self-organisation where international collaborations 'are formed through the individual interests of researchers seeking resources and reputation'.

Several studies of collaborations have suggested that scientists are sceptical about policy-makers' attempts to steer and intervene in collaborations—which scientists sometimes see in the EU's requirement that the Framework Programme projects include participants from several EU member states—that could lead to 'artificial networks' lacking coherence and the ability to produce high quality research (Engels and Ruschenburg 2008, p. 355). Drawing on his results, Melin (2000, p. 39) recommends that 'the scientists themselves should choose with whom they would like to cooperate, and under which forms. Initiatives and directives from politicians and funding agencies are not welcomed by the scientific community and can lead to the establishment of contacts with people other than the scientifically most interesting ones'. Wagner (2008) suggests that science policy should seek and encourage bottom-up, self-organising global networks but mentions that 'these networks cannot be managed; they can only be guided and influenced' (Wagner 2008, p. 105).

Against the background of some of the key research collaboration studies reporting a strong sentiment from scientists in favour of the freedom to self-organise collaborations and against political steering, concerns raised by Hagstrom in 1965 about the impact of funding agencies on free collaboration (Hagstrom 1965) are particularly relevant because the role of funding agencies has increased. To empirically study the role of funding agencies in steering collaborations, it is of crucial

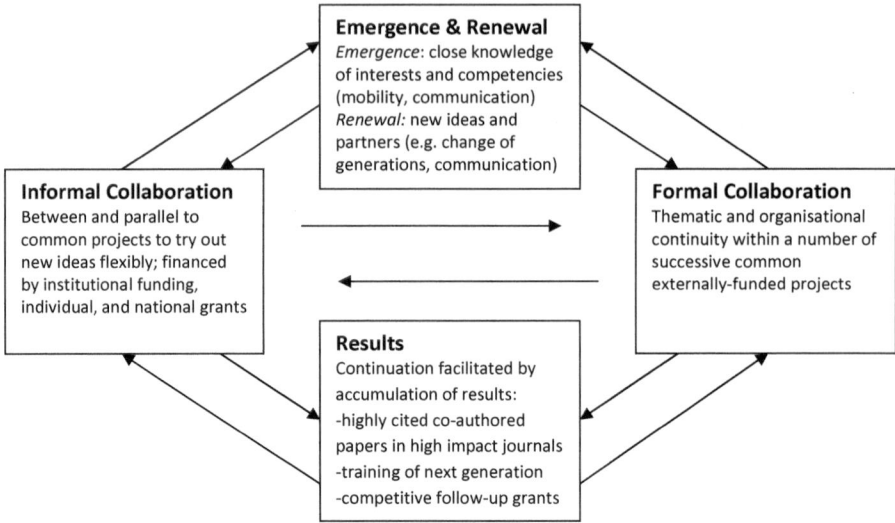

Fig. 5.1 Stylised model of long-term international collaboration process in nano S&T (Ulnicane, 2015, p. 440)

importance that collaborative funding schemes differ considerably. This can also be seen in the FP, where different funding schemes for collaboration vary greatly in terms of thematic guidance (from specific topics elaborated in the calls to thematically open schemes) and other requirements (e.g. related to the type of activities and composition) with which collaborative projects must comply. Thus, externally funded collaborative projects can entail different degrees and types of self-organisation and steering that must be studied empirically.

While the above-mentioned studies do not provide precise definitions of self-organisation and steering, they indicate a number of characteristics of both. Key features of self-organisation in international collaboration include initiation in a bottom-up manner by researchers themselves based on their individual interests and free choice of collaborators and modes of interaction. In contrast, steering is associated with initiatives and directives from politicians and funding agencies and artificial collaborations based on requirements to collaborate with people from specific countries, leading to contacts that might not be the most scientifically interesting ones. This study aims to further clarify relevant characteristics of self-organisation and steering by undertaking in-depth empirical research on international collaborations.

To thoroughly study the role of self-organisation and steering, this paper uses a process model of international collaboration (see Fig. 5.1) that focuses on four elements: emergence and renewal of collaboration, informal collaboration, formal collaboration, and results (Ulnicane, 2015). The emergence and renewal element explores how collaboration starts (selection of partners and topics) and how it later renews itself with new ideas and collaborators. An important ingredient of the emergence and renewal of collaboration is the diverse scientific, institutional, and social motivations of partners to launch and continue collaboration. Collaborations can develop over a longer time, combining informal and formal interactions. Whereas informal collaboration takes place outside joint externally-funded projects, formal collaboration is organised within common externally-funded projects. Both formal and informal collaboration can lead to variety of results, including co-authorships, training of researchers, and joint follow-up projects. The collaboration process can take different paths through these four elements: it can start informally and then lead to formal projects, or it can evolve through a number of successive formal projects with some informal interactions between or parallel to projects. There can be diverse feedback loops between different elements with results from earlier collaboration leading to follow-up projects and the renewal of collaboration.

While many definitions are used (e.g. Chen et al. 2019; Katz and Martin 1997; Laudel 2001) in extensive literature on (international) research collaboration, the process model used for the purposes of this study understands it as a joint research activity with a common aim or shared objective among scientists based in public research institutes in different countries (Ulnicane 2015, p. 434). Such joint research activity often starts because of international mobility (Jons 2009; Melkers and Kiopa 2010; Sugimoto et al. 2017; Wang et al. 2019) and afterwards can occur within a formal, common externally-funded project or be organised informally among collaborative scientists and funded from pooling together their institutional, national, individual, or industrial grants.

5.3 Research Methods and Data

To answer the research question and conduct an in-depth analysis of the role of self-organisation and steering in international collaborations, this paper combines multiple research methods and data sources. Two comparative case studies of international research collaborations are undertaken to obtain a detailed understanding of and rich insights into (George and Bennett 2005) self-organisation and steering. These collaborations were selected from the data on the main non-university research institutes in Germany (Heinze and Kuhlmann 2008) that collaborate with scientists in France, UK, and the Netherlands in the emerging field of nanosciences (Heinze 2010; Noyons et al. 2003).

According to the sociology of science studies (Crane 1972; Merton 1973), emerging sciences, also known as 'hot fields', are particularly relevant for studying collaborations because they are characterised by intense interaction, communication, and rivalry. Moreover, nanosciences are interdisciplinary. The two collaborations in the case studies bring together research groups with complementary expertise in chemistry, physics, and life sciences. Each research group involves scientists at different career stages, from professors and institute directors to post-doctoral researchers and PhD candidates. Both collaborations have a strong focus on training early career researchers.

The diverse case method (Gerring 2008) was used to choose two very different collaborations: one has a long experience of informal interaction, while the other has, so far, been largely limited to a formal externally-funded project.

The within-case analysis follows the method of retrospective process tracing (George and Bennett 2005), using substantial evidence from multiple data to reconstruct the emergence and development of collaboration according to the stylised model of collaboration presented above. To undertake a careful tracing of events, the cases draw on publication, citation, organisational, project and CV data, semi-structured interviews, and site visits to the institutes of the collaborators in the four countries.

Seventeen semi-structured interviews (seven for the first case study and ten for the second) were an important source of information. Several collaborators at each institute at different career stages were interviewed. The general interview protocol focused on the emergence of collaboration; motivations; the choice of topics, partners, and forms of interaction; and the evolution of the collaboration and its results. Each interview included extensive preparation of studying interviewee's publications, CV, projects, and organisational websites.

To compare both case studies, this paper follows the approach of contextualised comparison (Locke and Thelen 1995), looking for analytical parallels in different contexts. While the two collaborations differ in terms of their topics, length, and forms of interaction, the stylised model of collaboration allows for comparing the role of self-organisation and steering in each and looking for similarities and differences.

5.4   Self-Organisation and Steering in Two International Research Collaborations

For a better understanding of the key differences between self-organisation and steering, two contrasting cases are examined. While self-organisation of scientists is expected in the first 'long informal collaboration' case study, steering from a funding agency is predicted in the second 'novel project' case.

The first 'long informal collaboration' started with a researcher doing his PhD abroad with supervisors in two countries. The international links triggered by the joint PhD led to an informal trilateral international collaboration (involving researchers based in Germany, the Netherlands, and UK) lasting over ten years, which was structured around regular research seminars. Due to the international mobility of the two main collaborators, the tradition of regular seminars ended after more than ten years. However, several of the main collaborators continue to cooperate as a core-group within two larger applied FP projects. In their joint research activities, the collaborators combine different experimental techniques. The three

Table 5.1 Key characteristics of the two case studies of international collaboration

Case study 'Long Informal Collaboration'	Case study 'Novel Project'
Informal international collaboration launched by a PhD candidate wanting to do his PhD in a neighbouring country and extended by an internationally mobile postdoc	An internationally mobile postdoc acquires and coordinates an FP project focusing on interdisciplinary training and fundamental research
Strong focus on training	Project has a strong focus on training PhDs and postdocs
After ten years of informal collaboration and international mobility of two key collaborators, partners acquire three joint applied FP projects	After the project, some collaborators continue informal interactions but cannot find a coordinator for a follow-up project application
Formal and informal collaboration leads to co-authored papers	Successful project leads to co-authored papers

institutes where the collaborators are based are referred to here as Institutes A, B, and C.

The second collaboration 'novel project' started in the form of a four-year FP-funded project. While the project involved eight institutes from six countries, this case study focuses on the four institutes constituting the core of collaboration. These four institutes based in Germany, the Netherlands, UK, and France are referred to here as Institutes A, B, C and D. In the project, they combined different experimental competencies of synthesis, characterisation techniques, and expertise from life sciences. The leading role in this collaboration was played by a researcher at Institute A who developed a novel interdisciplinary topic for a collaborative project and brought together a new network.

While the two cases summarised in Table 5.1 have major differences regarding the focus on either predominantly formal or informal collaboration, they also have some important similarities. Both collaborations were launched by internationally mobile early career researchers—PhD and postdoc—and have a strong focus on training PhDs and postdocs who play an important role in laboratory sciences such as nanosciences. Over time, an interesting change happens in both cases—informal collaboration leads to formal projects, while the formal project is followed by a number of informal interactions.

To establish what role self-organisation and steering play, the emergence and evolution of both collaborations will be compared.

5.4.1   Emergence

'**Long informal collaboration**' was initiated when a student wanted to do his PhD in a neighbouring country at Institute A. As a researcher recounts, this collaboration started 'when a young man walked through the doors of [our institute] and said "I want to do a PhD". And we thought it was so cool that we said "Wow, we have to take this guy".'

However, Institute A is a non-university institute and cannot award a PhD degree. The engineering degree the researcher had from his home country was not seen as sufficient to enrol him as a PhD at the local university near Institute A. Thus, in order to do his PhD at Institute A, he had to register as a PhD student in his home country. Institute B was chosen for that purpose. He was jointly supervised by the professors at Institutes A and B and did his PhD research in both institutes. This joint PhD initiated collaboration between Institutes A and B. His supervisor at Institute B explains that the joint PhD:

> Brought us together—professor [in the Institute A] and me and we started talking and said 'o, Jesus, very interesting because we can do here more engineering as a technical university'. At [the Institute A] they are more up-stream, a lot of fundamental things. So we can combine our strengths and this guy [common PhD] could be a glue between the two groups.

While this collaboration was initially bilateral, very soon it became trilateral when a former postdoc from Institute A, who had also previously visited and co-authored a highly cited paper with researchers at Institute B, got a permanent position at Institute C in his home country and joined the collaboration.

A leading role in initiating **'novel project'** was played by Institute A, which sees participation in and coordination of international projects as a sign of scientific quality and a way to be internationally recognised in its research field and integrated into international networks. Scientific leadership in preparing and coordinating the project was provided by a postdoc at Institute A. He developed an idea for this project while he was abroad. He states that the scientific idea for the network was based on an idea from local research which needed additional expertise and resources in order to be developed further:

> That is based on some ideas in in-house research and also patents which we did and also based on a local cooperation . . . Then it turned out that we need some more complementary research and we were trying to get some funding . . . [A]t that time I was postdoc [abroad] and I met one [scientist] there and so I decided to set up such a network.

He states that for a novel interdisciplinary topic he looked for potential international partners in a number of ways: using links from individual international mobility, looking for scientists with relevant competences and equipment on the internet, searching via publications and projects on their websites, and asking for advice within the scientific community:

> I did relatively naïve approach. I was sitting at my desk at [laboratory abroad] and trying to think or to make a plan what I would be interested in. So and then I checked what is needed and I made a plan and then I checked the internet for possible partners. So I knew what I wanted to have all those techniques and then I checked out and called people, introduced myself, trying to find others who knew people working in that respective field.

The relevant competences and equipment of researchers at University B were identified via an internet search of websites and previous projects. Other collaborators were found via individual international mobility and local links at the respective sites. The researchers involved emphasise that, compared to their other collaborations, the origin of this project was rather unusual as the partners had not collaborated before. Nevertheless, they state that this approach worked out well. One of the researchers involved explains that:

> [The core initiators of the project] approached a large group of people via e-mail and internet. And that is very curious because that would be a recipe for disaster, having non-motivated people in the project. An interesting thing is that it turned out to be an incredibly productive network. So this e-mail and internet procedure worked out well, and I think one of the reasons for that is that they did not ask friends, but they started looking for who is the best in this type of field, and they simply approached them. And if you start with the group of friends, you want to be friendly to everybody, and maybe you don't make the best choices in all cases. And in this case they were pretty lucky in the choices that they made, and they found very interesting consortium of really interested multidisciplinary people and research groups. From the beginning I was surprised how good the atmosphere in this network was.

However, the coordinator says that the choice of partners was based not only on scientific needs but also on his perception of EU rules. He explains that:

> The real problem is: you have to obey the European rules. You have to find people, let's say from good regions; you have to include women, and, irrespectively of their scientific quality, you have to obey these rules in order to get chances. And obeying the rules means decreasing the scientific quality. It is not only the science anymore which plays a major role. That is a price you have to pay.

He adds that in order to fulfil the EU requirements for partners from different countries, he invited collaborators who performed tasks that are also done at his own institute. He explains: 'we could also do the chemistry; therefore we, in some sense, would have needed much smaller network—we could have done much more ourselves. But this obviously contradicts the idea of such networks, and you get of course also new things if you go to other partners'.

The initiator of the collaboration chose to prepare an application within a thematically open FP funding scheme dedicated to research training through fundamental research. He explains his choice as follows:

> If you are waiting for a call in the normal European funding scheme then it never fits very well. If you really want to do fundamental research, so Framework Programmes are meant for industrial support, and what we are doing is really basic research. And in this sense it does not fit to usual schemes, and, in addition, in the calls topics are old in some sense. At least two, three years before people set them up, and then the call rises, and then you have to write something about that. If you are working on that it is five years behind more or less. And therefore that is always the problem if you really want to do state of the art stuff. And in this sense [a thematically open scheme] was the obvious choice.

The application was granted in this highly competitive scheme. The coordinator assumes that the main reasons for it succeeding were the application's interdisciplinarity, its thematic coherence, and the quality of involved groups.

The emergence of two collaborations confirms some expected contrasts of self-organisation and steering. The 'long informal collaboration' started as a result of the initiative of a young researcher who wanted to do PhD at a top institute abroad and was supported not only by that institute but also by a professor in his home country. The 'novel project' started on the initiative of a postdoc to apply for an externally-funded project to get the necessary resources to develop further findings from in-house research. The development of the project demonstrates a mix of self-organisation and steering. Self-organisation includes carefully selecting external funding scheme that allows work on a topic of one's own interest and also the initial selection of collaborators. However, the selection of collaborators is also partly based on the researcher's perceptions of EU rules preferring geographical and gender diversity. That led to including international collaborators for tasks that could be done in-house; however, international partners brought new knowledge about these tasks. Informal collaboration is smaller and includes three partners, while the project is bigger with eight partners, which is at least partly due to EU rules.

5.4.2   Formal and Informal Collaboration

The main elements of the **'long informal collaboration'** over ten years were regular seminars, short visits and joint experiments, and at least five common PhD students. The focal point of the informal collaboration among the three groups was regular seminars that took place at least once a year, lasted for two to three days, and were hosted by the collaborating institutes on a rotating basis. The aim of these seminars was to give PhD researchers an opportunity to present their work, learn from the presentations of others, and find opportunities for collaboration.

These seminars led to short visits (e.g. three days or a few weeks) by PhDs to collaborating institutes, where they learned complementary techniques and did joint experiments which led to chapters in their PhD theses and co-authored papers. This informal collaboration also involved a researcher from Institute C receiving a fellowship to do a postdoc in Institute A, after encouragement from the group leader who had previously done the same.

There have been at least five common PhDs. The collaboration was initiated by a common PhD between Institutes A and B and there have since been at least four joint PhDs between Institutes A and C. The reason for these joint PhDs was the same as for the initial joint PhD: Institute A, as a non-university institute, cannot award PhD degrees, and it is difficult to enrol researchers with an engineering degree as PhDs at the local university near Institute A. Thus, a number of PhD students at Institute A were registered as PhDs at Institute C. They did their PhD research at Institute A, fulfilled the requirements for a PhD degree at Institute C, regularly discussed their results with the co-supervisor at Institute C, and, finally, were awarded their PhD degree at Institute C. Afterwards, some of them did their postdoc at Institute C.

An important characteristic of this collaboration was communication and collaboration across all levels of hierarchy: senior professors and group leaders as well as PhDs and postdocs. Collaboration was largely possible due to the support from the department leaders in Institutes A and B. As explained by a researcher, a crucial precondition for this long-term informal collaboration has been 'support by the people who make decisions, support by the directors; they wanted it'. At the same time, one of the directors emphasise the importance of communication among PhDs for successful collaboration in experimental research, where laboratory work is done by PhDs: 'eventually the job gets done by PhD students. If the PhD students don't talk, we [professors] can talk, but if they [PhDs] don't talk and they don't know each other, then it is a dead end'.

The informal collaboration was financed either by institutional funds (Institute A) or income from contract research (Institute B). One of the collaborators explains that 'it was funded by internal money. We never had any specific funding to pay for this seminar programme and we did it as cheaply as possible'. The collaborators outline a number of benefits from such informal collaboration: the ability to be responsive to new research ideas, no need to write reports, the 'beauties of excluding and ruling out bureaucrats from such collaborations', and the ability to have 'the least possible administration'. One senior collaborator summarises informal collaboration as follows: 'We had a joint work; a lot of joint work but there was no reporting, no funding agencies, no administrations, no bureaucracy. Great thing. As long as we can afford it'.

To summarise, this long-term informal collaboration had a number of advantages and disadvantages. Among the advantages mentioned were intellectual independence, flexibility, low administrative burdens, and no requirement to comply with external rules. At the same time, it had disadvantages, such as limited funding and no room for expansion. In order to expand, external funding is important. The long-term informal collaboration allowed the collaborators to develop mutual knowledge that helps in common formal projects.

The main renewal of collaboration took place when, after the two main collaborators were no longer internationally mobile, the tradition of joint seminars stopped. At the same time, a new generation of research leaders formed a core-group within three FP projects with more applied research topics and additional partners.

After ten years of informal collaboration, the major collaborators decided, according to one researcher, that they 'should really try to get some serious money to do some work together'. Thus, acquiring funding for common FP projects was seen as a way to expand their research. All three FP projects have been coordinated by the project leader at Institute A, where the acquisition and coordination of such projects helps strengthen a scientist's position. The two main projects are the so-called small- and medium-sized research projects; each has funding of around 3,000,000 euros and involves around ten partners, including research institutes, small and medium-sized companies, and hospitals. The third project is a smaller one with funding of around 100,000 euros; it supports the international exchange of researchers.

The two bigger projects emerged as a response to specific calls with predefined topics. The topics of these projects fitted well with the collaborative work previously published in co-authored papers and with new interdisciplinary expertise developed at Institute C. Moreover, there are thematic links between the two projects, as one builds on the other. Thus, there is a lot of thematic continuity between the informal and formal collaboration as well as between the two formal projects. Both projects are of a more applied character than the research done during the informal collaboration, but the main collaborators are also interested in applied research. For them, these applied research projects involve some very fundamental questions and allow doing some 'crazy things for the future'. Key collaborators see positive effects of acquiring FP projects not only as a source of additional funding but also for enhancing their institution's visibility.

While the core collaborators have known each other for a long time, the formal projects involve a number of new partners who were found via the web and publications or through suggestions from their colleagues. Within these projects, particularly within the first, the core collaborators have experienced the problem of 'unreliable partners'. According to the core collaborators, it turned out that some new partners had gotten involved in the projects 'simply because they supply quite a lot of money'. The main strategy for dealing with unreliable partners has been to replace them in a follow-up project and to design the tasks for the successive project in such a way that the core-group can, if necessary, deliver the main project outputs independently. As explained by one of the core collaborators, in the second project, his group 'wanted to be able to proceed as fast as possible without having to be dependent' and has 'found it more efficient to be a bit more autonomous where collaboration gives a lot of value added but is not absolutely essential to the successful outcome of the project'.

The training of young researchers was a major element in the informal collaboration, and it is also important for the common projects, which provide funding to hire PhDs and postdocs who carry out the laboratory work. Within the projects, training seminars have been organised and carried out by PhDs and postdocs to teach each other specific techniques and conduct joint experiments. Moreover, PhDs and postdocs in the projects have had to present their research results in project meetings every six months. A PhD student in one of the projects explains that presentations in the project meetings provide strong additional incentives to work well. According to her:

> It helps being a part of the big project . . . when you are working in a bigger group you feel more responsible for doing your own work and standing out within the group and trying to get results. It is a big motivator . . . because we have to do these six-monthly meetings and progress reports. You have to have worked to present and to show that you are getting the work done.

To summarise, the decision to acquire European projects has been important in renewing and expanding the collaboration while simultaneously building on common topics and key collaborative links developed in previous informal collaboration.

The **'novel project'** received FP funding of around 3,000,000 euros. The key FP rules that affected the project focused on recruitment, regular meetings, training, and interdisciplinarity. According to the rules, PhDs and postdocs had to be recruited from abroad, and gender balance was desired. While some partners had a lot of experience in recruiting high quality international researchers, others took a long time to find suitable candidates, and some joint activities were delayed due to some late recruitments.

The project was funded by the research training network scheme, which supported short-term exchanges of PhDs and postdocs as well as training workshops. The thematic coherence of the project was facilitated by the regular network meetings every six months and by close interaction between junior researchers. The coordinator explains the role of exchanges of PhDs in ensuring coherence:

> [PhD] students which we have are extremely smart, they are really extremely good. They are doing research, and in a lot of groups they are also giving the direction which we also search. And, if they are meeting . . . they are very proud of hosting their partners, and that provides coherence.

The interaction among PhDs doing experimental research ensured active communication and facilitated joint research activities. A researcher explains: 'my PhD students . . . send an e-mail to [a] production person, also a PhD [student] or postdoc, and then they have to produce that material, and in general they come to an agreement and things are being sent'.

Within the regular network meetings, PhDs initiated collaborations themselves which later resulted in internationally co-authored papers. A researcher states that, for him, it is safer to hire a PhD within an international network that provides a lot of support for PhD research on a novel topic. A former PhD who prepared material explains how she started to collaborate with another PhD doing physical characterisation:

> It was my own initiative. I think, basically, I had something that she wanted, and she had something that I wanted, so we started talking at the meetings. She wanted me to supply her with material . . . so she would not have to do the chemistry. And I asked—can you analyse them for me? And we found it to work very well.

She adds that 'at the beginning I did [the characterisation] myself, and then I found it is quicker and nicer if I send [material to the partner abroad]'. Both PhDs also visited each other for two to four weeks, which helped them to do 'a lot of analysis together'. This collaboration between the two PhDs led to a number of internationally co-authored articles.

As the project according to the project requirements was interdisciplinary, the collaborators initially experienced typical challenges in communicating across disciplines. Researchers state that it took time to understand each other's language and that presentations at the regular project meetings helped with this. A researcher explains that:

> [at] the first, the second network meeting we were all new to each other; we are still very polite of course, you do not want to question what everybody else is doing, saying, but then the third and the fourth meeting people start to ask 'but what do you really mean when you talk about milligrams' or 'what do you really mean when you talk about moles or number of molecules'. And it was very eye opening to get to that point.

A PhD student mentions that it might have taken as many as seven presentations at the network meetings combined with visits to each other's labs for partners to understand each other's research: 'at the beginning may be there was a little bit like—woo, don't understand that, but after a while, like a seventh time, they finally knew what you were doing, and also when they visited your lab or you visited theirs it became more clear.' A senior researcher explains the need to understand a collaborator's research:

> You got to be able to communicate with the collaborators. I don't think collaboration will work very well if you have no idea about the other aspects of the work. You may have less expertise in it, you may have less capability in it, but you must at least have some idea about how it works, what its capabilities are. All collaboration requires a little bit of knowledge on both sides of both sides.

This fits with the notions of 'contributory' and 'interactional' expertise developed by Harry Collins and Robert Evans (2002). For them, contributory expertise means enough expertise to contribute to the specific science field, while interactional expertise implies sufficient expertise to interact with representatives from that field. Thus, each collaborator has contributory expertise in their own area and interactional expertise in partner's area.

During the last two years of the project, the collaborators actively discussed the possibility of preparing an application for a joint follow-up project. The major stumbling block in preparing a follow-up project turned out to be difficulties in finding a coordinator as, due to their workload or lack of institutional support, none of the partners could commit themselves to a very time-consuming coordination task for the next four years. However, a number of project partners continue to interact informally.

To summarise, while both collaborations evolved quite differently, there were also some important similarities. The focal point for both was interaction and short exchange visits among junior researchers as well as regular meetings, which occurred more often during the project than in informal collaboration. While informal collaboration provides more freedom to choose topics and forms of collaboration, external funding is needed for expansion. Many rules that come with external funding reinforce activities that researchers already do—for example, meet regularly, provide training, and collaborate across disciplines. The most negative aspect of EU-funded projects, as seen by all researchers, is heavy administrative tasks.

5.5   Results

To sum up, 'long informal collaboration' has resulted in co-authored publications, common PhDs, and, recently, joint projects. For an overview of this collaboration, see Fig. 5.2 with full lines depicting actual directions and dotted ones indicating potential future developments. The scientists involved deem that collaboration has given 'PhD students much broader perspective and appreciation of science going on in other institutions' and contributed to the 'quality of the students you educate in the long run'.

The 'novel project' was seen as successful by collaborators and the funder, which mentioned it among its success stories. It resulted in co-authored publications, training, and new contacts (see Fig. 5.3).

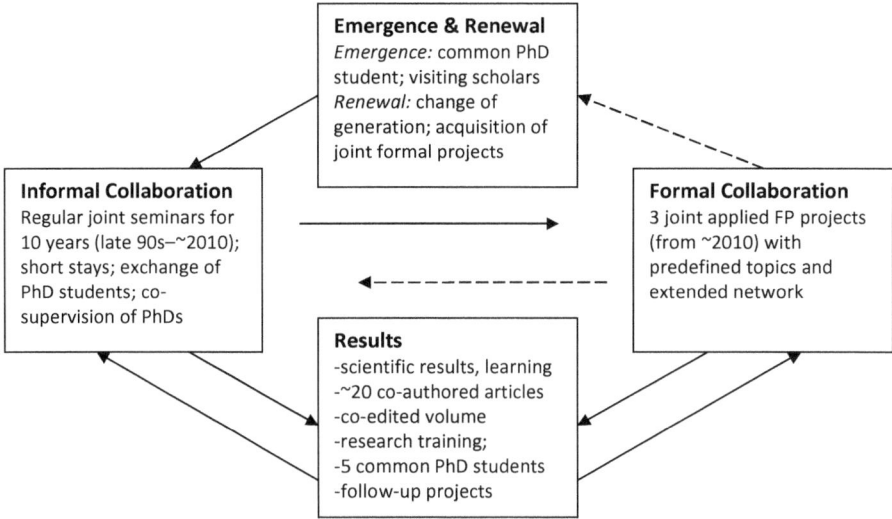

Fig. 5.2 Stylised model of 'long informal collaboration'

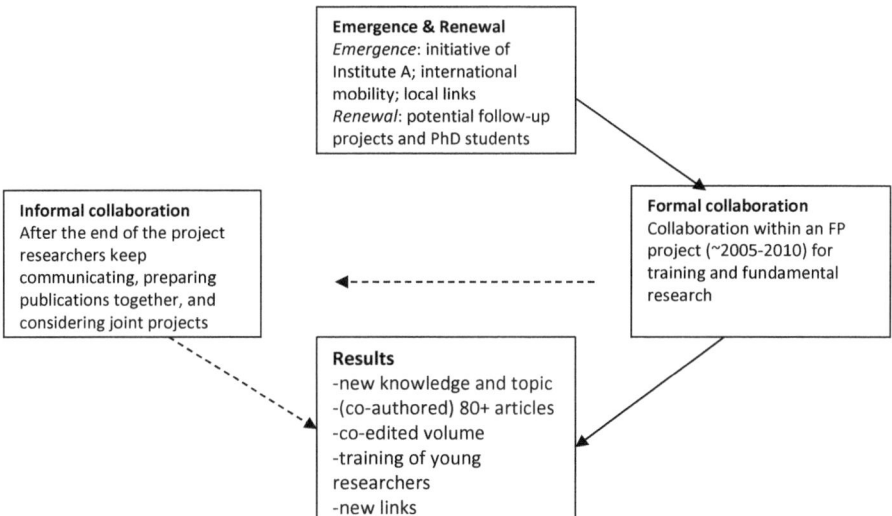

Fig. 5.3 Stylised model of 'novel project'

Thus, while the main types of outputs—co-authorships and training—are quite similar in both cases, resources provided by the externally-funded project can facilitate more productivity.

5.6   Conclusions

The important starting points of this article were two concerns raised in the literature. First, the suggestion from Hagstrom's 1965 work on the scientific community that, due to the increasing dependence on grant-giving agencies, free collaboration might be replaced by a more complex form of collaboration (Hagstrom 1965). Second, indications in studies of international research collaboration that self-organisation is the best way towards productive collaborations, while steering from funders can lead to 'artificial collaborations'.

Against this background, in-depth case studies of two collaborations demonstrate that when collaborating both within or outside externally-funded projects, scientists still have various degrees of freedom to choose their collaborators, common topics, and interactions. When applying for external funding, scientists carefully consider which funding scheme will best fit their research interests.

Rather than being purely self-organised or steered, the examples of successful and productive collaborations illustrated here present a mix of both. The elements of self-organisation and steering that were identified as relevant relate to choice of the topic, type of research (basic-applied, interdisciplinary, etc.), collaborators, recruitment of researchers, modes of interaction, and results. Interestingly, in the case of highly competitive EU funding schemes, not only formal rules but also perceptions of preconditions for success (e.g. geographic diversity) steer the design of collaborations. In contrast to literature that depicts self-organisation as supportive for productive collaborations and steering as problematic, insights presented in this article suggest that both can facilitate and limit collaborations (see Table 5.2).

One interesting result from comparing formal and informal collaboration is the many similarities between the two, including a focus on training, joint publications, and interdisciplinarity. Thus, project rules and requirements can largely reinforce the type of collaboration that scientists are interested in anyway. Moreover, all studied researchers and institutes are simultaneously involved in a number of diverse collaborations. Each specific collaboration—formal or informal—is a result of a number of factors, including research and resource needs, institutional strategies, and career stage of researchers. Thus, in the case of 'long informal collaboration', involved institutes had resources (institutional and business) and support from established professors to collaborate informally for ten years on a small scale. Externally-funded projects were sought later in order to expand the collaboration and advance the careers of the next generation of research leaders. In the second case, project funding was necessary in order to expand research. It was supported by institutional strategy and fit with the career development of project leaders.

Characteristics of self-organisation and steering identified in this study would benefit from testing on a larger sample of diverse cases from different disciplines (Heidler 2017), countries, and funding schemes. While both collaborations here were relatively small-scale and focused largely on interactions among scientists, in future, large-scale collaborations (Ulnicane 2020) involving society, stakeholders, industry, and policy-makers—which are increasingly relevant in the context of the

Table 5.2 Summary of elements and effects of self-organisation and steering

	Self-Organisation	Steering
Elements in international research collaborations	Freedom to choose topic and type of research (e.g. basic, interdisciplinary);	Support for certain topics and types of research;
	Freedom to choose collaborators;	Rules on specific types of collaborators (geographic, gender);
	Freedom to recruit any researchers;	Rules on recruitment of researchers (e.g. from abroad)
	Freedom to choose modes of interaction and results;	Requires certain modes of interaction (regular meetings, training events, short exchange visits etc.) and results;
Effects on international research collaboration	*Positive*	*Positive*
	Scientists collaborate on topics and types of research they are most interested in;	Provides resources to expand collaborations;
	Freedom to choose collaborators;	Facilitate inclusion of diverse collaborators;
	Limited administration;	Learning from additional collaborators;
	Trusted collaborators;	
	Potentially problematic	*Potentially problematic*
	No resources to expand collaborations;	In order to acquire funding, scientists adjust topics and types of research;
	Excludes collaborators from less developed regions and underrepresented groups;	Compromise on collaborators;
	Less opportunities to learn due to limited number of partners;	Heavy administration;
		Unreliable collaborators;

recent focus on supporting digital data infrastructures (Aicardi et al. 2018) as well as challenge- and mission-oriented research (Ulnicane 2016)—should be studied in the future as well.

Acknowledgments This work has benefited from comments on earlier versions presented at the 2015 Atlanta Conference on Science and Innovation Policy, Atlanta (US), the 2017 STS Austria workshop 'Community and identity in contemporary techno-sciences', Vienna (Austria), and the 2017 Eu-SPRI (European Forum for Studies of Policies for Research and Innovation) conference, Vienna (Austria). Valuable feedback from editors and reviewers is gratefully acknowledged. This research was done at the Science, Technology and Policy Studies (STePS) Department at University of Twente (Netherlands) and funded by the German Research Foundation DFG (FOR 517); it was written up at the Institute for European Integration Research EIF at University of Vienna (Austria).

References

Adams, J. 2013. Collaborations: The fourth age of research. *Nature* 497: 557–560.

Aicardi, C., M. Reinsborough, and N. Rose. 2018. The integrated ethics and society programme of the Human Brain Project: Reflecting on an ongoing experience. *Journal of Responsible Innovation* 5 (1): 13–37.

Beaver, D.D., and R. Rosen. 1978. Studies in scientific collaboration. Part I. The professional origins of scientific co-authorship. *Scientometrics* 1 (1): 65–84.

Bozeman, B., and J. Youtie. 2017. *The strength in numbers. The new science of team science.* Princeton: Princeton University Press.

Chen, K., Y. Zhang, and X. Fu. 2019. International research collaboration: An emerging domain of innovation studies? *Research Policy* 48 (1): 149–168.

Chou, M.-H., and I. Ulnicane. 2015. New horizons in the Europe of knowledge. *Journal of Contemporary European Research* 11 (1): 4–15.

Collins, H.M., and R. Evans. 2002. The third wave of science studies: Studies of expertise and experience. *Social Studies of Science* 32 (2): 235–296.

Crane, D. 1972. *Invisible colleges. Diffusion of knowledge in scientific communities.* Chicago: The University of Chicago Press.

Crawford, E., T. Shinn, and S. Sorlin, eds. 1993. *Denationalizing science. The contexts of international scientific practice.* Dodrecht/Boston/London: Kluwer.

Engels, A., and T. Ruschenburg. 2008. The uneven spread of global science: Patterns of international collaboration in global environmental change research. *Science and Public Policy* 35 (5): 347–360.

George, A., and A. Bennett. 2005. *Case studies and theory development in the social sciences.* Cambridge: MIT Press.

Georghiou, L. 1998. Global cooperation in research. *Research Policy* 27 (6): 611–626.

Gerring, J. 2008. Case selection for case-study analysis: Qualitative and quantitative techniques. In *The Oxford handbook of political methodology*, ed. J.M. Box-Steffensmeier, H. Brady, and D. Collier, 645–684. Oxford: Oxford University Press.

Hagstrom, W.O. 1965. *The scientific community.* New York/London: Basic Books.

Heidler, R. 2017. Epistemic cultures in conflict: The case of astronomy and high energy physics. *Minerva* 55 (3): 249–277.

Heinze, T. 2010. The research field of nanoscience & -technology (Nano S&T). In *Governance and performance in the German public research sector. Disciplinary differences*, ed. D. Jansen, 173–183. Dordrecht: Springer.

Heinze, T., and S. Kuhlmann. 2008. Across institutional boundaries? Research collaboration in German public sector nanoscience. *Research Policy* 37 (5): 888–899.

Jons, H. 2009. 'Brain circulation' and transnational knowledge networks: Studying long-term effects of academic mobility to Germany, 1954–2000. *Global Networks* 9 (3): 315–338.

Katz, J.S., and B.R. Martin. 1997. What is research collaboration? *Research Policy* 26 (1): 1–18.

Laudel, G. 2001. Collaboration, creativity and rewards: Why and how scientists collaborate. *International Journal of Technology Management* 22 (7–8): 762–781.

Locke, R., and K. Thelen. 1995. Apples and oranges revisited: Contextualized comparisons and the study of comparative labor politics. *Politics and Society* 23 (3): 337–367.

Melin, G. 2000. Pragmatism and self-organization—Research collaboration on the individual level. *Research Policy* 29 (1): 31–40.

Melkers, J., and A. Kiopa. 2010. The social capital of global ties in science: The added value of international collaboration. *Review of Policy Research* 27 (4): 389–414.

Merton, R. 1973. Behaviour patterns of scientists. In *The sociology of science. Theoretical and empirical investigations*, ed. R. Merton and N. Storer, 325–342. Chicago: University of Chicago Press.

Noyons, E., R. Buter, A. van Raan, U. Schmoch, T. Heinze, S. Hinze, et al. 2003. *Mapping excellence in science and technology across Europe. Nanoscience and nanotechnology*. Leiden: University of Leiden.

Shrum, W., J. Genuth, and I. Chompalov. 2007. *Structures of scientific collaboration*. Cambridge: MIT Press.

Steen, J.V. 2012. *Modes of public funding of research and development: Towards internationally comparable indicators*, OECD Science, Technology and Industry Working Papers, 4. Paris: OECD.

Sugimoto, C., N. Robinson-Garcia, D.S. Murray, A. Yegros-Yegros, R. Costas, and V. Lariviere. 2017. Scientists have most impact when they're free to move. *Nature* 550 (7674): 29–31.

Ulnicane, I. 2015. Why do international research collaborations last? Virtuous circle of feedback loops, continuity and renewal. *Science and Public Policy* 42 (4): 433–447.

———. 2016. 'Grand challenges' concept: A return of the 'big ideas' in science, technology and innovation policy? *International Journal of Foresight and Innovation Policy* 11 (1–3): 5–21.

———. 2020. Ever-changing big science and research infrastructures: Evolving European Union policy. In *Big science and research infrastructures in Europe*, ed. K. Cramer and O. Hallonsten, 76-100. Cheltenham: Edward Elgar.

Wagner, C.S. 2008. *The new invisible college. Science for development*. Washington, DC: Brookings Institution Press.

Wagner, C.S., and K. Jonkers. 2017. Open countries have strong science. *Nature* 550 (7674): 32–33.

Wagner, C.S., and L. Leydesdorff. 2005. Network structure, self-organization, and the growth of international collaboration in science. *Research Policy* 34 (10): 1608–1618.

Wagner, C.S., H. Park, and L. Leydesdorff. 2015. The continuing growth of global cooperation networks in research: A conundrum for national governments. *PLoS One* 10 (7).

Wagner, C.S., T. Whetsell, J. Baas, and K. Jonkers. 2018. Openness and impact of leading scientific countries. *Frontiers in Research Metrics and Analytics* 3.

Wang, J., R. Hooi, A.X. Li, and M.-H. Chou. 2019. Collaboration patterns of mobile academics: The impact of international mobility. *Science and Public Policy* 46 (3): 450–462.

Chapter 6
The Project-ed Community

Béatrice Cointe

6.1 Introduction

Projects are now ubiquitous in scientific work, to the extent that they might appear self-evident. With their institutionalisation in science policy and funding, they have become 'a key organizing principle into science' (Felt 2016, p. 136). The 'projectification' of research (Felt 2016; Yliojoki 2016) has become an object of inquiry in its own right. Projectification goes along with a series of widely documented trends in the organisation of scientific research. These include the rise of 'New Public Management' in the governance of research, the focus on so-called 'excellence', the encouragement of academy-industry and interdisciplinary collaborations, and increased expectations for research to display its relevance to society (Gläser and Laudel 2016, p. 121). In this context, projects are simultaneously ways of organising day-to-day scientific work and instruments of science policy. They serve the thematic steering of research according to governmental science policy and the accountability of funding. They are also, more often than not, collective endeavours that bring together researchers across teams, countries or disciplines. All of this makes them privileged sites to investigate evolutions in the collective dynamics of science. To what extent do projects exemplify, and participate in, the emergence of new forms of scientific communities?

This chapter takes an ethnographic perspective to investigate how projectification translates in collective research dynamics. It draws on the participant ethnography of a large interdisciplinary project exploring potential avenues for the production of biofuel by microorganisms. Microbio-E was a two-year project bringing together a dozen research groups around Aix-Marseille, France. Its disciplinary core was

B. Cointe (✉)
Centre de Sociologie de l'Innovation, Mines ParisTech, Paris Cedex 06, France
e-mail: beatrice.cointe@mines-paristech.fr

© The Author(s) 2021 127
K. Kastenhofer, S. Molyneux-Hodgson (eds.), *Community and Identity in Contemporary Technosciences*, Sociology of the Sciences Yearbook 31,
https://doi.org/10.1007/978-3-030-61728-8_6

in microbiology, but it also involved disciplines from biophysics to sociology. While project-based funding is well established in France,[1] participants in Microbio-E often described this particular project as relatively unusual in interviews and meetings. Microbio-E stood out because of its very broad thematic and disciplinary scope and, to a lesser extent, its applied horizon. Both characteristics stemmed from Microbio-E's institutional embedding. The project was funded by the 'Aix-Marseille's Excellence Initiative' (AMIDEX), a scheme aiming to develop world-class research, innovation and higher education around Aix-Marseille University (AMU).

I took part in Microbio-E for 18 months as a post-doc. I visited the teams involved and interviewed 40 permanent researchers, PhD candidates, post-docs, and administrative staff related to the project.[2] Interviews covered everyday work, roles within the project and relationships with other teams and participants, experiences of interdisciplinarity, and perceptions of the bioenergy promise advertised in the project. I participated in project meetings and seminars and collected project documents: project proposal, intermediary project report, slides from meetings and presentations, documentation on research groups, and previous project proposals on related topics. I was based in the project's core research group and had regular informal discussions with project participants.

One practical and methodological challenge in conducting fieldwork was to define Microbio-E as a group. Whether the project constituted its own community (beyond the ad-hoc gathering of researchers) and, if so, what delimited it, became central questions. These had additional salience in this case, because one stated objective of Microbio-E was to foster a local research community around bioenergy. This chapter attempts to address them by tracing the contours of what I call the 'project-ed community'. This entails a more general reflection upon what defines a project as a collective venture. In line with the analytical programme presented in this volume's introductory chapter (Kastenhofer and Molyneux-Hodgson, Chap. 1 this volume), the ethnographic perspective interrogates the dynamic role of projects in shaping and transforming research collectives. It makes it possible to delve into the heterogeneities that make up daily scientific life and, thus, informs reflections upon the concepts that can equip the empirical analysis of collectivities. In the case drawn upon in this chapter, it allows for addressing the question to what extent 'projectification' can help us to account for contemporary forms of scientific communality.

---

[1]The share of projects in research funding in France increased from 11% in 1982 to 21% in 2002 (Thèves et al. 2007); projects became a dominant form of funding with the creation of the National Research Agency in 2005.

[2]Interviews took place either in French or in English and were translated by the author when needed.

6.2 Project-ed Communities

The notion of a 'project-ed community' is a play on the different meanings of 'project'. It is meant to underline that a project exists on several levels and serves several purposes. First, the 'project-ed community' refers to the community that is gathered to carry out a project. Moreover, this community is also 'projected' as an image, a display of relevance and quality: the project is meant to showcase and nurture research competences. In that sense, community-building itself can be seen as a project, as much as the strengthening and development of existing communities. Last, the planned nature of projects points to a third meaning: 'to project' is to plan for the future, to look forward. Since projects are temporary and goal-oriented, a 'project-ed community' is gathered by a vision of what it *will* achieve at least as much as by what it is doing at any particular moment.

To test out this conception, I trace how Microbio-E and its community materialise (1) in project documents, (2) in institutional arrangements, and (3) in daily research. All three takes provide different, but coherent, pictures of the project-ed community. This points to the versatile function of projects in science: Microbio-E, I argue, can be equally analysed as an argumentative device, as a strategic venture, and as an arena for scientific work.

This chapter contributes to the ongoing investigation of how new collective scientific structures affect the way research is lived and done. Recent works suggest a proliferation of new entities that are emerging to organise research collectively according to thematic, geographical, or science policy rationales. Besides projects, examples include local research fields (Merz and Sormani 2016), regional clusters (Vinck 2016; Merz and Biniok 2016; Robinson et al. 2016), thematic networks and centres (Strathern and Khlinovskaya-Rockhill 2013), and emerging fields such as synthetic biology (Meyer and Molyneux-Hodgson 2016; Molyneux-Hodgson and Meyer 2009) and bioenergy (Tari 2015). One of their common traits is that they are not primarily based on disciplines and often embed research in extra-scientific financial, societal and political concerns, echoing (and maybe institutionalising) Knorr-Cetina's 'transepistemic arenas of research' (1982). Knorr-Cetina introduced the notion as a critique of analyses focusing on "specialty communities", and used it to emphasise the entanglement of research practices in multiple relationships not contained in laboratories and specialties. She also stressed the importance of focusing on groupings that are meaningful for participants, which invites an inquiry into the relevance of these new communities for scientists.

When it comes to projects, the focus has been on how they affect the organisation of scientific work, especially within research groups (Barrier 2011; Hubert and Louvel 2012; Jouvenet 2011), in terms of the relations between the governance and the content of research (Tricoire 2006, 2011), and temporality-wise (Felt 2016; Ylijoki 2016; Schultz 2013). Projects enact their own temporality, or 'project time' (Ylijoki 2016), altering the perception, quantification and organisation of time and generating tensions. Research, in a project, becomes something that fits in a predefined time span (Felt 2016, p. 136). Here, I consider projectification from a different angle: the focus of this chapter is not on how research fits into projects, but rather on how one project takes its place within a research landscape. This perspective is informed by the work of Leonelli and Ankeny (2015), who show how some projects perpetuate into lasting communities, becoming 'blueprints for the way in which whole communities should do science' (Leonelli and Ankeny 2015, p. 705). Leonelli and Ankeny conceptualise as 'repertoires' the shared sets of norms, infrastructures, procedures and resources that allow these communities to perpetuate and adapt to their research and funding contexts. The case of Microbio-E, in which the community itself is a project, allows for the study of how potential repertoires are tentatively built as part of project research.

6.3   The Project as Argumentative Device

In a context dominated by project-based funding, projects start out as written proposals arguing for the allocation of money to specific research groups and questions. Documents such as project proposals and reports are central in defining, justifying, and ultimately enabling projects: at the outset, a project is a successful proposal. A project can then be conceived as an argumentative device that assembles on paper a set of competences, research questions, promises concerning output and impact, and funding constraints so as to demonstrate feasibility and relevance.

Two documents delineated the contour of Microbio-E in this way: the project proposal (Microbio-E 2014) and a midway report (Microbio-E 2015). They described the project as uniting '100 researchers and engineers from 13 different labs' around Aix-Marseille, listing their areas of expertise, resources, and expected contributions. They also stated the budget needed (around 1.5 million euros over 2 years) and accounted for its actual use. Most of the budget funded short-term contracts: 11 post-docs, 3 PhD candidates, and one research engineer, all listed in the progress report.

The bulk of the documents was devoted to laying out what the project was meant to achieve and how. Microbio-E's full title, 'BIOmass valorization by MICRObes for BIOEnergy Production', emphasised its main ambition: "to set up and promote innovative and original scientific projects involving interdisciplinary approaches from biology to process engineering that push back the frontier of knowledge relative to the great world challenges and that will eventually allow (sic) the emergence of new biotechnological processes and economy" (Microbio-E 2014, p. 3). The proposal claimed that Microbio-E would foster "the emergence of an internationally recognised task force aimed at remov[ing] biological constraints currently limiting the development of advanced biofuels" by combining the "strong expertise in microbiology, metabolism, lipidomics, bioinformatics, biophysics, bioprocess, chemical engineering, economy" available locally (Microbio-E 2014, p. 2). While referring to encouragements for scientists "to act for the development of alternative energy sources and new feedstock for chemistry" (Microbio-E 2014, p. 2), the project focused on fundamental questions, which made up the main part of the proposal. Interviews confirmed that most teams were devoted to basic research and considered their work as "really, really upstream" (Interview 31, lecturer in chemistry). The objectives were thus threefold: to further basic scientific understanding on topics related to advanced bioenergy; to stimulate cross-disciplinary networking and collaborations among local teams; and to create a regional 'task force' able to increase the University's academic impact through publications and patents.

The proposal then detailed how the resources put together in the project would be organised to achieve its goals. It divided the project in three 'tasks' and several 'subtasks'.

- Task 1 would gather biologists, biophysicists and electrochemists studying enzymes and enzymatic reactions involved in biomass formation and degradation at a molecular level;
- Task 2 would explore the potential of micro-algae for the production of biofuels and high-value compounds, associating plant biologists, bioinformaticians and bioprocess engineers;
- Task 3 would focus on the production of hydrogen by micro-organisms and its use in fuel cells. It would involve biologists, electrochemists, process engineers, and social scientists, who were supposed to study "challenges of the use of biomass in the bioenergy sector and [...] in the Hydrogen industry" from a socio-economic perspective (Microbio-E 2014, p. 27).

Project documents, especially the proposal, sought to assert the relevance of the project's ambition and organisation to the concerns of the funder and, through it, to the policy context more generally. Following a local call, Microbio-E was funded by AMIDEX,[3] whose purpose was to foster so-called 'excellent' local research on a set of priority themes. The proposal, reflecting the expectations from AMIDEX, sketches a project that appears emblematic of many of the trends identified by sociologists of science. This suggests that these trends have become incorporated in conceptions of how research should be presented. For instance, the project ticks several of the boxes of 'Mode 2 knowledge production' (Gibbons et al. 1994). Gibbons and colleagues have described Mode-2 research as externally funded, transdisciplinary, problem-oriented, occurring in applied contexts, and evaluated according to its social and economic utility, as opposed to the disciplinary, curiosity-driven, institutionally stable, and internally assessed 'Mode-1' research (Gibbons et al. 1994; Ylijoki 2016). Microbio-E was, indeed, designed as interdisciplinary—cutting across natural sciences to "break the locks that will allow innovative strategies to become economically viable" (Microbio-E 2015, p. 2), and across the natural and social sciences to "enhance the chances of contributing to an innovative project" (Microbio-E 2014, p. 27). It claimed to be problem-driven and to contribute to solving energy issues. On paper, Microbio-E is a striking

---

[3]AMIDEX is one of 10 'Initiatives d'Excellence' (IDEX). IDEXes originate in a national programme launched in 2010 to reinvigorate innovation after the 2008 financial crisis. They are meant to drive the emergence of world-class academic clusters in France, with the aim of improving French scores in global university rankings (Juppé and Rocard 2010).

example of 'strategic research': basic research expected to produce a broad base of knowledge from which solutions to practical problems may emerge (Irvine and Martin 1984; Rip 2004). The proposal interwove grand promises with detailed, mundane descriptions of highly specialised research: it was an explicit attempt at articulating scientific excellence and societal relevance, a characteristic of strategic research (Rip 2004). In the project rhetoric, bioenergy served as an 'umbrella term' (Rip and Voss 2013) under which various strands of research were packaged and related to a societal and political concern—namely climate change and energy. The project proposal then appears as an argumentative device aiming to demonstrate that a specific combination of scientific resources would serve the interests of the funder and that the objectives of involved researchers were compatible with those of the funder.

6.4   The Project as Strategic Convergence

The community displayed in project documents does not only exist on paper. The very process of jointly writing a project proposal fostered interactions across groups. It consolidated a convergence of strategies and interests. Microbio-E stemmed from two distinct strategies for navigating the science policy environment: that of basic biology laboratories and researchers, and that of Aix-Marseille University. Their convergence enabled the enrolment of additional partners. The project then worked as a device to keep these strategies together and ensure their synergy, at least for its duration.

6.4.1   Bioenergy to Sell Basic Microbiology

Though interdisciplinary, Microbio-E had its core in biology. Basic biology was already well established in Aix-Marseille, with a close-knit network of strong groups that published, recruited, had enough money to run, and were recognised nationally and internationally (Interviews 1 and 34, senior researchers in biology; fieldwork data). Yet, in interviews, researchers often stressed that basic biology was increasingly hard to fund in its own right, and that they constantly had to justify their work as relevant to society (Interviews 1, senior researcher in biology; 5, researcher in electrochemistry; 8, researcher in biophysics; 9, lecturer in process engineering; 20, researcher in biology). Bioenergy was one of the umbrella terms under which they could sell their expertise, they explained, so many framed their research as relevant to future progress in hydrogen production, algae-based biofuels, biomass degradation, or biofuel cells. They argued that this not only helped obtaining funding but also publishing in high-impact journals (Interviews 1, senior researcher in biology; 5, researcher in electrochemistry), which was an excellence criterion as defined by AMIDEX.

As one senior researcher explained, energy constituted a distinctive thematic niche in biology, where most applications relate to health (Interview 15, physics professor). Interviewed researchers quickly added that this positioning was not purely opportunistic but also helped choose among a multitude of interesting topics, and opened up fascinating fundamental issues (Interviews 1, senior researcher in biology; 15, physics professor). Referring to energy is at times an individual (Interview 38, researcher in biology) or a punctual strategy (Interviews 9, lecturer in process engineering; 17, researcher in bioinformatics; 34, senior researcher in biology), but it had also become deeply ingrained in some research groups, as group leaders explained in interviews.

The leading group in the Microbio-E consortium was led to the energy theme by its research on hydrogen. Since its creation in 1991, it had studied enzymes involved in hydrogen production or degradation in micro-organisms. In the early 2000s, the group realised this could be linked to energy issues (interview 15, physics professor). By 2005, French research funding was targeting hydrogen as an energy carrier. The group rode the wave, promoting its hydrogen-related skills to obtain money and equipment. "We have been living on it for 10 years", I was told. In an emphasis that energy might be an area of application but that it was not their main scientific interest, they also told me: "we do *not* work on energy" (Interview 1, senior researcher in biology). Energy-oriented projects fostered lasting research orientations and collaborations, for instance on biofuel cells or on hydrogen production from biomass (Interview 1, senior researcher in biology; project proposal archives).

One other research group, focused on microalgae and lipids, also benefitted from the hydrogen hype of the 2000s. This group was created in 2006 with an explicit focus on biofuels, building on longstanding local research in plant biology. Its positioning resulted from hydrogen-related projects, as its leader explained, especially a "programme called Biohydrogen [...] that coordinated research on biohydrogen to try to reinforce it, to make it more visible" and that "made [them] realise that it was in [their] best interest to develop this" (Interview 25, senior researcher in biology). The group then created a technological platform supported by the Region, France and the EU, thereby increasing visibility and capacity to attract collaborators and recruits. For this group, contributing to the development of biofuels is a guiding principle. This contrasts with the former group, where relevance to bioenergy as a potential area of application was conceived of as a way to obtain funding, but not as a scientific objective in itself.

6.4.2 Aix and Marseille's Excellent Adventure

The project brought these two versions of strategic research together and aligned them with the objectives of Aix-Marseille University (AMU). AMU was created in 2012 following national reforms increasing the autonomy of universities and encouraging rapprochements between academic institutions. It merged three local universities. A few months later, the 'initiative for excellence' AMIDEX was launched.

After a probation period, it obtained stable funding of 26 million euros a year from 2016 onwards.

AMIDEX, whose slogan is a straightforward "Towards more excellence with Aix-Marseille University", enacts a specific vision of excellence defined by performance in international rankings, the ability to attract international students and academics, interdisciplinarity, and the integration of scientific, economic and industrial actors. AMIDEX organised calls for projects to fund "top international level research and higher education projects (emergent, interdisciplinary and innovative)" on "five priority scientific themes [...] where Aix-Marseille University and its partners can become leaders at an international level within 10 years" (AMIDEX website). Energy featured as priority theme; it was, in fact, pushed forward by biology groups (fieldnotes).

To foster cross-disciplinary interactions, AMIDEX created interdisciplinary networks for each priority theme. The steering committee of the Energy network gathered biologists, physicists, engineers, lawyers, economists, and sociologists, who met every month. They have thus learnt to know one another. This is where the biologists met the non-biologists who would join them in Microbio-E and where they worked to align their strategies with those of AMIDEX.

6.4.3 The Making of Microbio-E

AMIDEX had identified energy as a high-potential theme in the region, but one that was scattered and poorly structured, except for bioenergy—an effect of the strength of the biology network and the strategic positioning of several groups in the field of bioenergy. Yet, initial efforts to organise bioenergy research failed to obtain funding.

As a result, by 2014, AMIDEX had funded very few projects on energy. It needed a flagship energy project before its evaluation in 2016. Those who had submitted energy-related projects were prompted to try again. One scientist central in the project recalls:

> The lab managers were summoned, and they told me: well, you need to write a project. I said, ok, but I already wrote two, I'm not writing a third one to get thrown out. So I met the vice-president, I asked: what would be needed? And he told me: you include mechanics, and you include social sciences. I include mechanics, I include social sciences. And we get the project. (Interview 1, senior researcher in biology)

The project was pieced together quickly, over "two or three week-ends" (Interview 32, senior researcher in biology), drawing on previously rejected proposals. Pressured by short deadlines, participants proposed input based on their ongoing or planned work. The need "to bring everyone together" (Interview 1, senior researcher in biology) accounts for Microbio-E's broad scope.

The inclusion of social sciences resulted from a top-down injunction informed by a rather abstract commitment to interdisciplinary research. Contrary to what the few paragraphs about social sciences in the proposal suggest, there was no clearly defined research agenda.[4]

When focusing on what brought the participants in Microbio-E together, the project appears quite different from the way it is formulated in project documents. In this version, Microbio-E appears as the extension of an already strong network of biologists. For its scientific leaders, Microbio-E was the continuation of a strategy to frame their basic research as relevant to the development of bioenergy. The proposal succeeded in enrolling participants from other disciplines who could join under the umbrella of bioenergy. It also aligned with the ambition of AMIDEX to mould local research according to its criteria for excellence. These strategies converged in the project's function as a device to strengthen a community sharing an affiliation (AMU) and a topic (bioenergy), and to display its promises. This version of the project-ed community does not stand alone: it is a stage in the evolution of a local research community and contributes to its perpetuation and performance at the scales of both research groups and the university.

6.5   The Project as Arena of Research

Ultimately, projects constitute arenas of research: they fund and organise scientific work. To describe them as such, we can follow them in day-to-day research, and ask how and where participants actually encounter the project as such. This means looking for projects in their daily manifestations. This version of Microbio-E is fuzzier than the previous two, which could be traced quite directly from a set of documents or from the accounts of group leaders. Yet, in the case of projects as large as Microbio-E, it is how most participants will relate to it: not everyone involved takes part in the writing of project documents or in long-term strategic decisions. This is where the ethnographic method proves most useful. I followed Microbio-E in two directions: first, searching for the project-ed community as a coherent whole; second, considering the individual collaborations that resulted from the project.

6.5.1   The Project as a Whole

In practice, Microbio-E as a whole turned out to be quite elusive. While researchers knew the project existed and what it was about, it was not always clear who/what was part of it and who/what was not. Researchers sometimes appeared uncertain about whether they were part of it when I contacted them. Doubts about the exact perimeter of the project surfaced during interviews:

---

[4]There were no attempts to conceal the fact that the main contribution expected of social sciences was to ensure proposal success.

> I am not even sure, actually, that I know exactly which projects are on the payroll, and which are not. Hydrogenases, for sure; molybdoenzymes, I think we're on the payroll too, in relations with the formate deshydrogenase [...]. So, I think I am related to this by two projects. (Interview 31, lecturer in chemistry)

A discussion on whether a team was included or not similarly illustrates the fuzziness of Microbio-E's boundaries:

> So, they're not directly funded, they could be in it without being in it, that was not very, very clear... they would have liked to be in but they were not. (Interview 25, senior researcher in biology)

So, where is the project-ed community to be encountered? Project documents seem like an obvious starting point. As outlined above, they framed the project in terms of research questions, links to bioenergy, and teams involved. They included lists of participants, hires, and publications. Yet, the version of the project presented in documents did not match its real-life enactment perfectly. The researchers listed and the actual participants in Microbio-E were not exactly the same, and the same holds true for research questions. Besides, project documents were not distributed among participants. Only a handful of participants contributed to the proposal, and, according to one of them, "no one had a final version" (Interview 38, researcher in biology). In contrast, several researchers I interviewed admitted having never read it. Project documents were not for internal use, but intended for displaying the project to external readers, mainly AMIDEX.

The budget was a second concrete manifestation of the project as a whole, and a crucial one. Its management was centralised under the authority of the project coordinator, resulting in extra work for the administrative staff, for whom the existence of Microbio-E was very concrete indeed, and occasionally overwhelming (Interview 40, administrative staff). It mainly funded short-term contracts for people from outside the region, without guarantees that they would be able or willing to stay. Half of them were to be co-supervised by two research groups: co-supervised postdocs and PhD embodied the promise of collaboration, but also its temporary character. Funding translated into very concrete personnel additions to research groups, as well as in the mention of AMIDEX in publications. Yet, the financial lens does not fully account for the project, either. Not all research groups involved got funding from Microbio-E, and connection to the project was not necessarily limited to the staff funded by it:

> I think that, beyond the PhD that is funded, it is a project in which we are involved in a more important way, because the very essence of the lab is linked to these energy issues, so... we could almost put the whole lab in the project, if we wanted. (Interview 25, senior researcher in biology)

One year into the project, a gathering of Microbio-E as a group took place. All participants were invited to a day-long meeting. It consisted of presentations displaying what each team had been doing and a lunch buffet for networking. The concluding session provided an overview of the project's ambitions, allowing for a discussion of the purpose and future of this project-ed community that was gathered in a single room for the first time. Yet, there was only one such meeting in two years, not allowing for an in-depth collective discussion of findings or for contacts beyond existing collaborations. Like the project documents, it was a display of strength rather than an arena for collective work and community-building.

Interestingly, my own work also participated in enacting the project as a community. As I circulated across teams, I brought the project-ed community into existence—or at least, made it aware of its own existence. Interviewees often asked questions about Microbio-E and its organisation. I acted as a facilitator for community-making by mapping collaborations, transmitting information, and constructing and spreading a narrative about the project. But I was an imperfect messenger: I did not meet each and every participant, and my grasp of the science was limited.

In addition to these situated manifestations of the project, its participants were connected by two shared elements. First, the project narrative and bioenergy label provided a unifying perspective. Everyone in Microbio-E became identifiable as related to bioenergy, and was encouraged to consider potential applications in this domain. Sometimes, this genuinely sparked new research orientations: one researcher explained how she was inspired by a collaboration between biologists and process engineers and wanted to try something similar (Interview 36, biology professor). However, not everyone bought into this new identity. Several participants considered the relation of their work to bioenergy as tenuous at best, explaining: "it is not going to guide what we do on a daily basis" (Interview 14, lecturer in physics), or that "there's nothing established, like a process or something too specific, so it's more the perspective that you work in the same direction" (Interview 19, postdoc in biology). Second, the very existence of the project, as a materialisation of the university's ambition to intensify local interactions, acted as an incentive for collaboration. "I think that it forces us—well, forces us, it encourages us to collaborate, that's not a bad thing" (Interview 28, researcher in biology), said one plant biologist, while the bioinformatician he collaborated with as part of Microbio-E stated that "when you join a consortium, you have to try to interact; so even if they have the resources at home, they look on the fact that we do it with a favourable eye, to create bonds" (Interview 17, researcher in bioinformatics).

Interviews suggest Microbio-E provided a frame for informal exchanges and revived old scientific connections; it led to the identification of potential new partners; and it allowed teams to showcase their research, and to secure involvement in a local bioenergy research field. For some, this was the main motivation for joining: they were looking for opportunities to join a new community and to develop their work in new directions (Interviews 17, researcher in bioinformatics; 24, R&D team manager).

All of these manifestations of Microbio-E as a coherent community played out only occasionally in the work of most participants. Besides being difficult to delineate, the project-ed community as a whole did not appear directly relevant to day-to-day research activities. Interviewed postdocs, especially, tended to feel remote from the project, even though their position and salaries directly depended on it (Interviews 21 and 33, postdocs in biology).

6.5.2   A Patchwork of Subprojects

The structuring impact of Microbio-E mostly played out in the smaller collaborations that were initiated or sustained by it. In terms of scientific activity, Microbio-E mainly worked as a patchwork of independent subprojects. Collaborations emerged and developed—or not—as part of the day-to-day pursuit of research, with the project working as a catalyst.

This loose, federalist structure was to an extent built into Microbio-E. As one coordinator explained, there was deliberately no top-down organisation of research activities:

> I don't have control over people's research. In terms of research, I have control only over my research, my team's research, but not the others' research. I am not going to say, what you're doing is not good, or you're wrong. Everyone is responsible for their own topic. (Interview 1, senior researcher in biology)

This gave way to various engagements with and within the project. For some, Microbio-E hardly changed anything, but merely provided resources to develop ongoing research and collaborations. The researchers already well inserted in the Microbio-E network and topics had no need to divert from their interests (Interviews 2, senior researcher in biology; 14, lecturer in physics; 34, senior researcher in biology; 38, researcher in biology). For them, the project-ed community was a mere extension of their own local network: "we were already married", one researcher told me after detailing the history of the connections among the research groups involved (interview 34, senior researcher in biology). Symmetrically, involvement in Microbio-E had little effect on groups that hitherto had few connections with other teams in the consortium. They had proposed subprojects that corresponded to their own ongoing work, and largely stuck to them, arguing that two years is too short a period to build meaningful new collaborations (Interviews 20, researchers in biology; 22, lecturer in mechanical engineering; 32, senior researcher in biology).

In contrast, four collaborations were strongly identified with Microbio-E, either because they were initiated by it or because they were in line with its philosophy of fostering regional interdisciplinary collaborations to connect basic research with bioenergy applications.

The first was a collaboration between microbiologists and process engineers which had been initiated *before* Microbio-E. It studied hydrogen production by bacterial consortia, seeking to scale-up results obtained on the lab bench. It involved three permanent researchers in addition to one Microbio-E postdoc and one PhD candidate. The collaboration took shape through the transfer of instruments and staff from the process engineering lab to the microbiology lab and through their necessary adaptation to different experimental conditions and lab practices. A process engineer spent two years as a visiting researcher in the microbiology lab, installed her experimental devices, and eventually supervised a PhD candidate based there. This collaboration served as a blueprint for Microbio-E, as can be traced in the project proposals that preceded Microbio-E. It came to significantly shape the activities of the two teams involved: Microbio-E sustained and reinforced a collaboration that was not limited to this single project. The project was part of the ongoing constitution of a small-scale shared 'repertoire' (Leonelli and Ankeny 2015) of experimental methods and devices, theories, meeting venues, potential funding sources, and strategies to present and frame research. This repertoire was embodied in experimental reactors as they moved from the process engineering lab to the microbiology lab and in the diverse set of people, competences, and research questions assembled around them.

The second example wa a collaboration *initiated for* Microbio-E. It involved microbiologists, biochemists, bioinformaticians working on genetic sequences, and a technology-transfer unit focused on renewable energy, none of whom had ever worked together before. Microbio-E paid two postdocs and one PhD student. They converged around a shared object—*Asterionella formosa*, a freshwater microalga that grows with a community of bacteria and is of fundamental as well as industrial interest. The collaboration led to the multiplication of the forms of manipulation and scrutiny that the alga and its suit of bacteria were subjected to as they travelled from lab to lab: they were cultivated in containers ranging from a few millilitres to 30 litres, observed and counted using several microscopy techniques, and genetically sequenced by an external facility so that bioinformaticians could study them. Combining all these techniques required regular interactions and meetings, making this the most collaboration-intensive part of Microbio-E (fieldnotes and interviews), and resulting in a paper co-signed by all participants. Skills were sought for outside of the Microbio-E perimeter—here too, the collaboration went beyond the frame of the project. However, it is hard to tell to what extent the resulting group will last, especially since the persons involved full-time held short-term contracts.

The third example emerged *on the margins of* the Microbio-E project, but developed into collaborations within the consortium. It grew out of a PhD thesis on the production of alkanes—a family of chemical compounds that can be used directly as fuel—in microalgae. The PhD student's team identified a light-activated enzyme catalysing the production of alkanes in microalgae and used the Microbio-E network to study the enzyme further. They collaborated with other Microbio-E teams to characterise the enzyme structure and to do spectroscopic analyses. This collaboration was not planned in Microbio-E, but was facilitated by it. It was also presented as a success story that could provide the backbone for future collaborations (meeting notes, March 2016): a "beautiful story that we hope to continue" (Interview 15, physics professor).

The inclusion of social science was another collaboration initiated because of Microbio-E. As opposed to the three previous examples, it was not built around a shared object of enquiry, but driven by constraints from the funders. As is often the case in such situations, we moved through different roles (Balmer et al. 2015). Yet, we did not enter a 'collaborative mode', mutually shaping the knowledge produced (Calvert and Martin 2009). The main objective was to foster mutual knowledge and exchanges about energy across the social and natural sciences. The collaboration existed through my circulation across teams and the resulting interactions and mutual observations (I too was subjected to curiosity, scrutiny, and questions about the role, outputs, and practices of sociology). The project itself constituted the common ground upon which to establish these interactions. It meant something for each participant; it was the motivation for our meetings; and it was an object of interrogation and investigation. Through the presence and findings of social sciences, Microbio-E appeared as an experiment in community-making: not a roadmap, but a nutrient broth, or an incubator (to use biological metaphors) in which some collaborations might thrive while others barely caught on.[5] Indeed, when we presented our study, some of Microbio-E's participants asked us if we had recommendations about what was done well or not and about what worked best in terms of project organisation.

6.6 Conclusion

Combining ethnographic observation, interviews, and document analysis, I have tried to understand projects in contemporary technoscience. In particular, I sought to characterise projects as collective ventures and to specify how they take their place among existing research institutions, communities, and practices. My analysis was based on the in-depth study of a large interdisciplinary project with a regional focus and an ambition to link basic research to potential applications relevant to bioenergy—Microbio-E. In defining my research object, I found that Microbio-E as a whole was loosely connected and hard to delineate. The closer I got to the day-to-day practice of research, the fuzzier the notion of the 'project' seemed to get.

In fact, I encountered several versions of Microbio-E. In project documents and general presentations, the project works as an argumentative device: it is a coherent narrative connecting research questions, available scientific competence and resources, funders' requirements, and societal concerns to demonstrate the relevance and excellence of a collective research plan. In retracing the genesis of Microbio-E and its institutional embedding, the project appears as a strategic convergence: it emerged where the long-term interests of those involved met, because it brought resources that they could use for their own agendas. Last, when observing daily scientific work, the project appears as a protean arena, a pool of resources that influences research collaborations and topics to an extent, but intervenes only punctually as a discrete entity. Thus, Microbio-E functioned as a boundary object (Star 2010). It was a flexible arrangement of material, procedural, and social resources that was ill-structured as a whole, but it was mobilised by different groups with their own references and objectives. In its co-existing versions, it served a range of strategic needs: those of researchers needing money or contacts; those of the

---

[5]Some participants understood the project in this way, explaining in interviews that not all collaborations were expected to fare equally well: Microbio-E was a test to identify which could yield interesting results and secure future funding (Interview 4, senior researcher in biology).

university promoting excellence; those of research groups and networks seeking increased visibility and prestige.

This has implications for how we conceive of projects, because it shows that projects are not just a way of organising research and research communities. Considering the project in terms of how it shapes the work and interactions of researchers only tells one part of the story. This is what I try to convey with the notion of a 'project-ed community' and the various forms of 'project-ing' that it implies. Certainly, projects create communities—these are provisional, more or less loose, and may not outlast the project, but they are communities nonetheless in that they are united by shared objectives, resources and constraints. Projects also contribute to shaping and reconfiguring existing scientific communities and groupings, for instance by drawing them together or introducing new areas of research (though, here again, this may only be temporary). But 'project-ed' communities are not just communities constituted by a project. Projects do more than dispatch resources. First, they are displays of strength and relevance: they constitute symbolic resources for research groups and academic institutions to justify their existence and emphasise their potential contributions to science and society. They also constitute projections of a desirable future: they show what a current grouping could become and achieve if resources keep flowing. In that sense, community-making can be a project in itself and an objective of projects—it was very much so with Microbio-E. From this conception of projects, we can draw conclusions along two lines.

First, from a methodological and analytical perspective, there is a tension between the relevance of projects as arenas of research and their artificiality. The study of Microbio-E confirms that projects affect research practices because they allocate resources, foster specific forms of collaboration, define temporalities, and frame scientific work. All the same, projects are a lot about what *will* or *could* be. One function of projects is to display research communities and their work to the outside world (starting with funders). In that sense, projects are strategic constructions to enable the pursuit of daily work, and do not fairly represent the reality of this work. Most of the subprojects that constituted Microbio-E reached beyond the strict frame of this project, be it in terms of topics, participants, or temporality. This warrants caution on the observer's part. While projects provide privileged settings in which to witness the entanglements that constitute research practices, one has to be careful not to confer on them more importance than they have for participants.

Second, from a conceptual point of view, my ethnographic take on Microbio-E contributes to the understanding of what glues scientific communities together, especially since community-making was one explicit objective of Microbio-E. While Microbio-E is not representative of all scientific projects (those come in too many shapes, sizes and contexts), it provides insights into how scientists work together when they are constrained by the timeframe, predefined objectives, and funders' expectations that come with a project. Given its stated aim to foster a strong local research collective, Microbio-E also exemplifies the 'promise of communities' that Kastenhofer and Molyneux-Hodgson point at in their introduction (this volume), and so its outcomes give us insights into how such promise fares in practice. Microbio-E did foster several collaborations, across many disciplines. I have shown that these were usually grafted onto pre-existing ties or resources and, most crucially, that they took material forms: adapted laboratory instruments, biological organisms or molecules scrutinised using different methods, collaborating researchers. In sharing very concrete matters, subgroups within Microbio-E tentatively developed their own small-scales 'repertoires' (Leonelli and Ankeny 2015)—sets of shared resources, practices, and (small-scale) infrastructures that structure the way they do science and help them navigate the research and funding context. Leonelli and Ankeny show that the constitution of repertoires allows projects to lastingly structure the way whole communities do science; but they also emphasise that not all projects lead to the development of a repertoire. What the study of Microbio-E shows is that projects can be designed to experiment with potential repertoires, not necessarily at the project scale, but by fostering interactions on specific topics and with an expectation of results. This is, however, at odds with their short temporalities and reliance on non-permanent staff.

Acknowledgements This work was funded by the AMIDEX project 'Microbio-E' (Aix-Marseille University). I thank the participants in Microbio-E and their colleagues who kindly answered my questions. I also thank Marie-Thérèse Giudici and Pierre Fournier for supervising my work. The phrase 'project-ed community' was suggested by the organisers of the STS Austria Workshop on Community and Identity in Contemporary Technosciences held in Vienna in February 2017. Karen Kastenhofer and Susan Molyneux-Hodgson and two anonymous reviewers provided precious feedback on previous versions of this manuscript.

References

AMIDEX [institutional website]. n.d. http://amidex.univ-amu.fr/en. Accessed 30 June 2017.

Balmer, A.S., J. Calvert, C. Marris, S. Molyneux-Hodgson, E. Frow, M. Kearnes, et al. 2015. Taking roles in interdisciplinary collaborations: Reflections on working in post-ELSI spaces in the UK synthetic biology community. *Science and Technology Studies* 28 (3): 3–25.

Barrier, J. 2011. La science en projets: financements sur projet, autonomie professionnelle et transformation du travail des chercheurs académiques. *Sociologie du travail* 33: 515–536.

Calvert, J., and P. Martin. 2009. The role of social scientists in synthetic biology. *EMBO Reports* 10 (3): 201–204.

Felt, U. 2016. Of timescapes and knowledge spaces: Re-timing research and higher education. In *New languages and landscapes of higher education*, ed. P. Scott, J. Gallacher, and G. Parry, 129–148. Oxford: Oxford University Press.

Gibbons, M., C. Limoges, H. Nowotny, S. Schwartzmann, P. Scott, and P. Trow. 1994. *The new production of knowledge*. London: Sage.

Gläser, J., and G. Laudel. 2016. Governing science. *European Journal of Sociology* 57 (1): 117–168.

Hubert, M., and S. Louvel. 2012. Le financement sur projet: quelles conséquences sur le travail des chercheurs. *Mouvements* 71: 13–24.

Irvine, J., and B.R. Martin. 1984. *Foresight in science. Picking the winners*. London: Frances Pinter.

Jouvenet, M. 2011. Profession scientifique et instruments politiques: l'impact du financement "sur projet" dans des laboratoires de nanosciences. *Sociologie du travail* 53: 234–252.

Juppé, A., and M. Rocard. 2010. *Investir pour l'avenir. Priorités stratégiques d'investissement et emprunt national*. Paris: La Documentation Française.

Knorr-Cetina, K. 1982. Scientific communities or transeptistemic arenas of research? A critique of quasi-economic models of science. *Social Studies of Science* 12 (1): 101–130.

Leonelli, S., and R.A. Ankeny. 2015. Repertoires: How to transform a project into a research community. *Bioscience* 65 (7): 701–708.

Merz, M., and P. Biniok. 2016. The local articulation of contextual resources: From metallic glasses to nanoscale research. In *The local configuration of new research fields. Sociology of the Sciences Yearbook 29*, ed. M. Merz and P. Sormani, 99–116. Cham et al.: Springer.

Merz, M., and P. Sormani, eds. 2016. *The local configuration of new research fields: On regional and national diversity. Sociology of the sciences yearbook 29*. Cham et al.: Springer.

Meyer, M., and S. Molyneux-Hodgson. 2016. Placing a new science: Exploring spatial and temporal configurations of synthetic biology. In *The local configuration of new research fields. Sociology of the Sciences Yearbook 29*, ed. M. Merz and P. Sormani, 61–77. Cham et al.: Springer.

MICROBIO-E. 2014. *Project proposal (unpublished document)*. Marseille: AMIDEX.

———. 2015. *Progress report. (unpublished document)*. Marseille: AMIDEX.

Molyneux-Hodgson, S., and M. Meyer. 2009. Tales of emergence: Synthetic biology as a scientific community in the making. *BioSocieties* 4 (2–3): 129–145.

Rip, A. 2004. Strategic research, post-modern universities and research training. *Higher Education Policy* 17: 153–166.

Rip, A., and J.-P. Voss. 2013. Umbrella terms as mediators in the governance of emerging science and technology. *Science, Technology & Innovation Studies* 9 (2): 39–59.

Robinson, K.R.D., A. Rip, and A. Delemarle. 2016. Nanodistricts: Between global nanotechnology promises and local cluster dynamics. In *The local configuration of new research fields. Sociology of the Sciences Yearbook 29*, ed. M. Merz and P. Sormani, 117–135. Cham et al.: Springer.

Schultz, E. 2013. Le temps d'un projet: Les temporalités du financement sur projet dans laboratoire de biophysique. *Temporalités* 18. [online, URL: http://journals.openedition.org/temporalites/2563].

Star, S.L. 2010. This is not a boundary object: Reflections on the origin of a concept. *Science, Technology and Human Values* 35 (5): 601–617.

Strathern, M., and E. Khlinovskaya Rockhill. 2013. Unexpected consequences and an unanticipated outcome. In *Interdiscplinarity: Reconfiguration of the social and natural sciences*, ed. A. Barry and G. Born, 119–140. London: Routledge.

Tari, T. 2015. *The quest for the Oily Grail. The emergence of a research area on bioenergy and its role in the production of knowledge* (PhD thesis). Université Paris-Est, LISIS.

Thèves, J., B. Lepori, and P. Larédo. 2007. Changing patterns of public research funding in France. *Science and Public Policy* 34 (6): 389–399.

Tricoire, A. 2006. Externaliser la contrainte. Le dispositif de pilotage d'un projet de recherche communautaire (enquête). *Terrains & Travaux* 2 (11): 122–139.

———. 2011. La structuration d'un projet européen. Du réseau scientifique au collectif de recherche. *Terrains & Travaux* 1 (18): 81–101.

Vinck, D. 2016. The local configuration of science and innovation policy: A city in the Nanoworld. In *The local configuration of new research fields. Sociology of the Sciences Yearbook 29*, ed. M. Merz and P. Sormani, 81–98. Cham et al.: Springer.

Ylijoki, O.-H. 2016. Projectification and conflicting temporalities in academic knowledge production. *Theory & Science* 38 (1): 7–26.

Chapter 7
The Epistemic Importance of Novices: How Undergraduate Students Contribute to Engineering Laboratory Communities

Caitlin Donahue Wylie

Scholars and practitioners have long viewed learners as works-in-progress, as somewhat empty vessels to be filled with appropriate knowledge and skills to become future practitioners. A variety of powerful critiques portray this mindset as inaccurate and a barrier to effective learning and identity formation (e.g., Jones and Brader-Araje 2002; Wilson and Peterson 2006; Duschinsky 2012). These authors argue for embracing learners' existing knowledge and identities to facilitate their learning and also to enable their perspectives to inform a community's existing knowledge (e.g., Hakkarainen et al. 2004).

However, some communities continue to talk about novices according to the old model of passive receivers of knowledge. For example, academic research groups in engineering tend to assume that expertise aligns with hierarchy, such that higher-status members (e.g., faculty, graduate students) have more valuable knowledge than lower-status members (e.g., undergraduate students). This study challenges that view by presenting evidence that social and technical knowledge does not only flow from supposed experts to supposed novices; rather, it circulates and changes as it travels between them. What, then, do the identities of "expert" and "novice" mean in practice?

Based on participant observation and interviews in two engineering research laboratories in an American university, I argue that it is more productive to define a community based on mutual learning rather than a hierarchy of experts. Similarly, sociologists Grit Laudel and Jochen Gläser define scientific identities based on whether the person is learning or teaching: "An *apprentice* learns to conduct research while working under the direction of others; a *colleague* conducts independent research and contributes the results to the community's knowledge; a *master is a colleague* who additionally acts as a mentor for apprentices" (2008, p. 390, original

C. D. Wylie (✉)
Program in Science, Technology, and Society, University of Virginia, Charlottesville, VA, USA
e-mail: wylie@virginia.edu

© The Author(s) 2021
K. Kastenhofer, S. Molyneux-Hodgson (eds.), *Community and Identity in Contemporary Technosciences*, Sociology of the Sciences Yearbook 31,
https://doi.org/10.1007/978-3-030-61728-8_7

emphasis). However, while these roles hold true in some situations, I found that lab members regularly swap the roles of learner and instructor, regardless of their status as an undergraduate student, a graduate student, or a faculty member/principal investigator (PI). Status does matters, though, because identity relies on how individuals perceive themselves as well as how others treat them; identity is therefore "double sided" (Stevens et al. 2008, p. 357, drawing on concepts of identity and the self from Holland et al. 1998). For example, lab members might be surprised if undergraduates demonstrate expertise about a topic because they assume students are primarily learners, even if those students consider themselves experts about that topic. Likewise, PIs are likely to consider themselves experts in their discipline, though of course they continue to learn. Thus, the identities of "PI" and "student" do not always predict which members teach and learn.

In particular, I found that undergraduates play a wider variety of roles in research communities than scholars and practitioners realize. Lab members' perceptions of undergraduates grant these students unique opportunities to shape knowledge construction. Specifically, lab members perceive undergraduates, as a category of both ascribed and achieved identity, as low-status, low-stakes learners as well as interdisciplinary, open-minded scholars. These roles enable undergraduates to actively contribute to the construction of knowledge and communities by serving as local experts at lab tasks and as non-expert outsiders who challenge the knowledge that lab members take for granted. Thus, students—and all lab members—play the roles of expert and novice depending on the situation. These roles are not strictly tied to members' more stable identities as undergraduates, graduate students, or PIs. One benefit of this fluidity is that students can more easily develop an identity as a member of a research community if they are actively contributing as well as gaining knowledge.

Novices, such as apprentices, construct their professional identity through experiences with a community (Lave and Wenger 1991). Novices bring their own backgrounds and future aspirations, which must be included in a learning process to create a "trajectory"—a sense of context and purpose for learning that then shapes novices' identities (Wenger 1998, p. 154). But communities often reject novices' knowledge as "marginal" and irrelevant to their own practices and beliefs (Wenger 1998, p. 216). Novices look like empty vessels indeed when the community doesn't recognize their past experiences as valuable. Novices then may struggle to develop a sense of trajectory and thus identity in the community. Education scholar Etienne Wenger criticizes this approach as not only discouraging potential future members but also overlooking "a wisdom of peripherality"—the insights about a community and its practices that are available only to relative outsiders (1998, p. 216). By revising communities' view of knowledge to be diffuse rather than centralized in the most experienced or highest-status members, we gain novices' important perspectives while also supporting their formation of trajectory and identity. Communities will still marginalize novices, due to their insufficient expertise or what Wenger calls "competence" (1998, p. 216). But embracing novices' view from the periphery would promote their identity formation and success in the community and improve learning for all members, thereby strengthening the community's current practices and future population. By shifting between the roles of expert (e.g., instructor) and novice (e.g., learner), all members develop a sense of the knowledge that they can teach and learn from others.

To represent the epistemic exchanges between novices and experts, I propose a model of community interactions as a system of transfers of ideas between community members with different identities (e.g., PIs, graduate students, and undergraduates). These exchanges simultaneously build identities and communities. Research communities typically define their primary output as knowledge, not new practitioners. But Wenger (1998, p. 214) suggests that a "community of practice"—in which novices participate alongside experts—is a powerful site of knowledge creation *because* of the dynamism of novice-expert interactions. I suggest that applying Wenger's view of learning as the central activity of research communities will help us understand how these communities produce new members as a mechanism of creating new knowledge. To acquire expertise, after all, everyone must learn and continue to learn; expert skills and knowledge are not static. As novices learn, they help other members learn too. Therefore, studying novices, as the community members who focus the most explicitly on learning, provides a window into how people change their own and their community's knowledge. First I situate this study among theories about initiation into technoscientific communities, then I discuss how novices' expertise and inexpertise both benefit these communities.

7.1 The Roles of Novices in Communities

Existing models of how communities incorporate novices focus on the reproduction of the community more than on novices' existing attributes or values, by assuming that successful novices will perpetuate established ways of doing and knowing (Mody and Kaiser 2008, pp. 379–81). Learning these norms relies on the process of socialization, i.e., adapting a newcomer to a group's practices and behavior. For example, physician and philosopher Ludwik Fleck focuses on a student's mastery of a community's thought style as the key part of indoctrination into a scientific community, such that "[a thought style] becomes natural and, like breathing, almost unconscious, as a result of education and training as well as through his participation in the communication of thoughts within his collective" (1979, p. 141). Fleck does not consider novices' existing identities or beliefs, except by assuming that they will be replaced with that of the collective.

According to Fleck, learning what a collective believes and how members think is more important for a novice than learning hands-on practices or, implicitly, merging a novice's own beliefs with the collective's. For example, Fleck presents textbooks as the primary mode of learning a thought style and thereby joining a thought collective (1979, p. 112). He defines a thought style as a result of "authoritarian" instruction (1979, p. 104), in that "any didactic introduction to a field of knowledge passes through a period during which purely dogmatic teaching is dominant" (1979, p. 54). There is no mention of how students align this "dogmatic" education with their existing knowledge and identities. Philosopher Thomas Kuhn similarly argues that students must learn a scientific field's "paradigm," which includes a thought style and accepted knowledge as well as research methodologies (1996, p. 109). Like Fleck, Kuhn credits textbooks as powerful agents of transferring a paradigm to students, though he criticizes the "narrow and rigid education" they provide (1996, p. 166). He omits interacting with colleagues as part of a paradigm or as a way of learning a paradigm. However, textbooks and unquestioned dogma are insufficient for building novices' identities or preparing them to join a research community.

In particular, textbooks cannot convey a community's tacit knowledge, meaning knowledge that is not articulated (Collins and Evans 2007; Kaiser 2005). Research skills, such as the appropriate use of tools and ideas, and social norms rely on tacit knowledge. This is one reason why engineers historically learned their profession through apprenticeship (Calvert 1967). Apprenticeship provides direct instruction and hands-on practice, alongside camaraderie, informal interactions, and feedback when a learner breaks a social or technical norm. Anthropologist Jean Lave and Wenger (1991) argue that group interaction is so embedded in learning that the two cannot be separated. Only through "legitimate peripheral participation" in a "community of practice" can learners gain both the social knowledge and the practical skills required to be accepted as members (Lave and Wenger 1991, p. 29).

Fleck and Lave and Wenger present novices as blank slates by not acknowledging their existing knowledge and experiences. Fleck's thought collective is unidirectional, in that novices are transformed to see through the collective's eyes. However, a thought collective relies on bringing together many people's views, experiences, and questions as they try to make sense of new ideas and evidence. Fleck (1979, p. 144) even claims that thought collectives are egalitarian: "All research workers, as a matter of principle, are regarded as possessing equal rights." These ideas value multiple perspectives and identities, but novices do not seem to qualify as "research workers" to Fleck. Lave and Wenger's idea of a community of practice is similarly top-down, in that novices are immersed alongside expert practitioners. A community of practice pursues "the generative process of producing their own future" by preparing learners as future members (Lave and Wenger 1991, pp. 57–8). Lave and Wenger and their research participants do not consider what novices bring to the community; instead, they portray learners' experiences of being taught to carry on the community's status quo.

In contrast, Kuhn argues that newcomers, such as students, are a common and important cause of paradigm shift, thanks to their incomplete indoctrination into the existing paradigm and thereby their different worldview from more experienced researchers (1996, p. 90). But Kuhn's celebration of novices as an injection of creativity and outside perspectives is only mentioned briefly, and it is not shared by Fleck or Lave and Wenger. Ethnographies have shown the epistemic value of including a variety of scientific—and non-scientific—identities in research communities (e.g., Traweek 1988; Galison 1997; Doing 2009; Wylie 2015a, b, 2016, 2018a), but they do not explore the impact of learners.

A few studies note the contributions of graduate students to research communities. Sociologist Robert Campbell (2003) argues that managing graduate students is an important everyday task of being a scientist, though it is not explicitly taught or highly valued. Students therefore provide scientists with the opportunity to learn how to mentor. Working with students is an example of the "articulation work" of making scientific knowledge (Fujimura 1996), alongside organizing resources, communicating with administrators and funders, and other undervalued tasks that make research possible. Graduate students typically do hands-on research work while PIs plan projects and train the students (Felt et al. 2013; Delamont, Atkinson, and Parry 2000; Walford 1981). Education scholar Geoffrey Walford (1983) argues that because graduate students in science perform so much of a research group's work, decreasing their number would reduce most groups' productivity. Thus, student labor outweighs even the significant time and effort invested in training them. Vincent Lariviere (2012), a scholar of information science, found that graduate students who are involved in multiple projects and collaborations tend to be more productive, i.e., they publish more papers and finish their degrees more quickly. These studies consider graduate students as an important workforce but not necessarily as contributors of ideas or community-building. These studies—and most others (e.g., National Academies 2017)—do not mention undergraduates as members of labs or of the broader context of a research community's "epistemic living space" (Felt 2009).

Omitting undergraduates from studies of knowledge production and laboratory life overlooks a segment of the research workforce as well as a source of technical skill and interdisciplinary ideas. For example, graduate students report many personal and professional benefits of mentoring undergraduates, including more productivity, confidence, enjoyment, communication and mentoring skills, and improved comprehension of field-specific knowledge (Dolan and Johnson 2009, pp. 491–2). Teaching undergraduates in courses also improves graduate students' research skills, such as their ability to develop good research questions and study designs (Feldon et al. 2011). These benefits are striking in their variety and importance to graduate students' professional abilities, and they deserve further study.

This paper investigates how novices shape an expert technoscientific community as an exchange rather than a transfer of experts' knowledge and values to novices. This view is especially important now, as movements for interdisciplinary research, citizen science, and technology entrepreneurship celebrate novices' contributions to technoscience. Therefore, I suggest that we think about novices and experts interacting in a community as an interconnected ecosystem, with feedback loops and shifting, situated relationships. Because undergraduates' experiences as lab workers resemble apprenticeships (in terms of active learning, social immersion, identity-building, and mentoring), studying students offers valuable insights into epistemic exchanges between novices and experts.

7.2 Methodology

This paper draws from an ethnography of two engineering laboratory communities that include undergraduate workers in a medium-sized public research university in the United States (Wylie 2018b; Wylie 2019; Wylie and Gorman 2018). One PI, whom I call Kate,[1] studies the properties of materials. Kate's lab has one postdoc, nine graduate students, and three undergraduates. Including Kate, nine lab members are women and five are men; this proportion of women is unusually high.[2] Two lab members belong to racial or ethnic groups that are underrepresented in engineering in the United States, which is a lower proportion than the national averages.[3] The other lab's PI, whom I call Dan, develops electronic sensor systems. Dan's lab has one postdoc, 11 graduate students, and two undergraduates. Two lab members are women and 13 are men, which is an unusually low proportion of women. One lab member belongs to an underrepresented group, which is also an unusually low proportion. The five undergraduates work for hourly pay and had worked in their labs for at least one term (and up to three years) before this study began. I attended group meetings, shadowed each undergraduate in the lab to observe how they interacted with other members, and interviewed members during the academic year of 2016–2017.

Grounded theory and inductive analysis guided the qualitative data analysis of my observation fieldnotes and interview transcripts (Creswell 2007). I indicate which data are from interviews with citations; all other quotations are from my fieldnotes. The study examines what and how undergraduates learn from working in labs, a question that also yields opportunities to observe the formation of identities and communities in action. In observations I was looking for moments of learning, when knowledge passed between interlocutors explicitly or implicitly. I found significant exchange and construction of knowledge between supposed experts and novices. I sorted these moments of epistemic exchange—i.e., events and statements from fieldnotes and transcripts—by context to understand the circumstances in which they occurred. These data fit two broad categories, explored in the following sections of this paper: situations in which undergraduates are experts and decidedly not experts. These findings capture the epistemically rich benefits of members' movement between the roles of expert and novice in technoscientific communities.

7.3   Undergraduates' Expertise

Despite their low status as novices, undergraduates can be research groups' local experts about techniques. The opportunity to share their knowledge grants undergraduates a sense of belonging in the community and a clear contribution to the group's research. According to Kate's graduate student Samuel, "we ask each other

---

[1] All names are pseudonyms.

[2] In 2016 in the United States, women earned only 21% of engineering undergraduate degrees, 25% of master's degrees, and 23% of doctoral degrees (Yoder 2016, pp. 15, 23, 27).

[3] In 2016 in the United States, students from underrepresented racial or ethnic groups earned only 18% of engineering undergraduate degrees, 16% of master's degrees, and 13% of doctoral degrees (Yoder 2016, pp. 15, 23, 27).

for help a lot because everyone's an expert in something" (interview 10). The community benefits from members' sharing of their diverse skills. Psychologists call this collective awareness of each member's specific expertise transactive memory, which is best established through training with a community (Wegner 1986; Gorman 2002). For example, Dan's research group depends on undergraduates Will and Rick to build websites and 3D-print materials, respectively. Both students learned these skills on their own, not from working with Dan's group, and no other member had these skills. Similarly, Kate named undergraduates Jessie and Gretchen as her group's experts with a certain machine thanks to their frequent use of it. Graduate students first trained the two undergraduates to operate the machine, and the undergraduates then honed their skills through trial and error.

The most valuable knowledge and skills undergraduates can have are those that other members lack. For example, undergraduate Elena shared her knowledge of physiology learned from her biomedical engineering major with Dan's research group of electrical and computer engineers. During a meeting, a graduate student was presenting about how electronic sensors measure people's stress. Dan asked whether cortisol levels could indicate stress, and the graduate student didn't know. Dan asked another graduate student, who also didn't know, and clearly neither did Dan. Then Elena spoke up, explaining that cortisol levels change with circadian rhythms and exercise, so they are a nonspecific, problematic indicator of stress. This information inspired a lively discussion among Dan and the graduate students about how to use cortisol level as a "ground truth" for stress, including how to exclude other variables and measure changes with noninvasive sensors. The group needed the undergraduate's knowledge, which she had learned from another discipline. Making connections between disciplines is an important and common contribution from undergraduates, at least in education systems that prioritize general education alongside discipline-specific courses. In the United States, undergraduates take courses in various fields, while PIs' and graduate students' most recent education is specific to their chosen field.

When undergraduates have abilities that others don't, PIs encourage them to share. Undergraduates spend less time in the lab than graduate students do, due to their heavy course load and shorter degree time, so disseminating their skills is important for a research community. For example, Dan asked Rick to present his 3D printing method to the group "so we don't have to rely on just a couple of people for our 3D modeling." Dan asked several questions during Rick's presentation, revealing his own minimal knowledge as well as his eagerness for himself and others to learn. It surprised me that both Will and Rick wanted to teach their skills so that they could stop doing those tasks. Will tried to hand off the group's website management to all new undergraduates, as yet unsuccessfully. Rick had already delegated most 3D-printing work to newer undergraduates because "I've probably learned what I've needed to." Both undergraduates told me they were bored with those tasks and wanted to focus on learning new skills. They did not revel in their mastery, implying that this expert status may not align well with their identities as learners. This desire to learn is important for novices' success in communities. Aversion to mastered topics could impede a community's efficient division of labor, though, if members don't want to work on tasks they already know.

Inviting undergraduates to teach recognizes their skills and defines them as valuable community members, thereby building their double-sided identity and their sense of belonging in the research group. For example, Kate asked Gretchen and Jessie to write instructions for the group about experimental procedures they knew well because "I'm always looking for people to pass down their expertise on skills and techniques." She meant "down" to posterity and future lab members, not "down" the hierarchy from PI to graduate student to undergraduate; after all, in this case undergraduates were the instructors. Gretchen and Jessie were happy to oblige, perhaps because they were being treated like experts. Kate even included them on a document she emailed to the group listing each technique next to the names of lab members who know how to do it. Kate is thereby codifying the group's transactive memory and tacit knowledge, making it more explicit and accessible. Accordingly, Kate advised a new graduate student to ask Jessie how to work with a material they both study, indicating Jessie's skill and making Jessie proud. Unlike Rick and Will, Gretchen and Jessie aren't trying to hand off their responsibilities to others. Their skills are newer—acquired in the past year—so it's possible they are not as bored with them as Rick and Will are, after three years each in their research group.

It is interesting that these lab members express no competition. After all, scientists talk about competition with other research groups in their field as a defining feature of their struggle to create an identity for their research group and for themselves as PIs (Hackett 2005). But perhaps members don't compete as strongly within a research group. In these two groups, no one hoards skills for individual benefit or feels threatened that someone might surpass them in these skills. Instead, they appreciate everyone's efforts to acquire skills and share them. Swapping skills makes everyone into learners, not just the novices (who are sometimes the experts).

Another context in which undergraduates are considered experts is regarding the specific tasks they have done. Dan's graduate student Edward said, "Will is teaching me today how to use the system to test the app" that Will had built. Will is arguably the singular expert on this smartphone application because he programmed it. Likewise, Kate's undergraduate Jessie told me about a future paper she would write with graduate student Laurie:

> I would have to have a big role in writing that [paper] just because I did the experiment. The theoretical stuff and all of the discussion, the more conceptual stuff, goes over my head a little bit more, but the actual work? That I'm good with. (interview 13)

Jessie believes that Laurie needs her to write about the experiment, as the experimenter. Separately, Laurie told me that undergraduates are "maybe even more practical than some graduate students or [people with] PhDs ... They're just not getting so heavy-handed in the details of the scientific understanding. Sometimes you can be more focused on output and results and progress" as a graduate student, to the detriment of the research (interview 11). Laurie values Jessie's "practical" skills of doing experiments and not getting lost in theories or concerns about research productivity. Laurie can handle the "conceptual stuff," as can any expert in the field, but only Jessie knows about "the actual work" of these particular experiments. This local, situated, one-of-a-kind expertise often belongs to undergraduates, as the most hands-on lab workers. Of course, the "theoretical stuff" can't stand alone; it relies on the practitioner's experienced reality of the research.

7.4   The Epistemic Value of Inexpertise

In addition to sharing their expertise, novices enable other kinds of epistemic interactions thanks to their identity as *non*-experts. For example, PIs and graduate students perceive undergraduates as low-stakes learners and broadly-educated, interdisciplinary scholars. Undergraduates take more diverse courses than graduate students do, and they have less experience in the research group's field, making them less indoctrinated. These attributes are why Kuhn (1996, p. 90) credited most paradigm shifts to fields' relative newcomers. Likewise, undergraduates lack graduate students' deep knowledge, although their flexibility and open-mindedness make them capable learners and thinkers. Dan's postdoctoral researcher James explained this difference in terms of identity: "Entering graduate school, [students] may have a self-definition about 'I'm an electrical engineer,' 'I'm an engineer,' 'I'm a doctor.' The undergraduates don't have any definition about themselves. They are open for any kind of things" (interview 2). As a result, for James, "sometimes I prefer more undergraduates instead of grads" in a research group, thanks to undergraduates' broad interests that are not yet restricted by a field-specific identity. Also, lab members assume (often correctly) that undergraduates know little about the group's work. As a result, they offer explanations, invite questions, and listen to undergraduates' sometimes unorthodox ideas. In comparison, PIs and undergraduates expect graduate students to have significant field-specific knowledge. This role grants undergraduates more freedom to learn than graduate students, whose reputation in the community as knowledgeable and professional is more significant for their academic and career success.

7.4.1   How Low-Stakes Learning Encourages Epistemic Exchange

Undergraduates' double-sided identity as research novices means that they are encouraged to ask questions, and they often do. PIs also invite questions from graduate students but expect them to know more than undergraduates; therefore, graduate students' questions are more risky for their reputations than are undergraduates'. For example, graduate students might be admonished for asking a "dumb" question, while undergraduates are likely to be forgiven. This grants undergraduates a privileged position as question-askers, which simultaneously provides opportunities for graduate students to hear answers to questions they may share but don't ask for fear of looking ignorant. In addition, undergraduates' questions and lack of field-specific knowledge create a demand for PIs and graduate students to explain ideas and instructions, including contextualizing them in terms of what the undergraduates already know. This combination of expertise about a topic and about an audience, known as pedagogical content knowledge, is crucial for effective teaching and communication (Magnusson et al. 1993). Thus, answering novices' questions can improve the answerers' own understanding of concepts and inspire them to think about their knowledge from different perspectives.

Undergraduates can challenge PIs' and graduate students' assumptions by asking seemingly basic questions. Kate's graduate student Laurie, for example, was showing undergraduate Jessie how to use a machine that measures chemical bonds. A graph of measurements popped up on a computer screen, and Laurie rejected it because it had no "peaks." She adjusted a parameter on the machine, explaining to Jessie: "I know from experience there's a lot of noise [i.e., meaningless results]. You want the [graph] lines to be smoother, so upping the power can smooth it out." She ran the analysis with higher power. The new results showed one tall, thin peak which Laurie dismissed, saying, "That's cosmic." Jessie asked, "What was wrong with that one?" Laurie answered absently as she raised the machine's power again, "It's called a cosmic ray. It's not data." She contrasted the cosmic "spikey" peak with a desirable "smoother," "broad" peak. This is a typical example of undergraduates' frequent requests for justification. Explaining tasks requires practitioners to think about how and why they do that work and how to communicate those reasons to non-experts. Philosopher Donald Schön (1987) argues that such "reflection-in-action," i.e., questioning one's own assumptions and routines while enacting them, enables professionals to better identify problems and adapt their practices accordingly. Undergraduates' in-the-moment demands for explanations can thereby challenge assumptions that underlie "normal science" (Kuhn 1996) by inviting graduate students and PIs to recognize their assumptions and potentially revise them (Wylie et al. 2019). For example, which results are data and not data is obvious to Laurie but not to Jessie. Explaining the difference means that Laurie thinks through her judgments, creating an opportunity for error-spotting and new ideas. When not teaching an undergraduate, Laurie probably does not think about why she ignores cosmic rays.

Some research communities value undergraduates as knowledgeable enough to understand the field yet not specialized enough to be mired in isolated conversations with experts. Accordingly, Kate relies on undergraduates to judge the accessibility of graduate students' presentations in lab meetings, precisely for their lack of discipline-specific knowledge. For example, after graduate student Samuel presented a practice conference talk to Kate's research group, he commented that it is targeted at the conference's audience so "it's tough to understand this talk" if you're not an expert. Kate asked Jessie, "How'd it go for you?," thereby making Jessie a test case for whether non-experts could understand Samuel's talk. Jessie rarely speaks in lab meetings, but she did not hesitate to respond to Kate's question. She told Samuel that she understood his results, but the experimental setup "went over my head." She also asked why he used the oxygen isotope O_{18} instead of the more common O_{16}. Kate, impressed, said, "That's a really valuable question. Thank you." The graduate students then discussed where to add the answer to Jessie's isotope question into Samuel's presentation. It's possible an undergraduate could have been scolded for questioning a graduate student's methodological choice. Instead, the non-expert's confusion served as feedback to improve the expert's communication skills. This situation resembles interactional expertise, in that Jessie can talk competently about a field without necessarily contributing to it (i.e., contributory expertise as defined by Collins and Evans [2007]). Graduate students benefit from knowing when their presentations are too specialized and "go over [the] head" of educated generalists such as undergraduates. In response, graduate students acquire widely-applicable communication skills in the form of pedagogical content knowledge—i.e., they learn to recognize concepts that only experts know and to explain them effectively to non-experts.

Although undergraduates' questions can be important for a research community, many go unasked. This is evident from Dan and Kate's ongoing emphasis that students should ask questions, which suggests that they know students sometimes hesitate to ask and thereby publicly reveal their lack of understanding. For example, at one meeting for Dan's group, students discussed a journal paper they had read. Afterwards I asked Will and Rick about a recurring word in the group's discussion: "stochastic." I assumed it was yet another ubiquitous engineering term that I, as a non-engineer, didn't know. But neither undergraduate knew what it meant. Both said they had wondered about it during the meeting, but neither had asked. Will used context to decode the word: "He opposed it to 'deterministic,' so it can't mean that." He mused that it must be about predictability because the paper was about statistics. Rick posed some guesses while Will googled the word on his smartphone, then reported that it means there is some random variability. Will then asked Rick a question about the paper's equations and Rick looked embarrassed, admitting, "I didn't really get the math." This was a missed opportunity. Based on Rick's reaction, I imagine that Will and Rick didn't ask about "stochastic" because they feared being seen as ignorant. Neither wanted to admit that they "didn't really get the math," to the detriment of their learning and of other lab workers' reflection on their own thinking processes.

7.4.2 Teaching as Community-Building

Another way in which undergraduates create learning opportunities for lab members is by serving as mentees, thereby enabling graduate students to learn how to mentor (as Campbell [2003] found that faculty learn how to mentor by advising graduate students). Mentoring is a crucial part of community-building as it guides the socialization of novices while also granting leadership to the mentors, thereby solidifying both groups' double-sided identity as community members. Most studies assume that undergraduates' research mentors are faculty; however, most undergraduate lab workers interact more often with graduate students than with PIs, and they describe graduate students as more approachable than PIs (Dolan and Johnson

2010). Education scholar Jennifer Good and coauthors (2000) found that undergraduates from underrepresented racial and ethnic groups who served as peer mentors reported benefits to their own learning and professional skills; it's logical to assume that graduate student mentors from underrepresented groups, and perhaps all graduate students, earn similar benefits. Teaching undergraduates also helps graduate students hone their own research mindset and skills (Feldon et al. 2011). Undergraduates give graduate students valuable opportunities to learn how to communicate with and manage others. Dan's postdoc James argued that this is the most important reason to include undergraduates in a research group: "I think [teaching undergraduates] is not very helpful for the project, but it's very helpful for graduate students to learn how to teach the junior students" (interview 2). He believes that undergraduates' presence is more valuable for graduate students' development than for the group's research. Thus, even by doing simple tasks, novices can activate important learning mechanisms for other community members.

Few research communities formally teach experts how to work with non-experts. Accordingly, graduate students learn to work with undergraduates primarily through experience and by observing their PI's approach (Dolan and Johnson 2010). Dan regularly checks in with undergraduate and graduate students during meetings and sometimes coordinates how they work together. In one typical conversation, Dan asked a new undergraduate, Barry, what he was interested in working on. Barry answered that he'd like to analyze data from the group's sensor systems. John asked who was working on that, and graduate student Edward said he was and the work was "straightforward," recommending it for an inexperienced undergraduate. Dan said to Barry, "Edward will be busy with [a project] for a few days but you guys can get together after that" to plan the task and teach Barry to do it. Edward and the other graduate students witnessed how Dan assigned partnerships based on people's interests and relevant expertise. Dan deemed Edward's work with Barry lower-priority than Edward's other tasks by postponing their meeting; this implies that Edward was in charge of the partnership. In this exchange, Dan modeled how to create research collaborations. He was subtly preparing all the students to someday create these partnerships themselves, through a valuable real-world example provided by an undergraduate. Embedded in this everyday project management conversation was a lesson on how to build a team, especially one that includes novices.

Generally, experienced graduate students perform more effective mentoring and teaching than newer graduate students, suggesting that students acquire or at least improve these community-building skills during graduate school. For example, as senior graduate student Alison helped undergraduate Frank set up an apparatus in Kate's lab, she gently quizzed him about the procedure and accepted or corrected his answers. She created opportunities for him to practice, reinforce, and show off his understanding and for her to assess how well he knows the procedure. In one instance, I asked both of them what results the apparatus produces, and Alison said to Frank, "You can answer that." Frank answered, then asked Alison if he was right. She agreed and elaborated by explaining what can be learned from those results. Alison artfully boosted Frank's double-sided identity as a researcher by inviting him to demonstrate knowledge she knew he had, and then she built on his knowledge by sharing her own. She probably acquired these skills by observing

Kate's mentorship methods and by working with undergraduates in the lab for several years. In comparison, Kate's graduate student Kenny is about halfway through graduate school and is still learning how to work with undergraduates. In one instance, when undergraduate Gretchen was telling me how a tool worked, Kenny interrupted her to provide the explanation himself. He preferred the role of information-provider to standing back to let the undergraduate demonstrate her knowledge. Gretchen allowed herself to be interrupted, probably in deference to Kenny's higher status as a graduate student and his presumed greater expertise. There could also be a gendered component to this exchange, in that men tend to interrupt more often than women and women tend to acquiesce to being cut off more often than men (Tannen 1994). A graduate student's eagerness to showcase his knowledge over an undergraduate's suggests that he hasn't yet learned how to create space for novices to practice participating in a research community. Luckily, working in Kate's lab alongside undergraduates gives Kenny the opportunity to improve his communication and leadership skills.

7.4.3 How Broad Education and Outsider Status Encourage Creativity

Undergraduates bring ideas to research groups from their courses, previous experiences, and hobbies. They are, in a sense, embodied "trading zones" (Galison 1997): they carry knowledge from various fields into their research community, more so than other members due to their ongoing broad education. Dan's graduate student Edward credits this ability to undergraduates' use of social media to follow new trends in technology:

> Will and Rick know stuff that are more on the tech news, not on the textbooks . . . So they have better ideas . . . I start thinking from, like, textbook style, "Is it possible to do?" But they say, "Yeah, it has been done, maybe not feasible in some cases," based on a [specific] project's perspective. It's nice to have those ideas. (interview 4)

Edward rejects Kuhn's and Fleck's portrayals of students as reservoirs of textbook knowledge (except for himself); instead, he admires undergraduates' ability to suggest cutting-edge approaches for the group's research. For example, Edward credits Rick for improving how the group builds their sensor system: "We would just protect [the sensors] somehow, using some casing. We never thought really about 3D printing. But then Rick said, 'Yeah, we can 3D-print,' and the professor said, 'Yeah, that's a good idea'" (interview 4). The undergraduate brought the graduate students a novel idea (as well as the skill to achieve that idea), which they implemented. Dan's approval no doubt contributed to the decision to 3D-print casings, but everyone credits Rick for the idea. It seems that not knowing how the group typically protected sensors enabled Rick to suggest an unconventional way to solve that problem.

Undergraduates' ideas vary in practicality. Edward recounted Will's innovative solution to a problem with the group's sensor system: "We had this issue with the wifi routers … Will was saying, 'Why not use powerline data communication [instead]?'" (interview 4). Edward was impressed that Will knew about this emerging technology, about which Edward then read several research papers. He was less impressed with the technology's limitations: "It is the trend, but it's not really established and there is a lot of noise in the powerline data." He didn't adopt Will's suggestion; nonetheless, an undergraduate's open-mindedness and broad knowledge influenced a graduate student's learning and decisions.

Undergraduates can serve as a research community's windows on the world beyond the lab, as Rick and Will did above. For example, at a lab meeting Rick presented his idea for a new purpose for the group's sensor systems: monitoring environmental factors that affect public health. His ability to understand the group's systems and situate them in new uses is impressive, reflecting undergraduates' strong connection to fields outside the group's and perhaps also to the "real world" of users because they have not yet become specialized researchers. Rick summarized a paper about a method of monitoring air quality as "It's [our] system on top of a stoplight," thereby using his personal interest in public health to broaden his coworkers' thinking and potentially the impact of their research. Likewise, undergraduate Gretchen's interest in mechanical engineering inspired Kate to think more broadly about her research. Kate told Gretchen about a grant proposal she was writing about alloys. She detailed, in technical terms, the experiments she wanted to include, then added, "I'm trying to make it relevant to the navy" to improve the proposal's chances of being funded. Gretchen asked whether the navy could use these alloys to build engine turbines. Kate answered thoughtfully, "Yes. Oh, maybe I'll put a picture of a ship turbine blade in there." While the PI was thinking about lab-based specifics, the undergraduate asked about general applications, thus creating an opportunity for the PI to consider a new perspective. The PI appreciated this angle and incorporated it in her proposal. Perhaps non-experts can more comfortably play roles of innovators and outsiders than experts can. Without deep knowledge in the field, non-experts draw instead on their wide-ranging experiences and education.

7.5   Conclusion: Epistemic Exchanges

Communities depend on newcomers as future members. As those newcomers learn and become socialized, they also make impressive contributions to the community's generation of knowledge, methods, and social norms. For example, undergraduates import concepts and worldviews from other disciplines to lab workers who have deep specialized knowledge and less recent experience with other disciplines. By valuing undergraduates as broadly-educated outsiders, the community benefits from their knowledge and also makes them feel included in the community and its work. In addition, thanks to their role as novices, undergraduates have the privilege and responsibility to improve their own understanding. As a result, they ask questions that can inform other members and even inspire them to rethink their assumed knowledge and routine practices. By challenging experts to be aware of how they make knowledge, this reflection encourages more attentive work and more open-minded thinking. Undergraduates' laboratory labor is in itself legitimate and valuable; in addition, it creates opportunities for other lab members to learn how to mentor and collaborate. Thus, students provide influential injections of creative and diverse thinking while also enabling social relationships that couldn't happen without them, such as exchanges between experts and novices.

The identities of "novice" and "expert" are relative and context-dependent. Undergraduates are novices in the specific work of a research community, and they are experts at the work and role of learning. Graduate students have (or are acquiring) expertise in a specific field, though they are often novices at collaboration, leadership, and professional communication. PIs are experts in their disciplinary knowledge and professional skills but can be novices in other fields compared to undergraduates who have studied those fields more recently. These identities assign appropriate roles for people to play in different situations, such as question-asker-vs. answerer, task-focused doer vs. big-picture-focused interpreter, and visionary vs. pragmatist. Lab members shift fluidly between these roles during everyday social interactions, illustrating the concept of double-sided identity as they both perform and are treated as experts or novices in different situations.

This study shows that a technoscientific community is a dynamic network of epistemic exchange, not a purely top-down hierarchy. Community members trade, adapt, and produce knowledge and skills up and down the hierarchy as they change contexts and roles. Focusing on how researchers interact as experts and novices broadens our understanding beyond the work of producing new knowledge to include how communities produce new methods, mindsets, and practitioners. The laboratory is a potent place to observe identity work and community formation in action because it hosts unscripted interactions between various kinds of people around complex tasks. Beyond the lab, these trends continue, though more subtly, in other spaces of scientific interaction, such as publications, conferences, and research governance.

This model of community members' mutual epistemic influence offers a more nuanced and accurate portrayal of today's technoscientific communities. It revises the identity of "novice" to recognize the epistemic value of peripheral wisdom and even inexpertise, as well as of novices' power to call attention to a community's paradigmatic assumptions and actions. This model therefore also implies that communities should include novices, for the benefit of all members.

References

Calvert, M.A. 1967. *The mechanical engineer in America, 1830–1910: Professional cultures in conflict.* Baltimore: Johns Hopkins University Press.

Campbell, R.A. 2003. Preparing the next generation of scientists: The social process of managing students. *Social Studies of Science* 33 (December): 897–927.

Collins, H., and R. Evans. 2007. *Rethinking expertise.* Chicago: University of Chicago Press.

Creswell, J. 2007. *Qualitative inquiry and research design: Choosing among five approaches.* 2nd ed. Thousand Oaks: Sage.

Delamont, S., P. Atkinson, and O. Parry. 2000. *The doctoral experience: Success and failure in graduate school.* New York: Falmer Press.

Doing, P. 2009. *Velvet revolution at the synchrotron.* Cambridge, MA: MIT Press.

Dolan, E.L., and D. Johnson. 2010. The undergraduate-postgraduate-faculty triad: Unique functions and tensions associated with undergraduate research experiences at research universities. *Cell Biology Education* 9: 543–553.

Dolan, E., and D. Johnson. 2009. Toward a holistic view of undergraduate research experiences: An exploratory study of impact on graduate/postdoctoral mentors. *Journal of Science Education and Technology* 18 (6): 487–500.

Duschinsky, R. 2012. Tabula rasa and human nature. *Philosophy* 87 (04): 509–529.

Feldon, D.F., J. Peugh, B.E. Timmerman, M.A. Maher, M. Hurst, D. Strickland, and C. Stiegelmeyer. 2011. Graduate students' teaching experiences improve their methodological research skills. *Science* 333 (6045): 1037–1039.

Felt, U., ed. 2009. *Knowing and living in academic research: Convergence and heterogeneity in research cultures in the European context.* Prague: Institute of Sociology of the Academy of Sciences of the Czech Republic.

Felt, U., J. Igelsböck, A. Schikowitz, and T. Völker. 2013. Growing into what? The (un-)disciplined socialisation of early stage researchers in transdisciplinary research. *Higher Education* 65 (4): 511–524.

Fleck, L. 1979. *Genesis and development of a scientific fact.* Chicago: University of Chicago Press.

Fujimura, J.H. 1996. *Crafting science: A sociohistory of the quest for the genetics of cancer.* Cambridge, MA: Harvard University Press.

Galison, P. 1997. *Image and logic: A material culture of microphysics.* Chicago: University of Chicago Press.

Good, J.M., G. Halpin, and G. Halpin. 2000. A promising prospect for minority retention: Students becoming peer mentors. *The Journal of Negro Education* 69 (4): 375–383.

Gorman, M.E. 2002. Types of knowledge and their roles in technology transfer. *The Journal of Technology Transfer* 27 (3): 219–231.

Hackett, E.J. 2005. Essential tensions: Identity, control, and risk in research. *Social Studies of Science* 35 (5): 787–826.

Hakkarainen, K.P., T. Palonen, S. Paavola, and E. Lehtinen. 2004. *Communities of networked expertise: Professional and educational perspectives.* Amsterdam: Elsevier.

Holland, D.C., W. Lachiocotte, D. Skinner, and C. Cain. 1998. *Identity and agency in cultural worlds.* Cambridge, MA: Harvard University Press.

Jones, M.G., and L. Brader-Araje. 2002. The impact of constructivism on education: Language, discourse, and meaning. *American Communication Journal* 5 (3): 1–10.

Kaiser, D., ed. 2005. *Pedagogy and the practice of science: Historical and contemporary perspectives.* Cambridge, MA: MIT Press.

Kuhn, T.S. 1996. *The structure of scientific revolutions.* 3rd ed. Chicago: University of Chicago Press.

Larivière, V. 2012. On the shoulders of students? The contribution of PhD students to the advancement of knowledge. *Scientometrics* 90 (2): 463–481.

Laudel, G., and J. Glaser. 2008. From apprentice to colleague: The metamorphosis of early career researchers. *Higher Education* 55 (3): 387–406.

Lave, J., and E. Wenger. 1991. *Situated learning: Legitimate peripheral participation*. Cambridge: Cambridge University Press.

Magnusson, S., J. Krajcik, and H. Borko. 1993. Nature, sources, and development of pedagogical content knowledge for science teaching. *Examining Pedagogical Content Knowledge*: 95–132.

Mody, C.C.M., and D. Kaiser. 2008. Scientific training and the creation of scientific knowledge. In *The handbook of science and technology studies*, ed. E.J. Hackett, O. Amsterdamska, M. Lynch, and J. Wajcman, 377–402. Cambridge, MA: MIT Press.

National Academies of Sciences, Engineering, and Medicine. 2017. *Undergraduate research experiences for STEM students: Successes, challenges, and opportunities*. Washington, D.C: National Academies Press.

Schön, D.A. 1987. *Educating the reflective practitioner: Toward a new design for teaching and learning in the professions*. San Francisco: Jossey-Bass.

Stevens, R., K. O'Connor, L. Garrison, A. Jocuns, and D.M. Amos. 2008. Becoming an engineer: Toward a three-dimensional view of engineering learning. *Journal of Engineering Education* (July): 355–368.

Tannen, D. 1994. *Talking from 9 to 5: How women's and men's conversational styles affect who gets heard, who gets credit, and what gets done at work*. New York: William Morrow and Company, Inc.

Traweek, S. 1988. *Beamtimes and lifetimes: The world of high energy physicists*. Cambridge, MA: Harvard University Press.

Walford, G. 1981. Classification and framing in postgraduate education. *Studies in Higher Education* 6 (2): 147–158.

———. 1983. Postgraduate education and the student's contribution to research. *British Journal of Sociology of Education* 4 (3): 241–254.

Wegner, D.M. 1986. Transactive memory: A contemporary analysis of the group mind. In *Theories of group behavior*, ed. B. Mullen and G.R. Goethals, 185–208. New York: Springer.

Wenger, E. 1998. *Communities of practice: Learning, meaning, and identity*. Cambridge: Cambridge University Press.

Wilson, S. M., and P. L. Peterson. 2006. *Theories of learning and teaching: What do they mean for educators? Best practices, new research*. Working paper for the National Education Association.

Wylie, A. 2015a. A plurality of pluralisms: Collaborative practice. In *Objectivity in science*, ed. F. Padovani, 189–210. Springer International Publishing.

Wylie, C.D. 2015b. "The artist's piece is already in the stone": Constructing creativity in paleontology laboratories. *Social Studies of Science* 45 (1): 31–55.

———. 2016. Invisibility as a mechanism of social ordering: Defining groups among laboratory workers. In *Invisible labour*, ed. J. Bangham and J. Kaplan. Berlin: Max Planck Institute for the History of Science. Retrieved from https://www.mpiwg-berlin.mpg.de/sites/default/files/migrated/book_invisibility_and_labour_in_the_human_sciences_preprint_.pdf.

———. 2018a. Trust in technicians in paleontology laboratories. *Science, Technology and Human Values* 43 (2): 324–348.

———. 2018b. "I just love research": Beliefs about what makes researchers successful. *Social Epistemology* 32 (4): 262–271.

Wylie, C.D., and M.E. Gorman. 2018. Learning in laboratories: How undergraduates participate in engineering research. In *Proceedings of the American Society for Engineering Education*. Salt Lake City. https://peer.asee.org/29966.

Wylie, C.D. 2019. Socialization through stories of disaster in engineering laboratories. *Social Studies of Science* 49 (6): 817–838.

Wylie, C.D., S.J. Kim, I. Linville, and A. Campo. 2019. Graduate/undergraduate partnerships (GradUP): How graduate and undergraduate students learn research skills together. In *Proceedings of the American Society for Engineering Education*. FL: Tampa. https://peer.asee.org/32879.

Yoder, B. L. 2016. *Engineering by the numbers*. American Society for Engineering Education. Retrieved from https://www.asee.org/documents/papers-and-publications/publications/college-profiles/16Profile-Front-Section.pdf.

Chapter 8
Tracing Technoscientific Collectives in Synthetic Biology: Interdisciplines and Communities of Knowledge Application

Alexander Degelsegger-Márquez

8.1 Introduction: Technoscientific Communities?

Scientific work is collective work. As scientists, we are trained in collective environments. We work and live in departmental and laboratory settings. We collaborate in teams, locally and remotely. We participate in conferences and publish in journals related to our field of work. We relate to colleagues in a specific discipline or an interdisciplinary field of research. We feel part of a scientific community. Or do we?

Scientific communities (in plural) have been defined as collectives that are based on knowledge, more concretely on a common way of producing scientific knowledge (Böhme 1974; Gläser 2006). Mathematicians have a different way of going about scientific knowledge production than biologists, molecular biologists a different one from zoologists. Acquiring the appropriate ways of producing knowledge is an integral part of a young scientist's education and socialisation. Belonging to a scientific community, one could argue, is part of a scientists' identity.

The relevance of such a concept of scientific community has long been contested, however, in parts of STS literature. Already at the end of the 1970s, the role of scientific communities in knowledge production as well as their importance for scientists' subjectivities was questioned (Whitley 1978; Knorr-Cetina 1981). It was suggested, for instance, that the more immediate laboratory and collaboration environment plays a much more decisive role than the diffuse layer of scientific communities.

The question of the relevance of scientific communities gains new momentum in an environment where research funding is increasingly mission-driven and application-oriented, trying to identify the 'next big thing' in terms of economic

A. Degelsegger-Márquez (✉)
Gesundheit Österreich GmbH, Vienna, Austria
e-mail: alexander.degelsegger@goeg.at

© The Author(s) 2021
K. Kastenhofer, S. Molyneux-Hodgson (eds.), *Community and Identity in Contemporary Technosciences*, Sociology of the Sciences Yearbook 31,
https://doi.org/10.1007/978-3-030-61728-8_8

163

potential. Decision-makers do not only choose certain thematic areas over others when distributing funding. They increasingly aim at establishing or nurturing scientific communities in a strategic manner (Kastenhofer 2018). How researchers and their practices relate to these efforts is an unresolved question.

One field that received significant policy-level attention in recent years is synthetic biology. Synthetic biology is about bringing engineering principles (like abstraction, modularity, or standardisation) into biology (cf. European Commission 2005). Synthetic biologists think about living organisms as devices that can be designed to perform specific functions, e.g. produce drugs or detect toxins. Its proponents and critics alike argue that this profoundly changes the way biological research is done.

According to social scientists following and reflecting on the field, synthetic biology spans visions of understanding nature and engineering life (Kastenhofer 2013) or, in other words, of comprehension and construction (O'Malley et al. 2007). Synthetic biology does not aim at representing nature but at remaking it. In doing so, it potentially changes our understanding of what 'natural' means (Calvert 2010).

While both the label and the practices of synthesising or engineering living organisms go back to the early twentieth century, contemporary synthetic biologists themselves relate their work to practices in the early 2000s located in interdisciplinary laboratories in the US (cf. Benner and Sismour 2005; Endy 2005). Since then, synthetic biology has received significant funding in the US and in parts of Europe and Asia.

Three aspects make synthetic biology a particularly interesting case to restate the question of the relevance of scientific communities. First, synthetic biology can be described as a technoscientific field of research (cf. Schmidt et al. 2010) where boundaries between scientific knowledge production and technological applications are blurred (Nordmann 2011; Kastenhofer 2009). The case helps to better understand a potential related blurring of boundaries between scientific communities. Secondly, in integrating biology and engineering, synthetic biology engages researchers with very distinct epistemic orientations. The notion of interdiscipline (Jacobs and Frickel 2009) has been proposed to study the emergence of disciplinary hybrids like synthetic biology. The case helps us to understand how we can conceive of interdisciplinary scientific communities in contrast to disciplinary scientific communities.[1] Thirdly and finally, synthetic biology can be considered a field in ongoing emergence (Morrison 2012) with slightly divergent origin stories, competing visions regarding its present condition, and the continuous construction of promissory narratives. Given that the field is not stabilised, the mechanisms of community constitution are empirically accessible and can be productively researched in contexts of emergence.

---

[1]I propose to differentiate disciplines from scientific communities in that the former are a specific form of the latter: disciplines are, in concrete terms, disciplinary scientific communities. This definition builds on Marcovich and Shinn (2011) who understand disciplines as 'well-defined and differentiated epistemological and organizational units … and institutional … bodies that are separated by recognized boundaries' (p. 586).

I thus set out to study the role scientific communities play in synthetic biology as a technoscientific, interdisciplinary hybrid. What are the communities that matter? Is there something distinctive about the synthetic biology case that points to specificities of technoscientific communities? I address these questions from two angles: first, I review community-making efforts in synthetic biology. Both the top-down side (i.e. the public funding mobilised) and the bottom-up initiatives from researchers are important here. Secondly, I take challenges in synthetic biology research collaborations as an empirical entry point to studying community dynamics. Before approaching my empirical material, I will review key concepts in the sociological study of scientific communities.

8.1.1 From Community to Scientific Community

In sociological theory, the concept of community is linked to authors like Ferdinand Tönnies, Max Weber, or Talcott Parsons. Tönnies (1887/2001) first defined community as such and contrasted the concept with that of society. Communities are characterised by social ties that build on personal social interactions, roles, values, and corresponding beliefs (ibid., pp. 22ff). Decision-making in communities is consensual. Individuals are born into communities and the latter encompass all their actions and relations. In Weber's words, the orientation of social action in 'communal' social relationships is 'based on a subjective feeling of the parties, whether affectual or traditional, that they belong together' (Weber 1921/1978, p. 40). The conceptual opposite of community ('Gemeinschaft') is society ('Gesellschaft'). 'In Gemeinschaft [groups of people] stay together in spite of everything that separates them; in Gesellschaft they remain separate in spite of everything that unites them' (Tönnies 1887/2001, p. 52).

Following Tönnies and Weber, Talcott Parsons also distinguished between community and society. This distinction results from five so-called pattern variables (cf. Parsons and Shils 1951) between which, according to Parsons' structural functionalism, each actor must choose in a given situation of agency. According to these dichotomous variables, communities are characterised by affectivity (vs affective neutrality), functional diffuseness in the definition of situations (vs functional specificity), particularism (vs universalism), ascription (vs achievement), and collectivity orientation (vs self orientation). Applying these traditional sociological approaches to conceive of scientific communities is not without tensions.

Concepts of scientific community are typically connected to characteristics that, according to sociological tradition, resemble societies more than communities. Merton (1942/1993) delineated a joint scientific ethos and thereby focused on the norms shared within the scientific community at large, taking the latter for granted as both result and source of this joint ethos. Science as a field of collective human activity is, in his perspective, characterised by universalism, communalism (in the sense of common ownership), disinterestedness, and organised scepticism (ibid.).

Interestingly, all but one (communalism) of these Mertonian values stand against the traditional sociological definitions of community.

With the practice turn in STS, however, the relevance of Mertonian norms in practices of scientific knowledge production has been questioned. Laboratory studies (e.g. Knorr-Cetina 1981) bring scientific practices closer to traditional ideas of community, pointing to, for instance, the relevance of place (e.g. specific laboratory settings) or affectivity. Values are not something external to scientific practices and discourses. They do not inform or orient these practices but are generated within them (Swidler 2001). Maybe it is because of these tensions with traditional sociological accounts that the notion of scientific community made its way into the sociology of science through the backdoor.

8.1.2 Scientific Communities

The notion of scientific communities was introduced into the sociology of science by Thomas Kuhn (1962/2002) and Karl Polanyi (1962).[2] It was somewhat secondary to their work, however, and therefore undertheorised (cf. Gläser 2006, p. 46). Böhme (1974) was the first to offer a more thorough definition. He defines scientific communities as being based on knowledge (as opposed to values). Scientists coordinate their action through a common object (the research subject, topic or method, etc.). At the core of scientific communities is the organising principle of argumentation: scientific communities constitute arenas for scientists to discuss knowledge claims. Members of a scientific community are not defined by the norms they share (as they might be in a Mertonian view of science) but by the way in which they produce knowledge.

The understanding of scientific communities as defined by shared practices rather than shared values is taken up by later theoretical work. The shared practice described as the basis of scientific communities is the production of scientific knowledge (Holzner and Marx 1979). Gläser (2006) conceptualises communities of scientific knowledge production as the mode of social organisation that allows scientists to produce the specific kind of knowledge that is scientific knowledge. Collective concerted production is structured through a common object and rests on autonomous, decentralised decisions of the producers working the object. Production communities hence solve the problem of motivation, information, and integration in different ways than markets, organisations, or networks would. They are, thus, a distinct type of social order. With this characterisation, Gläser refines the argument that scientific communities are defined by the way in which they produce knowledge: they are characterised by work on a common object and forms of organisation that allow for decentralised knowledge production.

---

[2]Of course, Fleck's (1935/1981) concept of thought collectives and their role in the construction of scientific facts already pointed in the direction of scientific communities.

There are, however, different approaches to refining our ideas of scientific communities. Meyer and Molyneux-Hodgson (2010), for instance, draw our attention not only to the production of knowledge but also to its dissemination. Elzinga (1993) distinguishes disciplinary epistemic communities from hybrid epistemic communities. While the former are organised according to traditional patterns of academic behaviour, the latter are driven by outside rationales from the realm of policy-making or commercial applications. Interdisciplines are characterised by porous boundaries and a collective interest in problem-solving (Frickel 2004)—it is possible to think of problems coming from realms outside academia. All this suggests that there is more to scientific communities than the production of knowledge.

As we shall see later, the analysis of collectives in synthetic biology also raises doubts as to whether it is only the *production* of knowledge that is constitutive for technoscientific communities. The case also helps to specify the relationship between communities and collaborative networks.

8.1.3 Communities and Collaboration

If scientific communities are characterised by a decentralised production of knowledge (Gläser 2006), inter-institutional research collaborations (cf. Katz and Martin 1997) represent one kind of this decentralised production. This variety seems to be particularly important in fields of research that bridge traditional disciplinary boundaries.[3] In interdisciplinary settings, organisationally less integrated modes of production might not be effective (e.g. because researchers in one field do not attend conferences or refer to literature from the respective other field). Collaboration in inter-institutional teams is thus not only relevant for maintaining existing communities of decentralised knowledge production (in a context of growing internationalisation and projectification), but it is also a prime mode of knowledge production for interdisciplinary communities-in-the-making. What remains unclear is whether there is a distinct point where interdisciplinary collaborations spanning several institutions and communities turn into a newly consolidated disciplinary community.

8.1.4 Interdisciplinary Communities

In the literature on interdisciplinarity and field-creation, the concept of the interdiscipline (Frickel 2004) provides a useful conceptual angle to think about the

---

[3]Rhoten et al. (2009) identify team collaboration as one among four types of interdisciplinary practices, the others being cross-fertilisation, problem-orientation, and field-creation.

links between interdisciplinary collaborations and scientific communities. Interdisciplines are stabilised disciplinary hybrids that result from the collective action of scientific and intellectual movements (SIMs). They are characterised by the perforation of existing epistemic and institutional boundaries or the invention of new porous ones. They 'tend to exhibit ... organisational, economic and epistemological variability. Disciplines tend to be anchored in university departments and maintain tight control of ... the production and employment of Ph.D. students. ... Interdisciplines are more likely to be located in less powerful (and thus less stable) institutes, centres, or programmes' (ibid., p. 273). Recurring formal and informal inter-institutional collaborations are an example of these less stable interdiscipline-like arrangements.

The activities of proponents of a field like synthetic biology (cf. Bensaude-Vincent 2013) provide the collective action framing required for interdiscipline formation. Researchers promote synthetic biology as a new and engineering-inspired way of thinking about organic systems. Policy-makers decide to set up funding schemes with the synthetic biology label. Both researchers and research funders engage in community-building.

The notion of interdiscipline provides an alternative to asking whether synthetic biology is or will ultimately become a disciplinary community. Its focus on porous boundaries, frame alignment efforts, and temporal arrangements seems particularly helpful to conceptualise collectives in the technosciences. However, as we shall see, the concept of interdiscipline is not enough to understand what is going on in synthetic biology.

While interdiscipline-like structures emerge in some places, in others there is no indication for community beyond what has been called community-making devices. '[C]ommunity-making devices ... help make, or at least articulate the need for, community' (Molyneux-Hodgson and Meyer 2009, p. 139). They can include workshops, conferences, or journals. The specific community-making devices in synthetic biology will be my first empiric entry-point for conceptualising collectivity in synthetic biology. Before that, however, I will briefly introduce the methodological background of the following analyses.

8.2   Method

My engagement with the field of synthetic biology started in 2008 when I was a research intern at the Austrian Academy of Sciences' Institute of Technology Assessment (ITA). ITA had a project (COSY) that was investigating the communication and perception of synthetic biology in a variety of publics. In COSY and a related EU project (SYNBIOSAFE), we were focussing on accounts of the ethical and social implications of synthetic biology. My document analyses of relevant publications and the informal discussions with the project PI brought me closer to the discourses around the label of synthetic biology. Ever since this project work, I have been following the field, first loosely through contract work for ITA, then as part of my PhD work (2011–2017).

Apart from document analyses and the involvement in social science reflections on synthetic biology, the data basis for the present article comprises 15 semi-structured, theory-generating expert interviews (Bogner and Menz 2009). They were conducted between 2014 and 2016 with synthetic biology researchers (PIs and postdocs) in Austria, Germany, Singapore, and the UK. They have been selected to cover institutional, national, and supranational (European) funding environments for synthetic biology. Most interviews took place in the researchers' offices. The interview strategy I chose was to appear as an interested layperson with regard to the science (the synthetic biology research) and a co-expert with regard to science organisation (funding schemes, policies, etc.), which was possible because of prior research. Using co-authored publications as a conversational anchor at the beginning of the interviews, I invited the interviewees to speak about their experiences in inter-institutional collaborations.

These collaborations, heuristically speaking, are an intermediate layer connecting individual with collective practices. They are one site to observe the significance of collectives. Interviews are a suitable method for the research question at hand because the narratives they evoke are part of the discursive-material practices in a given field. 'An interview is not a window on social reality but it is a part, a sample of that reality' (Czarniawska 2004, p. 41).

The interviews were transcribed and analysed following Charmaz' (2006) constructivist version of grounded theory methodology that sees data and the analysis as created in a shared experience of interviewer and interviewee. As became clear in the coding process, it is particularly the interviewees' accounts of collaboration challenges that can serve as a heuristic to investigate the relevance of communities for technoscientific research.

8.3 Synthetic Biology: To Be or Not To Be (a Community)

8.3.1 Community-Making

Synthetic biology has been in the focus of policy-making and funding right after the label re-surfaced in influential review articles in the mid-2000s (e.g. Endy 2005; Benner and Sismour 2005). In line with early success narratives (e.g. the anti-malaria drug artemisinin), synthetic biology became a field that research policy decision-makers in a variety of countries aimed to nurture strategically. Large-scale public funding was first made available in the US in the form of the multi-university consortium SynBERC (starting in 2006). In Europe, the UK was the first country to mobilise significant public funds. Synthetic biology researchers were closely involved in the design of these public funding instruments. Since 2013, and after the publication of a national synthetic biology roadmap (Technology Strategy Board

2012), six synthetic biology centres were established at different university locations.

The fact that researchers identified with the label synthetic biology were invited to advise programme development in the UK shows that such community-making already relies on some sort of collective (before the funding expands the community). In the case of synthetic biology, there were indeed some community-making devices that can be considered researcher-driven and that were in place before large-scale public funding set in.

The origin stories of the field regularly refer to the Synthetic Biology x.0 conference series that began at MIT in 2004 and internationalised in 2006. The SynBio x.0 conferences are supported by the BioBricks Foundation. The people behind the BioBricks Foundation are also the driving force behind the Registry of Standard Biological Parts and another very visible event series in synthetic biology: the International Genetically Engineered Machine (iGEM) competitions.

The iGEM competition brings together undergraduate student teams to compete for the best design of biological systems. iGEM started in 2003 as a local MIT initiative. In 2004 the competition had participants from five US universities, and in 2005 it became international. The 2016 edition had 5600 participants from 42 countries. iGem has become a kind of community hub that not only brings undergraduates closer to synthetic biology but, according to my interviewees, also serves as a platform for senior researchers in the field to meet

The above-mentioned Registry of Standard Biological Parts and the Standard Biology Open Language (SBOL) used to describe these parts, can both be understood as community-making devices in synthetic biology. SBOL provides a vocabulary, visuals, and a model that help to define the design of DNA components. With the help of SBOL, one group of synthetic biology researchers can define an abstract design of a biological component (e.g. a switch where parts might be left undefined), while another group continues completing unspecified DNA sequences (cf. Galdzicki et al. 2014, p. 548). The designed component might then be 'plugged' into a standardised chassis provided by yet another group of researchers. SBOL and related efforts towards standardisation can be seen as devices that make it easier for a community of researchers to go about knowledge production-oriented collaboration.

iGEM and SBOL are community-making devices specific to synthetic biology. In addition, there are also synthetic biology journals and an increasing number of higher education curricula. These community-making devices specific to synthetic biology raise expectations of a tightly knit community of knowledge production. Is it that simple, however?

8.3.2   Community, Communities, Interdiscipline?

The proponents of synthetic biology in the early 2000s can be seen as a scientific and intellectual movement (SIM) establishing an interdiscipline (Jacobs and Frickel 2009). The existence of synthetic biology institutes, centres or programmes at certain universities, especially in the UK, further suggests a characterisation as an interdiscipline—i.e. a disciplinary hybrid of less stability than traditional disciplines. The fact that there are synthetic biology MA and PhD programmes at, for instance, UK universities, indicates that the ongoing institutionalisation of synthetic biology might actually point towards a discipline rather than an interdiscipline. However, despite the existence of a SIM, different degrees of institutionalisation, and a number of community-making devices, further evidence suggests that there is no single coherent biology community—there are communities.

While community-making devices like iGEM continue to be popular, some of the early synthetic biology journals have already been shut down.[4] More importantly, studies on publication output in synthetic biology suggest that divides in epistemic orientations in the field are continually reflected in knowledge production (Raimbault et al. 2016). The most prominent epistemic divide is between more biology-oriented researchers, aiming to understand biological systems, and engineering-oriented researchers, aiming to build biological systems (Kastenhofer 2013). Both groups tend to disagree on the degree of complexity involved. Interviewee F is a postdoc in a synthetic biology group in Austria:

> I mean, there has been quite a lot of critical voice around what one can do and what one actually cannot do, what is too complex. . . . I mean the whole iGEM and BioBricks idea is. . . . As a biologist I was always a bit skeptical of the simplicity of the idea. (Interviewee F)

Synthetic biology, thus, can be seen as a contested umbrella term (cf. Rip and Voß 2013). It serves as an anchor for promissory narratives, imaginaries of application and utilisation (in biofuels, pharma, etc), and funding schemes. It is regularly evoked in influential review articles that typically outline research agendas rather than reporting on research findings (cf. Oldham et al. 2012). Thus the label loosely unites researchers whose knowledge production activities, however, are not integrated in terms of the epistemic objectives, ways of producing knowledge and institutional formations.

Synthetic biology researchers might associate with the label in order to access career and funding opportunities. They might take part in some of synthetic biology's community-making devices (publish in journals, participate in iGEM). But they do not usually collaborate in actual research work. With the exception of iGEM, which seems to attract undergraduate teams and supervisors from all sub-communities of synthetic biology knowledge production, devices like the SBOL and the BioBricks Foundation pertain to an engineering-oriented interpretation. My own empirical analysis confirmed a divide in knowledge production activities: those synthetic biologists whose primary epistemic goal was to understand biological systems referred to the engineering-oriented colleagues as an inspiration, as the following passage from my interview with a Germany-based synthetic biology PI shows:

---

[4]*IET Synthetic Biology* only published one edition in 2007. *Systems and Synthetic Biology* stopped publishing in 2015.

Well, they polarise. They are engineers ... from the bottom of their heart. They approach biology with a refreshing ingenuousness, and this is certainly important. This is also what stirred up things. ... They have been important to get things going, but it is certainly not my corner. (Interviewee B)

There was no evidence, however, that the two groups collaborate with each other at the project level. What unites the field was, thus, the umbrella term and the community-making devices. Synthetic biology is in ongoing emergence (Morrison 2012) because it does not exist as a community beyond dedicated community-making activities.

Synthetic biology researchers relate to the umbrella term out of their own interest and self-orientation. The term allows for a rationally motivated adjustment of interests, e.g. in regard to career opportunities before funders or policy-makers. In terms of the traditional sociological theory introduced above, this resembles societal forms of coordination more than communal ones: members of a society "remain separate in spite of everything that unites them" (Tönnies 2001/1887, see above).

This is not to say that communities do not matter in synthetic biology. They do, and we can now specify in more detail how: knowledge production practices of synthetic biology researchers are linked to a variety of disciplinary and interdisciplinary communities instead of one bounded synthetic biology community. What distinguishes the various synthetic biology communities of knowledge production is not exclusively geography or disciplines: each of the communities is international, and each has its way of establishing an interdiscipline. However, not all researchers that associate with the label of synthetic biology share a common object of investigation. They have different ways of producing knowledge (trial and error synthesis, modelling, complexity reduction, etc.), are motivated by different objectives (building life, understanding life), and share information by different means (via the Standard Registry or more traditional academic channels, for instance).

As elsewhere in the life sciences, research in synthetic biology is international and collaborative. However, there are different strands of synthetic biology that do not integrate their knowledge production activities in actual collaborations. Nevertheless, it is enlightening to have a closer look at research collaboration practices in the field. Accounts of collaboration offer a window into the kinds of communities that play a role in synthetic biology research. As we shall see in the next section, accounts of obstacles that researchers encounter in collaborations challenge the sociology of science's focus on knowledge production when thinking about scientific communities.

8.4 Challenges in Synthetic Biology Collaboration: Investigating Communities that Matter

8.4.1 Challenged Collaboration

Working together with others on a research problem involves practices that have been analysed in laboratory studies in great detail (e.g. Latour and Woolgar 1986): building an experimental setup, getting equipment to work, documenting data, connecting results to theories, producing text, etc. Different types of inter-institutional collaboration have been described, ranging from mutual stimulation to collaboration involving a division of labour (Laudel 2002). The collaboration I am interested in involves structured or unstructured interactions over space and time, where different researchers engage in similar practices, interact with human and non-human actors, etc. Collaborations are structured interactions between human actors. They are mediated by and mobilising both human and non-human actors.

In what follows, I want to focus on the former type of resistances. I refer to them as collaboration challenges. Researchers' narratives on their practices include accounts of such challenges.

Above, I have alluded to the different epistemic orientations or communities of vision relevant in synthetic biology—one focused on understanding nature, the other on engineering life (Kastenhofer 2013). Challenges from synthetic biology's inter-disciplinary setup also arise from differences in the way research results are reported.

One interviewee described differences in publication strategies between biologists and engineers that evoke the image of differing disciplinary communities of scientific knowledge production:

> [People] in biology will [pre-]announce their result, and by announcing the result they claim that they are the ones that work on that . . . saying 'This is my turf' . . . [I]n engineering most of the time you don't announce the result before you have it. You don't say 'I plan to do this', you say [that you] have done this and this is the result'. (Interviewee M)

This account exemplifies challenges that arise in an interdisciplinary field like synthetic biology. Not only are the disciplinary languages and the epistemic orientations (understanding vs building) different. The publication cultures in biology and engineering are also different.

While such differences between biology and engineering might be one of the reasons behind the divides in synthetic biology knowledge production, challenges also exist in actual projectified collaborations. Interviewee I is a mathematician who joined a synthetic biology collaboration at his university as an in-silico modeller. He gives the following account of the work involved in preparing a joint publication with his biology colleagues:

> Well, I have produced a draft, and there I have put everything which I found noteworthy, you know, including all the mathematical curiosities. And the biologists meant that this is not interesting to anybody, that it detracts from the message of our studies, that we have to shorten it'. (Interviewee I)

The researchers engaged in a collaborative epistemic effort had produced results that they then wanted to publish. In this effort to jointly convey to a scholarly audience the new knowledge they had produced, they encountered challenges. The mathematician interviewee felt the need to include detailed information on mathematical procedures and the mathematically most interesting results. He wanted to see these aspects in the main body of the joint article. Biologists argued that this distracts from the core argument, which in their view is biological. The parties ultimately agreed to put the formulae into the annex. This was the solution the interdisciplinary collaborators found in order to have the same article speak to (at least) two different communities. At the root of this challenge are ways of talking about research results that are defined not inside the collaboration but outside it: in communities the researchers belonged to apart from their potential identity as synthetic biologists.

We have seen that after more than a decade of top-down and bottom-up community-making, there is still no single synthetic biology community. The field is not transitioning into a disciplinary community (cf. Cain 2002). It instead can be described as an interdiscipline characterised by porous epistemic boundaries and temporal forms of institutionalisation. Even this character as an interdiscipline is limited to specific science policy environments (the UK, in particular). Researchers relating to the umbrella term synthetic biology have different objectives and forms of knowledge production.

Not only is the way of synthetic biology knowledge production contested. Interview data on research collaborations also reveal that the contexts where (and the forms how) synthetic biology knowledge is applied vary and are contested.

8.4.2 Communities of Knowledge Application?

The accounts of synthetic biology researchers presented above point beyond the production of scientific knowledge. The challenges between the biomathematician and the biologists were not so much related to the production of knowledge, but to its dissemination through a joint publication. As we shall see, beyond dissemination, the data also demonstrates how contexts of knowledge application become relevant for technoscientific communities.

What we observe in synthetic biology is a disconnect of communities of knowledge production, communities of dissemination and communities of application.[5] Researchers come together around synthetic biology-related topics to carry out research, but they report their results to other, more traditional and disciplinary scientific communities. They all attend iGem, but they have very different

---

[5]In line with earlier discussions in the philosophy of science one might speak of the context of discovery (knowledge production), the context of justification (dissemination) and the context of application instead (Reichenbach 1938 for the first two, Meyer-Abich 1988 or Gibbons et al. 1994 for the latter).

application contexts for engineered biology in mind. This is not to say that knowledge production, knowledge dissemination and knowledge application are practices that can be clearly separated from each other. STS research has convincingly shown that they are deeply interrelated. Application perspectives, for instance, affect knowledge production and viceversa (cf. de Laet and Mol 2000; Gibbons et al. 1994). The production and application of knowledge can also be conflated in one and the same collaboration:

> I have not had the interaction . . . with the modellist that I wanted to have. . . . [T]here is certainly some drive in synthetic biology . . . to create new models . . . but that does not seem to be what modellers want to do. They want to continue modelling the organisms that they have the . . . I guess that's what they have the data for'. (Interviewee G)

Interviewee G is a UK-based synthetic biology PI trained in biochemistry. While she wants to engage in knowledge production, the other parties necessary for the collaboration are interested in finding opportunities to apply knowledge they already produced in another context. Interviewee G herself is interested in knowledge application in an industrial context.

> I have worked with [company x] on a . . . project. [Company x] is different. . . . They have certain synthetic biology tools that they have commercialised. . . . There is a lot of industry . . . meetings that I have had . . . that we collaborate with, but we really are providing them with . . . a service or knowhow that applies to them in a specific situation. But [company x] is quite different. Basically, they are a company that operates almost in the same way as an academic, who is industrially focused'. (Interviewee G)

The application of Interviewee G's knowledge took a specific shape (industry-oriented, close to knowledge production) because of a specific local environment (the company having close relationships to her university) and her own trajectory.

My choice of the word application is deliberate. I use it because it encompasses more than the dissemination of research results in journals or the reutilisation of prior work in new research projects. It points beyond the cycles of scientific capitalism as discussed by Knorr-Cetina (1982). Instead, it connects epistemic work to non-epistemically oriented rationales. In synthetic biology, it can entail, for instance, the use of synthetic compounds or processes in the production of biofuels, drugs, or flavours. Institutionally, application can be linked to industry, start-ups, community-specific environments (e.g. the BioBricks movement) as well as policy. Both the local and institutional environment (university policies regarding industry collaboration, availability of partners, etc) and the individual career paths of researchers play a role in knowledge application. They in turn also shape knowledge production. It is the local community and innovation system that influence how synthetic biology materialises as a field in this specific environment.

Societal applications that can be derived from research results are another possible application context. One interviewee talked about a synthetic biology-based biosensor for arsenic detection in drinking water in rural areas in Bangladesh:

> [T]he business plan was that NGOs would essentially use money that's been devoted to this purpose to buy these devices . . . and then distribute them. . . . And another part of the

prototype that we've been talking about is designing an instruction sheet with pictures and graphics and things explaining how to actually use it'. (Interviewee H)

Researchers most often choose to publish, but they can also delay publication in order to first file a patent. Or they can look at societal impact and orient publication channels and types accordingly. What researchers choose depends, among other things, on the kind of communities they are part of. Thus, although the practices of knowledge production and application are often interlinked, the communities wherein these activities are embedded are not necessarily the same. Not only do communities of knowledge production affect application (e.g. getting researchers to publish in specific disciplinary journals), but the communities of knowledge application can affect knowledge production.

Proposing knowledge application as a relevant analytical anchor point mirrors the observation in technoscience-oriented STS literature that scientific research and technological development are no longer separated. Similar observations have been made regarding societal impact or commercialisation: observers of developments in higher education speak about the 'entrepreneurial university', particularly in the area of biotechnology (Yi 2015). Societal impact is a relevant dimension in research as evidenced by large-scale institutionalised performance assessments. Researchers have to think about applications and impact when applying for funds and, increasingly, when advancing their careers. In light of these developments, it seems necessary to consider *communities of knowledge application.*

Borrowing from, but expanding definitions of epistemic communities,[6] communities of knowledge application could be defined as work communities concerned with the application of certain types of knowledge. Communities of knowledge application might be closely related to communities of knowledge production (e.g. when researchers' publications in journals are referred to in the production of new knowledge), but they can also be separate (e.g. when researchers join collectives of technology transfer professionals at the university and beyond).

I am not claiming that the work researchers undertake with regard to the application of knowledge is always organised in communities. On the contrary, parts of my data show that they sometimes actively refuse to access certain communities and their practices, for instance when it comes to the commercial exploitation of research results:

Back in [x]'s lab I was working also with someone to patent . . . and then we realised it's just so much effort. . . . It takes so much of our time . . . and it will delay our publication'. (Interviewee F)

Others have also shown that researchers sometimes see the private sector as simply another venue for knowledge production (Fochler 2016). However, there are discussions on new types of scientists that can be related to communities these

---

[6]Continuing the use of the label community as it was introduced in the sociology of science (different from traditional sociological theory); this choice is justified by two reasons: one is that it allows the concept to remain visibly related to the notion of communities of knowledge production; the other reason will become clear below.

scientists belong to. Lam (2010), for instance, describes 'entrepreneurial scientists' and 'entrepreneurial hybrids' that embrace commercial application of knowledge results.[7] In his research on entrepreneurial scientists, Gulbrandsen (2005) argues that these are not part of the academic *and* the entrepreneurial world, but of *none* of these worlds. They are what he calls 'liminal scientists' with a certain distance to both worlds. He does not address the question as to whether there is a community of application made up of liminal scientists. Further research is required to clarify whether the commercial application of results of technoscientific research is driven by individuals who identify with academia, entrepreneurship, both, or none of these worlds.

Knowledge production, dissemination and application are entangled in everyday practices. Nevertheless, it is useful to sharpen our heuristic and conceptual tools to be able to distinguish communities oriented towards knowledge production from those more oriented towards application. Technoscientists are not necessarily 'only' knowledge producers.

8.5   Conclusion

In the present article, I have used my own empirical material from the case of synthetic biology to critically reflect on the notion of scientific community. The case proves Meyer and Molyneux-Hodgson (2010) right when they state that '[s] cientific communities matter' – with a strong emphasis on the plural form: researchers who identify with the field of synthetic biology do not necessarily have a common way of producing knowledge. Instead of one consolidated and bounded community, there are multiple communities of synthetic biology knowledge production and application. Synthetic biology thus is an umbrella term, advertising innovative research paradigms and collaboration modes at the intersection of the life sciences, chemistry, physics, mathematics, and computer science. It also designates a set of community-making devices, reflecting strategies of a purposeful steering of scientific development.

But how can we further specify the character of this 'communality in the plural'? We have seen that parts of the synthetic biology landscape can be described as an interdiscipline (Frickel 2004). The field is institutionalised in a temporary fashion in centres, institutes, and degree programmes. As Kastenhofer and Molyneux Hodgson write in the introduction to this volume, it is a field that seems to be stabilised by forces external to the epistemic enterprise. It is stabilised by funding, by the iGem competitions, and by promissory narratives. The transgression of professional identities that was at the foundation of synthetic biology – being an engineer, but doing biology –does not exclude researchers with more traditional disciplinary identities to engage with the label. Synthetic biology keeps maintaining porous boundaries

---

[7]We find several of these hybrids among the early proponents of synthetic biology.

between different disciplines and identity options. However, the boundaries between the different techno-epistemic orientations (understanding nature and engineering life) as well as the different ways and means to produce knowledge under the label of synthetic biology seem surprisingly stable. This suggests that the field, despite all community-making efforts, is not transitioning towards a consolidated disciplinary community.

The analysis of the synthetic biology case illustrates two further aspects that are not sufficiently accommodated in Frickel's notion of an interdiscipline. First, contexts of knowledge application also have to be accommodated in technoscientific interdisciplines. Their relation to knowledge production has to be negotiated. Porous boundaries are not only needed between disciplines, their methods, concepts, and theories. They also have to be established and maintained between academic and non-academic actors. In this sense, the local (organisational, geographical) contexts matter: it makes a difference whether a synthetic biology interdiscipline takes shape in an environment of potential private sector investors, in a virtual community of biohackers, or in development cooperation networks.

Under the umbrella term synthetic biology, we observe researchers sharing views on knowledge application while differing in their outlook on knowledge production. At the same time, researchers with the same epistemic orientation in their knowledge production can differ in their outlook on knowledge application. Being aware of this diversification in communities of scientific practice is important because of the increased relevance of knowledge application for science policy, research funders,[8] and science's public image.

The special relevance of application within synthetic biology brings us to the second aspect of the notion of interdiscipline that my analysis problematised. Frickel stated that interdisciplines 'are more likely to be located in less powerful ... institutes, centres, or programmes and do not enjoy control of internalized markets' (ibid., p. 273). The significant funds invested in synthetic biology, together with broader trends towards third-party funding, mission-orientation, and projectification, call our attention to a paradox: interdisciplines linked to umbrella terms are more fragile in an academic institutional environment, but at the same time, within contemporary innovation regimes, these same interdisciplines become important to obtain the very resources the academic environment is built upon.

The case of synthetic biology demonstrates that both, contexts of knowledge production and contexts of knowledge application, should be taken into account in an analysis of technoscientific communities. Communities can form around epistemic practices as well as around practices of application. The identities of researchers in general and in the technosciences in particular can even centre on the application of the knowledge they produce. Researchers might increasingly be part not only of one or several (inter-)disciplinary communities but of communities oriented towards

---

[8]E.g. in the context of the EU Research and Innovation Framework Programmes that concede specific importance not only to the dissemination, but, as it is currently called, the exploitation of knowledge.

knowledge application. While this might still be an exception, developments in research funding suggest it could become a rule in the future. Thinking about communities of both knowledge production and application allows us to grasp potential changes in the collectives (and related identities) relevant to technoscientists.

Acknowledgements The author wants to thank Karen Kastenhofer and two anonymous reviewers for valuable feedback.

References

Benner, S.A., and A.M. Sismour. 2005. Synthetic biology. *Nature Reviews Genetics* 6: 533–543.
Bensaude-Vincent, B. 2013. Discipline-building in synthetic biology. *Studies in History and Philosophy of Biological and Biomedical Sciences* 44: 122–129.
Bogner, A., and W. Menz. 2009. The theory-generating expert interview: Epistemological interest, forms of knowledge, interaction. In *Interviewing Experts*, ed. A. Bogner, B. Littig, and W. Menz, 32–80. Basingstoke/New York: Palgrave Macmillan.
Böhme, G. 1974. Die soziale Bedeutung kognitiver Strukturen: Ein handlungstheoretisches Konzept der scientific community. *Soziale Welt* 25 (2): 188–208.
Calvert, J. 2010. Synthetic biology: Constructing nature? *The Sociological Review* 58: 95–112.
Cain, J. 2002. Epistemic and community transition in American evolutionary studies: The Committee on Common Problems of Genetics, Paleontology, and Systematics (1942–1949). *Studies in History and Philosophy of Biological and Biomedical Sciences* 33 (2): 283–313.
Charmaz, K. 2006. *Constructing grounded theory*. London/Thousand Oaks/New Delhi: Sage.
Czarniawska, B. 2004. *Narratives in social science research*. London/Thousand Oaks/New Delhi: Sage.
De Laet, M., and A. Mol. 2000. The Zimbabwe bush pump. Mechanics of a fluid technology. *Social Studies of Science* 30 (2): 225–263.
Elzinga, A. 1993. Science as the continuation of politics by other means. In *Controversial science. From content to contention*, ed. T. Brante, S. Fuller, and W. Lynch, 127–152. Albany: State University New York Press.
Endy, D. 2005. Foundations for engineering biology. *Nature* 438: 449–453.
European Commission. 2005. *Synthetic biology. Applying engineering to biology. Report of a NEST high-level expert group. Technical report.* Brussels: European Commission Directorate-General for Research.
Fleck, L. 1981. *Genesis and development of a scientific fact*. Chicago: University of Chicago Press. (Original work published 1935.).
Fochler, M. 2016. Beyond and between academia and business: How Austrian biotechnology researchers describe high-tech startup companies as spaces of knowledge production. *Social Studies of Science* 46 (2): 259–281.
Frickel, S. 2004. Building an interdiscipline: Collective action framing and the rise of genetic toxicology. *Social Problems* 51 (2): 269–287.
Galdzicki, M., et al. 2014. The Synthetic Biology Open Language (SBOL) provides a community standard for communicating designs in synthetic biology. *Nature Biotechnology* 32: 545–550.
Gibbons, Michael, et al. 1994. *The new production of knowledge: The dynamics of science and research in contemporary societies*. London: Sage.
Gläser, J. 2006. *Wissenschaftliche Produktionsgemeinschaften. Die soziale Ordnung der Forschung*. Frankfurt/New York: Campus.

Gulbrandsen, M. 2005. 'But Peter's in it for the money'—The liminality of entrepreneurial scientists. *VEST Journal for Science and Technology Studies* 18 (1/2): 49–75.

Holzner, B., and J. Marx. 1979. *Knowledge affiliation: The knowledge system in society.* Boston: Allyn and Bacon.

Jacobs, Jerry A., and Scott Frickel. 2009. Interdisciplinarity: A critical assessment. *Annual Review of Sociology* 35 (1): 43–65.

Kastenhofer, K. 2018. Community and identity in contemporary technosciences: Conceptual issues and empirical change. *EASST Review,* 37(2). https://easst.net/article/community-and-identity-in-contemporary-technosciences-conceptual-issues-and-empirical-change/. Accessed 28 Apr 2018.

———. 2013. Two sides of the same coin? The (techno)epistemic cultures of systems and synthetic biology. *Studies in History and Philosophy of Biological and Biomedical Sciences* 44: 130–140.

———. 2009. Debating the risks and ethics of emerging technosciences. *Innovation: The European Journal of Social Science Research* 22 (1): 77–103.

Katz, J.S., and B.R. Martin. 1997. What is research collaboration? *Research Policy* 26: 1–18.

Knorr-Cetina, K. 1982. Scientific communities or transepistemic arenas of research? A critique of quasi-economic models of science. *Social Studies of Science* 12: 101–130.

———. 1981. *The manufacture of knowledge. An essay on the constructivist and contextual nature of science.* Oxford: Pergamon Press.

Kuhn, T. 1962/2002. *The structure of scientific revolutions.* Chicago: University of Chicago Press.

Lam, A. 2010. From 'ivory tower traditionalists' to 'entrepreneurial scientists'? Academic scientists in fuzzy university-industry boundaries. *Social Studies of Science* 40 (2): 307–340.

Latour, B., and S. Woolgar. 1986. *Laboratory life. The construction of scientific facts.* Princeton: Princeton University Press.

Laudel, G. 2002. Collaboration and reward. What do we measure by co-authorships? *Research Evaluation* 11 (1): 3–15.

Marcovich, A., and T. Shinn. 2011. Where is disciplinarity going? Meeting on the boarderland. *Social Science Information* 50 (3–4): 582–606.

Merton, R.K. 1993. The sociology of science: Theoretical and empirical investigations. Chicago: University of Chicago Press. (Original work published 1942.)

Meyer, M., and S. Molyneux-Hodgson. 2010. Introduction: The dynamics of epistemic communities. *Sociological Research Online,* 15(2), 14. http://www.socresonline.org.uk/15/2/14.html. Accessed 30 Oct 2017.

Meyer-Abich, K.M. 1988. *Wissenschaft für die Zukunft. Holistisches Denken in ökologischer und gesellschaftlicher Verantwortung.* München: Beck.

Molyneux-Hodgson, S., and M. Meyer. 2009. Tales of emergence—Synthetic biology as a scientific community in the making. *BioSocieties* 4 (2–2): 129–145.

Morrison, M. 2012. Promissory futures and possible pasts: The dynamics of contemporary expectations in regenerative medicine. *BioSocieties* 7 (1): 3–22.

Nordmann, A. 2011. The age of technoscience. In *Science transformed? Debating claims of an epochal break*, ed. A. Nordmann, H. Radder, and G. Schiemann, 19–30. Pittsburgh: Pittsburgh University Press.

O'Malley, M.A., et al. 2007. Knowledge-making distinctions in synthetic biology. *BioEssays* 30: 57–65.

Oldham, P., S. Hall, and G. Burton. 2012. Synthetic biology: Mapping the scientific landscape. *PLoS One* 7 (4): e34368.

Parsons, T., and E. Shils, eds. 1951. *Toward a general theory of action.* Cambridge, MA: Harvard University Press.

Polanyi, M. 1962. *Personal knowledge. Towards a post-critical philosophy.* London: Routledge.

Raimbault, B., J.-P. Cointet, and P.-B. Joly. 2016. Mapping the emergence of synthetic biology. *PLoS One* 11 (9): e0161522.

Reichenbach, H. 1938. *Experience and prediction. An analysis of the foundations and the structure of knowledge.* Chicago: University of Chicago Press.

Rhoten, D., E. O'Connor, and E.J. Hackett. 2009. The act of collaborative creation and the art of integrative creativity: Originality, disciplinarity and interdisciplinarity. *Thesis Eleven* 96 (1): 83–108.

Rip, A., and J.-P. Voß. 2013. Umbrella terms as mediators in the governance of emerging science and technology. *Science, Technology & Innovation Studies* 9 (2): 39–59.

Schmidt, M., et al., eds. 2010. *Synthetic biology. The technoscience and its societal consequences.* Dordrecht: Springer.

Swidler, A. 2001. What anchors cultural practices. In *The practice turn in contemporary theory*, ed. T.R. Schatzki, K. Knorr Cetina, and E. von Savigny, 83–100. London/New York: Routledge.

Technology Strategy Board. 2012. *A synthetic biology roadmap for the UK.* https://webarchive.nationalarchives.gov.uk/20130302042701/http://www.innovateuk.org/_assets/tsb_syntheticbiologyroadmap.pdf. Accessed 31 Jan 2019.

Tönnies, F. 2001. *Community and civil society.* Cambridge: Cambridge University Press. (Original work published 1887).

Weber, M. 1978. *Economy and society.* Berkeley/Los Angeles: University of California Press. (Original work published 1921).

Whitley, R. 1978. Types of science, organizational strategies and patterns of work in research laboratories in different scientific fields. *Social Science Information* 17 (3): 427–447.

Yi, D. 2015. *The recombinant university: Genetic engineering and the emergence of Stanford biotechnology.* Chicago: University of Chicago Press.

Chapter 9
Community by Template? Considering the Role of Templates for Enacting Membership in Digital Communities of Practice

Juliane Jarke

9.1 Introduction

The idea of 'communities of practice' was one of the most successful to travel from academic research into the world of business and management. Originally an analytical concept to investigate learning as a social and situated practice (Lave and Wenger 1991), it became a prescriptive term and desirable objective: managers came to view communities of practice as a 'supplementary organizational form' and were increasingly seeking 'to develop and support communities of practice as part of their knowledge management strategies' (Roberts 2006, p. 626).

The rising interest by management scholars was instigated by Brown's and Duguid's account (1991) of Julian Orr's PhD research and subsequent seminal book *Talking about Machines* (1996). Orr had demonstrated in his ethnography how vital informal knowledge was for getting work done effectively by following the knowledge sharing practices of a community of XEROX photocopy repair technicians. Based on Orr's insights, Brown and Duguid took the concept of 'communities of practice' into the business context and promoted it as a way to leverage what organisations knew. Subsequently, management and organisational researchers set out to study to what extent communities of practice could be 'cultivated' (e.g. Wenger et al. 2002). For example, following on his work with Jean Lave, Etienne Wenger moved towards a consultancy role and argued that the 'success of organizations depends on their ability to design themselves as social learning systems' (2000, p. 225). This claim related to organisations such as private businesses but also public sector organisations and civil society organisations.

J. Jarke (✉)
Institute for Information Management Bremen (ifib) & Centre for Media, Communication and Information Research (ZeMKI), University of Bremen, Bremen, Germany
e-mail: jarke@uni-bremen.de

© The Author(s) 2021
K. Kastenhofer, S. Molyneux-Hodgson (eds.), *Community and Identity in Contemporary Technosciences*, Sociology of the Sciences Yearbook 31,
https://doi.org/10.1007/978-3-030-61728-8_9

In such 'social learning systems' knowledge came to be seen as a 'thing' or 'shared object' that circulates within and between communities and needs to be organised and managed (Vann and Bowker 2004; Easterby-Smith and Lyles 2011). Key to enabling the cultivation of community was some kind of 'community knowledge pool' (Lesser and Storck 2001) in which members share documents, templates, and other knowledge artefacts. Such artefacts (or knowledge objects more broadly) may then be part of a 'common body of knowledge' (du Plessis 2008) along with tools, methodologies, and approaches. Hence, most of the attention in the business and management literature focussed on how communities of practice may be created and nurtured: community came to be regarded as an organisational form for enabling relationships between distant practitioners; 'knowledge objects' (as shared objects) needed to circulate through such communities and provide a 'container' for members' shared practices (Brandi and Elkjaer 2011; Østerlund and Carlile 2005; Cox 2005).[1]

Hence so far, knowledge objects have been understood as a means for community members to share their communal practice. What has received less attention is the role of knowledge objects in the performance of community membership. I argue in the following that the development and continuous reconfiguring of communal 'knowledge objects' can be understood as a way to grow community. The idea of circulating objects to perform community is prominent within anthropology. Miller (2010) describes the exchange of goods 'as a means to grow culture' (p. 10). Accordingly, in order to consider digital forms of togetherness, digital communities may not be regarded as a medium constructed for enabling relationships between distant practitioners, but exchange relationships between individuals are a means to grow community because, as Miller (2011, p. 207) argues, community is 'a series of increasingly expanding exchanges'. The work performed around the construction and negotiation of knowledge objects is a way for individuals to enact their membership and foster a sense of belonging and identity. Hence, community is not a medium for circulating knowledge objects; rather the configuring and circulation of knowledge objects enacts community. In this chapter, I argue that the creation and collection of knowledge objects are not simply an act of developing a 'common body of knowledge', and as such a by-product of the performance of community, but rather they are key to enacting membership in trans-local communities and hence perform community.

This chapter focuses on a specific and very common type of knowledge object in trans-local communities: templates. I demonstrate that templates for describing a community's shared practice are a way to enable 'exchange relations' between community members and foster their sense of belonging and identity. The chapter is based on a 3-year ethnographic study of a European Commission initiative: ePractice. ePractice set out to create (or nurture) a 'European eGovernment

---

[1]Different forms of such trans-local communities have been described and analysed, including concepts such as 'epistemic communities' (Haas 1992), 'virtual communities' (Cox 2005), or 'communities of interest' (Fischer 2001).

community of practice' in which eGovernment practitioners[2] would be enabled to share their experience and expertise across a number of topics related to eGovernment via online and offline means. This striving for community is well established throughout the European Commission and it follows this endeavour in the many different policy fields it works on (Jarke 2015).[3] Yet after establishing a European endeavour to strive for community, the question arises: what kind of community shall it be, as there are many ways of 'doing' community. In the case of ePractice, the 'community of practice' concept provided a solution to the question of how to do a 'European eGovernment community' well. As such, it has informed the way in which the European Commission and other actors conceived of ePractice and configured it—how they aimed to cultivate ePractice. Templates for describing eGovernment practices in a structured way became one of the main tools to facilitate the sharing of practices amongst geographically distributed practitioners. Through templates, the ePractice team strived to create a common European knowledge base on eGovernment.

In order to discuss the role that templates (for capturing and circulating practices) may play in the performance of digital community, we need to consider how practice relates to the performance of community and membership. In the following, I provide a theoretical framing for this question. I then present the ePractice case study and analyse the role of templates for the cultivation of this community. Subsequently I discuss my findings and argue that knowledge objects (such as templates) are key for performing membership in trans-local communities.

9.2 Theoretical Framework

9.2.1 Performing Community Through Travelling Practices

Originally, the 'community of practice' concept provided a conceptual space to investigate learning as a social and situated practice. With their monograph, Jean Lave and Etienne Wenger (1991) challenged the traditional view according to which learning is primarily understood cognitively, as happening in the mind of the learner. In contrast to this traditional 'schooling model', Lave's and Wenger's approach conceptualises learning as taking place not in an individual mind but in a participation framework, as a mode of being in the world and everyday practice. *Legitimate peripheral participation* hereby refers to a framework describing how roles within a community are defined and how engagement may take place.

---

[2]eGovernment or electronic government refers to the use of electronic technologies within the public sector. It comprises inter-governmental communication and workflow-design as well as communication between government and citizens, and government and businesses. More information is provided in the methodology section.

[3]European community-building relates to policy fields such as climate change, biodiversity, or—as was the case for this ethnography—eGovernment.

According to this framework, community as well as practice were conceptualised as intrinsically local. However, both concepts have also gained enormous importance in considerations of digital forms of togetherness. Through the proliferation of digital technologies across social life, the performance of community has increasingly become a distributed accomplishment. With the rising entanglement of information and communication technologies (ICTs) in organising everyday life and work activities, scholars as well as practitioners set out to translate concepts such as 'communities of practice' into ICT-mediated, digital environments (e.g. Brown and Duguid 2001; Vaast and Walsham 2009): the 'communities of practice' concept, developed as an analytical tool to research learning as a situated and 'legitimate' participation in a communal practice, came to be translated into a prescriptive term that designated a desirable objective within distributed, digital settings (e.g. Lave 2008; Amin and Roberts 2008; Cox 2005).

A concept originally intended as a 'way of looking' at learning as a situated practice became 'a thing to look for' (Lave 2008, p. 290). Suddenly organisations were seen as assemblages of communities that could be cultivated in order to leverage the benefits and opportunities associated with the idea of 'communities of practice'. Amin and Roberts (2008) note that

> [a]s the race for survival in the knowledge economy intensified, so too seems the desire to exploit the potential for creativity and innovation offered by CoPs [communities of practice], ever wishful of articulating and harnessing the intangible, the tacit and the practiced. (p. 354)

For example, Swan, Scarbrough and Robertson (2002) have described the appropriation of the concept in the business context in their analysis of a company where managers purposefully use '"communities of practice" as a rhetorical device to enrol key professionals and to mobilize and legitimize changes in work practices' (p. 477).

One of the very first initiatives to establish or build a trans-local community of practice was the EUREKA-project at XEROX, an initiative that aimed at enabling and enhancing knowledge sharing amongst XEROX' photocopy repair technicians. The starting point of the project was that much of the informal or 'tacit' knowledge 'remains embedded in practice' within small circles of colleagues and work groups (Bobrow and Whalen 2002, p. 47). Insights from studies such as Orr's and Lave's and Wenger's led to considerations about how organisations may facilitate and further knowledge sharing beyond local groups:

> Organizations face the challenge of somehow converting this valuable but mainly local knowledge into forms that other members of the organization can understand and perhaps most important, act on. (Bobrow and Whalen 2002, p. 47)

The EUREKA project was meant to accomplish this. In EUREKA, technicians wrote tips for other technicians that were moderated and approved by 'expert field technicians'. What a project such as EUREKA struggled with was the extent to which informal and local talking and telling could be formalised and even centrally managed in a 'tip data base' (Bobrow and Whalen 2002, p. 52). For example, Orr (2006) highlighted that one of the problems that was never anticipated was that '[t] echnicians did not all find it natural to write, nor did they find it natural to abstract segments of their experience' (p. 1807). Overall it was anticipated (or aspired to) that the EUREKA-system could serve as a representation of the informal relationships between community members. Instead of stories that the technicians exchanged during their breaks, knowledge as part of an expert help system (in the form of tips and tricks helping technicians with problems arising from their repair work on photocopiers) was meant to circulate through the wider 'community'. Knowledge hereby came to be regarded as a thing (or entity) that can circulate in a community rather than as something embedded in the stories that were part of the problem-solving skills of the technicians whom Orr (1996) had reported on. How the templates for describing tips were developed and negotiated was not reported by Bobrow and Whalen (2002). EUREKA-tips were simply depicted as a container or vehicle for the knowledge and experience of the technicians.

Reviewing such projects, Amin and Roberts (2008) pointed to a turn 'towards communities of practice as a driver of learning and knowledge generation across a variety of different working environments' (p. 353). In such frameworks, it was argued that knowledge could be shared following the 'routes prepared by practice' and made possible through 'common embedding circumstances' such as similar tasks and work contexts (Brown and Duguid 2001, p. 203). However, others argued for a shift of attention to a community's shared practice(s) rather than community itself:

> I prefer to relate the concept of community of practice to that of community of practitioners [because] a practice and the tradition of a practice do not respect organizational boundaries but instead traverse several organizations.... Practice, with its materiality, its technological knowledge and its transorganizational character [is what] organizes a community. (Gherardi 2006, p. 108)

With this argument, Gherardi (2006) reversed the causal relationship between community and practice and stated that 'the practice "performs" the community' (p. 108). In so doing she defined community as 'an effect, a performance, realised through the discursive practices of its members' (p. 110):

> The practice which ties a community, or several communities, together is what 'performs' the community or the constellation of communities. (Gherardi and Nicolini 2002, p. 422)

Hence, what scholars such as Brown and Duguid and Gherardi and Nicolini highlight is that the notion of community tended to dominate the importance of practice and that it was not an organisational construct called 'community' that needed to be focused on but rather a community's practices.

However, with the rising interest in building and performing digital communities, the notion of practice has undergone a profound translation. Practice became 'reinstrumentalized and reconfigured as a commercial object with specific uses' (Vann and Bowker 2004, p. 41). Labels such as 'good practice' or 'best practice' became a prominent strategy to tackle the abundance of participants and knowledge objects, to filter the informational overload as people often do not know (of) each other (Brown and Duguid 2001) and need some kind of guidance in order to determine relevance, reliability, and quality (Jarke 2017; Scott and Orlikowski 2012; Pollock 2012). What has received less attention is the work of configuring knowledge objects through which practices are meant to circulate in trans-local communities and their role for the performance of membership and community.[4] In order to understand this configuration work better, I now discuss the performance of membership as a practice of accounting and representing a community's shared practice(s).

9.2.2 Performing Membership Through Accounting and Representing a Community's Shared Practice(s)

There exists an intricate relationship between the performance of membership and the performance of community: membership (in a community of practice) is performed through shared practices and their accounts. What practice-based studies and practice approaches highlight are that the performance of community includes the performance of membership and what a community is about. The performance of subjects and objects of a community is co-terminus with the performance of the community itself and vice versa.

Munro (2001) considers accounts 'as a *display* of membership' (p. 474). This is because accounts are not radically distinct from but rather 'parts or tokens' of persons (Mosko 2000, p. 392). This means that the engagement of persons with (a) knowledge object(s) co-produces their identity as members of a community of practice and relates them to others:

> Objects exchanged . . . are inherently identified with the persons who give *and receive* them; hence, they are inherently connected with the relations between the persons who reciprocate them to the extent that a person is defined as the composite of relations. (Mosko 2000, p. 382, emphasis in original)

The production and consumption of accounts of communal practice(s) hence enact community membership. In other words, community members may be held accountable by others for their practice accounts. In addition, accounts tell us something about 'how things ought to be' (Law 1996, p. 295) and hence make visible the underlying assumptions (e.g. concerning the type of community) that are at work in a particular mode of ordering community. Accounts and accounting practices serve as ordering devices as they represent or stand for what is eligible and thereby belongs to a community and what is excluded. Yet these accounts are not just mere representations as they co-construct and co-produce what they are meant to represent.

---

[4]Latour has proposed the term 'immutable and combinable mobiles' (Latour 1987, p. 227) to account for such ordering devices. He was specifically interested in role of immutable mobiles for coordinating networks. The focus of this chapter is on the role of re-configuring such circulation devices for enacting/practicing membership in communities. The space of this chapter does not allow for a more detailed discussion on the differences and similarities of these two interests and concepts.

The process of accounting almost always aims to reduce complexity by categorising social reality. Yet categorisation is a process that leads to what Leigh Star (2010) calls 'intercategory problems' and 'intercategorical objects' which are those 'things that do not fit categories or standards' (p. 609). Over time, all standardised systems throw off or generate residual categories. These categories include 'not elsewhere categorized', 'none of the above', or 'not otherwise specified' (p. 614). They may shift over time and be refined as more and more objects come to inhabit a space. They move 'back-and-forth between ill structured and well structured' (p. 614). Therefore, we need to focus our inquiry on 'the practices or performances of representing, as well as the productive effects of those practices and the conditions for their efficacy' (Barad 2007, p. 49) rather than on accounts or representations as such. This means that we need to attend to the very practices that produce representations and accounts of community and membership in order to understand what they stand for and what the underlying assumptions of a community are.

In many cases, there will be a 'gap between formal representations ... and unreported "back stage" work' (Star 2010, p. 606). Even if community members share representations, there is often 'local tailoring' performed by some members which is invisible to the community as a whole. In particular, for trans-local communities, it is of interest to understand how shared representations come into being and make the work visible that produces such representations. I will argue here that knowledge objects (such as templates) are an ideal starting point for understanding what a community is (supposed to be) about and how participants continuously negotiate their understanding of community by refining them.

Bowker and Star (1999) have argued that membership can 'be described individually as the experience of encountering objects and increasingly being in a naturalized relationship with them' (p. 295). This means that as membership evolves, members become familiar with the specific accounts and representations that a community produces and that are acknowledged as valid. Subsequently knowledge objects (such as templates) become more stable.

To conclude this section: following Gherardi and Nicolini (2002, 2006) we need to focus on the practices that perform community rather than take community as a more or less stable assemblage of people sharing a practice. By attending to the production and consumption of knowledge objects as accounts and representations of members' practices, we can learn about how membership is performed in trans-local communities. In so doing, the invisible 'backstage work' becomes visible. In the remainder of this chapter, I analyse the work performed to configure templates for representing (supposedly) shared practices within ePractice, a digital assemblage striving to cultivate a European eGovernment community of practice.

9.3 Research Methodology and Methods

9.3.1 Empirical Site

The work presented here is based on a 3-year ethnographic study of ePractice: a European Commission project to build a community of European eGovernment practitioners. In June 2007 ePractice joined-up a number of legacy projects, and the European Commission aimed to establish it as the European meeting place for eGovernment practitioners and in so doing produce a European dimension to national eGovernment work.

eGovernment (or electronic government) is the use of information and communication technology (ICT) within public sector organisations in order to provide electronic services to citizens, businesses, and other public sector organisations. Overall eGovernment promised a citizen-centric approach that would bring huge savings and a better integration of public sector bodies by overcoming siloed organisational structures (cf. Bloomfield and Hayes 2009; Ahn 2012; Millard 2010; Rowley 2011). Across the Global North, eGovernment became mainly associated with online services such as driver licence renewal and income tax filing (cf. Ahn 2012).

In building and fostering a 'European eGovernment community of practice', the aim of ePractice was to facilitate the sharing of knowledge and experience amongst eGovernment practitioners across Europe. This was meant to be achieved through a Web portal comprising of a number of online and offline mechanisms such as Web 2.0 features (e.g. public member Web profiles, a community blog), a case study database, workshops, and conferences. By January 2012 over 140,000 people had registered as members of ePractice, over 1550 case studies had been submitted, and over 1800 events had been announced.

With an initial user base of just under 5000 individuals, ePractice grew exponentially throughout the period of my fieldwork. The service was free of charge and users typically worked within the realm of eGovernment—either as civil servants in the public sector, in the private sector (e.g. consulting firms), or academia. Using ePractice, they were able to set up a profile, connect with others, describe their projects as 'ePractice good practice cases', participate in theme-based communities, communicate and discuss via a blog, disseminate studies, activities and events, and also receive up-to-date information about eGovernment across Europe. Furthermore there existed the opportunity to participate in eGovernment-related workshops and conferences. These workshops provided a more formal place for coming together in working sessions but also left considerable room for more informal coffee-break networking interaction.

9.3.2 Methodology

Above, I argued that many studies on 'communities of practice' assume a view of communities as supplements of organisations; interest is directed to cross-community activities such as collaboration or knowledge transfer. Communities are taken to be composed of things that can be measured. This 'distal view' of organising is grounded in a 'sociology of being' because it is concerned with 'taken-for-granted states of being' (Cooper and Law 1995). A 'proximal view', in contrast, takes these taken-for-granted entities as outcomes of processes; it focusses on the relations between different entities and the processes that enact and re-enact them. With this approach, the researcher is interested in the performative work of ordering rather than order and looking at relations and recognising these relations as processes rather than durable entities. Hence, rather than being interested in measurable results and outcomes, proximal thinking is interested in the processes and doings that lead to these results (Cooper and Law 1995).

The interest of this paper requires a qualitative approach that attends to the practices, relations, and processes related to communal knowledge objects (Law 2004), and hence it will take a proximal approach. There are many interpretations of what practices may be and how they may be studied. I have followed Mol's (2003) compelling praxiography. For a praxiographer, attending to practices is crucial if wanting to understand how objects come into being, which in my case means attending to the 'tailoring work' (Star 2010) around communal knowledge objects. Following Mol, the identity of objects is never stable but fragile and ever changing; objects are not just brought into being and thereafter remain mute, but they are constantly enacted and re-enacted in practice and thereby multiply.

However, following the practices that bring knowledge objects into being poses methodical challenges in a trans-local setting such as ePractice. My research did not take place in a particular building, not even in a particular organisation. Nevertheless, I went out physically—mainly to Brussels—and attended workshops where I met people, listened to their presentations, and talked to them during breaks. I also went out virtually—mainly to the ePractice Web portal but also to other websites—and visited project websites, workshop and conference websites, LinkedIn profiles, Facebook, and Twitter. I conducted interviews face-to-face, via Skype, and via telephone. This mixed approach to researching ePractice was, however, not only specific to the research but also a common feature of the diverse communication channels employed by the people I observed and interviewed. The fieldwork was conducted from May 2007 to December 2010. An in-depth study of a specific eGovernment community (eGovMeasureNet[5]) was conducted from March 2009 to April 2010. In total, I conducted 73 semi-structured interviews with 58 interviewees and attended 23 events.

My involvement and roles in these events ranged from 'participant observer' to workshop 'rapporteur' to member of an Informal Expert Committee to community

---

[5]The real name of the community has been disguised.

presenter to project staff of the EU-funded thematic network eGovMeasureNet and community facilitator for eGovMeasureNet on ePractice. My participant observation also stretched to the online presence of ePractice: I was a registered user of ePractice and also a community facilitator for eGovMeasureNet. In this position, I created workshop events on ePractice, I posted in the eGovMeasureNet community blog, I uploaded documents, and I corresponded with the eGovMeasureNet community. Throughout my online and offline engagement, I presented myself as a researcher. My public profile on ePractice gave a brief summary of my CV as well as the research focus and interest of my study.

Additionally, I also set up Twitter, LinkedIn and Facebook accounts in order to connect with eGovernment practitioners whom I had met in the realm of ePractice. In all social networking spaces I followed their activities (tweets, groups, events) in order to understand some of the online communication dynamics beyond the ePractice portal. Yet participant observation has its limitations, and I felt that interviews were needed to complement my fieldwork and help me understand what I was encountering through the participants' own sense-making. Most interviews were voice recorded and transcribed. Participation in events was recorded in field notes, and photographs were taken. Further material includes workshop reports, the ePractice blog, and ePractice TV. In addition, the ePractice team made its own survey results and Web statistics available to me. Data was structured and analysed via codification software (ATLAS.ti). All names have been anonymised.

Overall, the long-time involvement in ePractice and the resulting longitudinal study allowed me to follow a proximal research approach that studied ePractice as emergent and becoming. Rather than take the existence of a European eGovernment community for granted, the methodology allowed for insights into how community was enacted. The research methods mirrored the ways of communicating and engaging in ePractice but also the materiality members of ePractice found themselves part of (for example the numerous journeys to Brussels, the attention to online dynamics, and the importance of communal knowledge objects).

9.4 Analysis

In the following, I present two vignettes that describe the 'tailoring work' required to produce 'shared representations' within ePractice and how exchange relations were established and performed as part of members' 'accounting practices' (Munro 2001). Key to this work was the development, use, and circulation of templates. In the first vignette, I report on a template that the European Commission developed in order to further 'good practice exchange' amongst European eGovernment practitioners. This is an account in which a coordinating institution developed and reconfigured a template to suit their objective (to build a European community). In the second vignette, I report on a template that was developed by a network of European eGovernment measurement practitioners to facilitate their sharing of practices and to develop a sense of belonging to a community of practitioners.

9.4.1   Vignette 1: Translating and Disentangling Practices Through Templates

Practices cannot simply circulate within a digital network and travel because they are local by definition. A study commissioned by the European Commission about transferability of good practices in eGovernment suggested study tours to the sites of the actual projects because 'the realities of innovation and change in the development of eGovernment on a wider scale represent a much more complicated mesh of events, activities and relationships' (European Commission 2007, p. 55).

Such study tours are an expression of the idea that practices are situated; they are part of evolving exchange relations within a (hopefully) evolving community. Through the participants' co-presence in space and time, an exchange relation is established around mutual attention, interest, and spent resources. An alternative to study tours that follows knowledge management initiatives and exploits the potential of ICTs is the creation of a digital 'knowledge base' to facilitate 'knowledge exchange'. In the case of ePractice, it featured 'ePractice good practice cases' which are written case studies of eGovernment projects. On the ePractice website, the following definition is put forward:

> ePractice cases are written summaries of real-life projects or business solutions developed by public administrations, entrepreneurs and corporations. Case studies included in our portal are based on actual experiences, and reading them provides a picture of the challenges and dilemmas faced by the professionals working in eGovernment.[6]

During the interviews, eGovernment practitioners declared repeatedly that knowledge practices may be indeed shared through 'ePractice good practice cases' if they are applicable and transferable to new environments. Good practice needs to translate into the reader's own work context; a reader needs to be able to take useful aspects of other practices and apply them to his or her own work. Hence, the ability of good practice cases to travel and be transferred is seen to depend on what Brown and Duguid (2001) have called 'common embedding circumstances':

> The other element is that the information has to be understandable and relevant. So that you can recognise it as something which could be applicable to your area and therefore you can translate that into what you need to know for that work. (Interview with Tom, local government, UK)
> Something only becomes a best practice if you can show that what you've done is transferable in some way. It is not like you go into a shop and purchase something; you've got to have the right environment in which whatever you purchase lives. (Interview with Brian, local government, UK)

How much of a project may be revealed in a good practice case if—like in the study cited above—eGovernment projects 'represent a much more complicated mesh of events, activities and relationships' (European Commission 2007, p. 55) than can be translated into a written text? In 2003, the notion of 'ePractice good

---

[6]Retrieved 9 March 2010 from http://www.epractice.eu/info/cases

practice cases' was introduced through the *eEurope Awards for eGovernment* (from 2005 onwards called *European eGovernment Awards*): a system of incentives for practitioners to compete and 'share' their eGovernment expertise. Talk and action, practical experience, and local knowledge were accounted for by writing a 'good practice case'. This translation from practice into text was accomplished quite differently from one case to another in terms of, for example, the level of detail, intended target audience, length, and truthfulness.

Within ePractice, a more structured approach was followed in the form of a template that needed to be completed in order to compile an 'ePractice good practice case'. Such a template allowed for comparisons amongst case studies and searches across them. It requested, amongst other things, a short description (abstract), the project's scope, its target audience, the technical and management approach, the policy relevance, and lessons learnt—aspects which the ePractice team 'believe[s] to be of high interest to the ePractice community' (case guidelines). From the beginning of ePractice, the template was provided as an HTML-based online text tool shown in the following screenshot (Fig. 9.1).

The structure or frame that a template provides was positively assessed by case authors and readers alike:

> What I found very useful is that the ePractice presentation of the projects is somehow divided in different sections, so that the people who are trying to promote their cases through this ePractice community can follow this pattern and this pattern then can, how to say, make them focus on what is relevant and what is important for a case presentation. So I found this quite useful. (Interview with Marko, NGO, Slovenia)

The template structure hence inscribed a certain reading of accounts and distinguished between what is understood to be 'relevant' or 'important' and what is not. In so doing, the template was more than a simple device that translates eGovernment projects into good practice cases (situated practices into knowledge objects in the form of written text). Rather the template also configured the production and consumption of accounts of good practices and hence the performance of membership itself.

The power of inscribing certain membership practices by providing a template was challenged by a number of ePractice members. They described 'good practice case'-writing as a process that required them to gather information from colleagues and collaborators throughout an organisation. Often MS-Word documents were created and circulated amongst these participants. This information then needed to be compiled, ordered, reviewed, and edited in order to comply with the online template provided by ePractice. Representing practices is not an easy task. One ePractice user suggested making a form available that could be easily circulated in one's own organisation in order to enable people to contribute to the 'case preparation':

> I had to do precisely this, because different people had to provide input, check that what I was writing was correct, etc. And I guess, that this is the standard procedure once you publish a case, esp. from public sector organisations. (Comment left on ePractice blog on 31 May 2010)

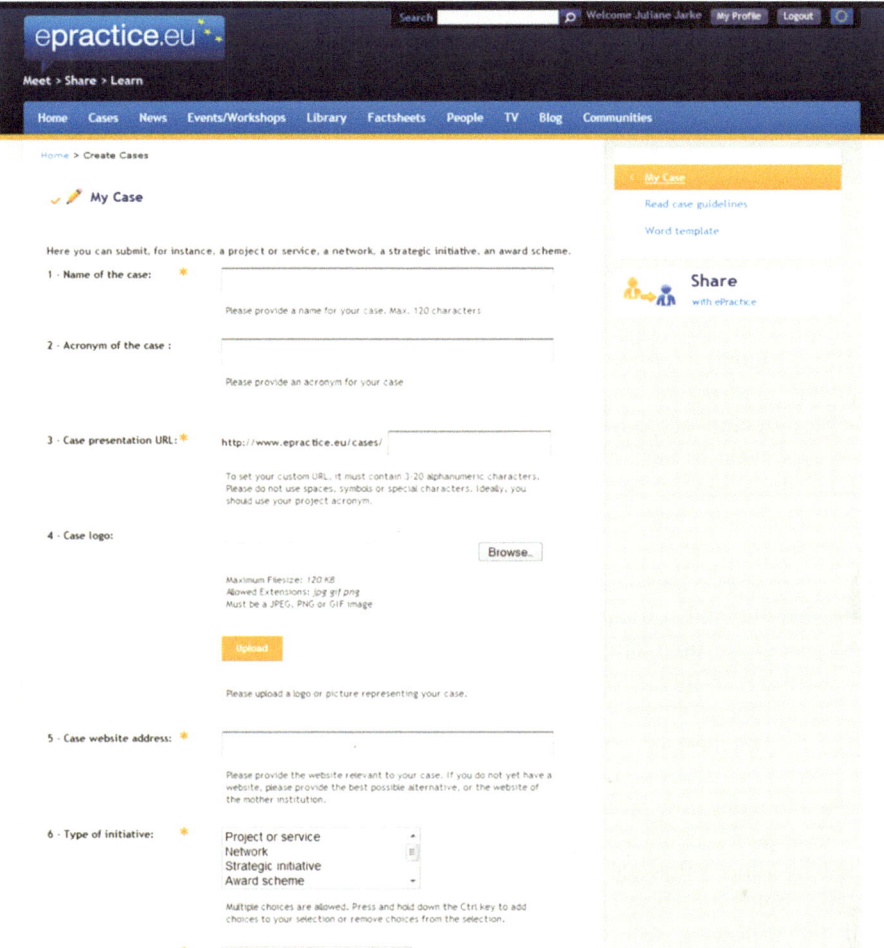

Fig. 9.1 Screenshot of ePractice case template. (Retrieved 12 Jan 2012 from http://www.epractice.eu)

One interviewee (Javier, civil servant in the Spanish regional government) referred to the ability of MS-Word documents to digitally circulate through an organisation in order for 'people to provide input' and 'check' that what was written was 'correct'. He also pointed to the portability of MS-Word documents beyond the workplace as well as their availability and accessibility continuously or in iterative steps. Hence, for ePractice case authors, good practice cases did not simply represent their project or their experience. They were rather an accomplishment, a project in themselves:

> Usually these cases are quite complex, they involve a lot of different organisations, and usually it's very hard to capture the whole totality of these complex projects in a one page summary, which is then put on the ePractice portal. (Interview with Marko, NGO, Slovenia)

The template, in its structured form, does not do justice to the complexity of the various practices. The abundance of people, places, data, objectives, and timings simply lead to overflows. As a result, many case authors, expressed their feeling that they needed to capture more, uploaded additional documents as part of the good practice cases (e.g. videos, slides, tutorials, flyers). This also served to overcome constraints such as word limits or rigid categories in which certain aspects of a 'case' did not fit.

'[C]aptur[ing] the whole totality of these complex projects' is an issue that Star (2010) describes as the 'problem of collecting, disciplining, and coordinating distributed knowledge' (p. 607). She points out that the challenge lies in the decision about which information is 'discarded as unimportant'. The case template does not fit all case authors and their respective projects. Rather they feel that the template 'imposes' a structure and constrains them. Practitioners subsequently created MS-Word documents to frame their content according to the template; they framed 'what they want to say' in order for it to fit the case template structure. Hence a question that needs to be asked is: 'How do [these] forms shape and squeeze out what can be known and collected?' (Star 2010, p. 607). And further, whose decision is it to determine what is to be regarded as relevant or important and what is not? To what extent does a template determine the production and consumption of accounts of membership and subsequently the ways in which community is performed?

As part of an 'upgrading strategy', some of the suggestions elaborated above as well as further features were introduced to ePractice after October 2010. In January 2012 the online template consisted of 33 different headings and allowed the inclusion of further material, including pictures (gallery), presentations via SlideShare, videos via YouTube, or any other files, such as project documentation, directly to the system. ePractice now provides a supporting PDF file with guidelines on how to fill in the case template and a downloadable template in MS-Word format. Cases may be tagged by other ePractice members; they can also be highlighted in one of the ePractice communities or shared via other social networking sites such as LinkedIn.

In the following section, I describe a further case in which a template was continuously reconfigured by a group of practitioners in order to capture their assumed joint practice and cultivate their community. What was striking was that the reconfiguring of such templates and their completion became a membership practice itself.

9.4.2 Vignette 2: Enacting Joint Practices through and with Templates

eGovMeasureNet was one of ePractice's biggest 'communities' and was focussed on eGovernment measurements and benchmarking (e.g. comparing eGovernment maturity), initially funded by the EU as a Thematic Network (2008–2010). Its domain of interest and what the community was meant to be about, however, was subject to continuous re-negotiation amongst participants. Generally, partners were interested in learning about 'good practices' in the rest of Europe. The heterogeneity of the participants was perceived as rather positive, enabling academics to better 'identify [the] real meaning of their work' and helping practitioners to reflect on what they do through a more 'research-based approach'. There was an aspiration of planting seeds in Europe of what good measurement practices entail.

One of the mechanisms employed by eGovMeasureNet was a 'Measure Template' that was developed to describe 'current practice in measurements of user satisfaction, and impact'.[7] It was meant to identify 'the common features, and the differences among the methods . . . in a collaborative way to support the converging practices' (ibid). The template was 'intended to support the description of eGovernment measurement methods to share the current practices and the experience [of] deploying them'. The term 'vehicle' was used, which ought 'to facilitate the discussion to promote take-up and harmonisation'.

The Measure Template was anticipated to be used as a 'procedural guide' that was going to help make sense of and get to know 'what the other network members were doing'. Simultaneously the Measure Template was seen as a means of supporting community-building through the joint definition of common work practices. It was developed collaboratively and was meant to reveal differences and similarities between, for example, the members' methodological approaches. This straightforward procedure turned out to be highly problematic and caused many controversies.

In the interviews, most interviewees stressed their surprise about the differences they had experienced 2 years into the project. Whereas one of the interviewees located these differences in different work cultures in public sector organisations across Europe, another placed eGovMeasureNet members' different professions (regional government, consultancy) at the fore. The interviews revealed the struggle of participants to produce a single 'good measurement practice' to accomplish successful coordination across sites and between the multiple enactments. The collaboratively developed template failed to coordinate between the various measurement practices; no 'shared representation' had been produced.

The design of the template was initially anticipated to be fairly simple with iterative updates as the project went along. This assumption, however, turned out to underestimate the differences between the measurement practices of eGovMeasureNet members. The jointly developed template had to be refined in several iterations. What was perceived as one of the biggest challenges was identifying the 'right questions'. This was due to the open structure of the template as no section was defined beforehand 'because we didn't really have an idea about what they would look like' (Fredrik, project coordinator).

Questions that arose during the collaboration were: (How) can measurement practices actually be compared? What might be different? What is local and specific, what can be made explicit and is transferable? These were not necessarily things

---

[7]Technical Annex of eGovMeasureNet project proposal, p. 37

members knew, and they were contested. For example, Clara, the project assistant, described in an interview how the template became increasingly detailed as partners negotiated these aspects.

This striving for ever more detailed and fine grained categorisation is what Star (2010) described as 'intercategory problems' which trigger ever more fine-grained categories. In the case of eGovMeasureNet, the resulting template was 20 pages long and very detailed. The template was sent out to eGovMeasureNet members in order to be completed with respect to either measurements they were carrying out in their own organisations or measurements that were taken at the national level of their respective countries.

Giorgio (academic, Denmark) mentioned that the 'real step' was to compare the Measure Templates and referred to a matrix that allowed comparison of 'concrete aspects'. Accordingly, practice exchange (or learning) did not take place by reading a single template but by comparing and analysing several of them. Only through this type of circulation and subsequent exchange was a 'learning community' enacted.

Such comparison allowed searching for patterns among approaches that 'work' in different contexts and identifying similarities and differences. In doing so, sharing practices shifted towards comparing templates and relating the information provided to each other. But no matter how detailed the template became and whether people had been involved at various stages of its development, the Measure Template remained open to misunderstanding and misinterpretation because practitioners related its different categories and questions to different local contexts: the practices of measuring and benchmarking were complex and involved multiple agencies, methods, approaches, and timings.

9.4.3   Configuring Templates, Performing Community (Membership)

The above vignettes describe how the European Commission set out to cultivate a community of eGovernment practitioners. Through ePractice they aimed to foster not just any type of European community or network but to establish a 'community of practice'. Their focus is not surprising, given the attention that the concept of community of practice received by managers as a successful way of coordinating distributed settings and leveraging practitioners' knowledge and expertise.

One way of facilitating this cultivation was through the use of templates to describe 'joint work practices'. The importance of practice had been demonstrated by Lave and Wenger (1991) when they introduced the term 'community of practice'. It was taken up by organisation studies and management scholars, who argued that for distributed forms of community, a coordination of practices was required. Coordinating practices through the templates is a specific way of attending to the challenge of facilitating the travelling of local and situated practices.

However, there is an important difference between the two vignettes: in the first vignette, the template was configured by the ePractice team; in the second vignette it was the practitioners themselves that negotiated the structure of their communal template. In the first vignette, the ePractice team reacted—at their own discretion—to feedback from registered ePractice members. The continuous reconfiguration followed a particular mode of ordering—one that was concerned with sharing knowledge and facilitating learning amongst members of ePractice. There was also another task assigned—to make the community and performance of membership explicit and visible by accounting for a joint practice. Hence, it was not just any ephemeral community that the ePractice team wanted to cultivate but a visible one. Visibility was in part achieved through documented accounts of performing community membership. Interviewees from the European Commission repeatedly stressed the importance of proving 'that exchange has happened'. The way these accounts were imagined to represent and stand for the envisaged community configured ePractice accordingly.

In the second vignette, the template was configured (or scripted) by eGovMeasureNet members themselves. Working on the template and amending it was experienced as a form of collaborative practice and communal sense-making (determining what eGovernment measurements are about). The 'sending back and forth', the circulation of the template throughout its development process, already established an exchange relation in which, for example, categories and legitimate practices were negotiated. Hence, despite the above mentioned complications and differences, eGovMeasureNet members felt a sense of belonging and were keen to mutually engage in joint activities. In order to reach a communal understanding and exploit different backgrounds, interviewees emphasised repeatedly that they had 'to learn to talk to each other'. This learning process was a very important part of community-building. Moreover, members not only needed to learn to talk to each other but also had to make their work visible to others. These very practices became part of performing membership. Overall, the Measure Template served as a frame for demarcating the eGovMeasureNet community and the understanding of its joint domain of interest. Hence the framing worked in two ways: (1) the framing of benchmarking and measurement projects to fit the scope of eGovMeasureNet, and (2) the adaptation of the boundaries of the Measure Template itself to the unanticipated breadth of activities and all the things that had remained 'intercategory problems' and 'intercategorical objects'. This means that the created 'good practice cases' were not merely a collection of the community's knowledge and shared practices but rather continuous reconfigurings of what valid representations of supposedly joint practices are.

9.4.4 Configuring Templates, Performing Exchange Relations

Both vignettes described practices of making work practices visible and of sharing practices digitally, which are distinctively different from learning and knowledge sharing in co-located encounters (as described by Lave and Wenger (1991) and Lave (2011)). Many of the associations that Orr (1996) described so vividly in his account of a community of photocopy repair technicians were absent. This is why digital accounts became so important: a shared repertoire of resources and accounts had to be developed, e.g. through the Measure Template that also included conceptual objects such as 'eGovernment measurements' and 'benchmarking'.

In both vignettes, a practice or a field of practice was at first passively shared, then actors driving community development efforts attempted to share experiences stemming from this practice in a more active way. In the first vignette, the active part of sharing was restricted to the exchange of experiences (filling out templates). In the second vignette, practitioners were actively involved in setting the rules of sharing (developing the templates). The first vignette demonstrates how interaction at the local level was fostered (through collecting and preparing the 'relevant data'), while the second vignette demonstrates how templates may foster trans-local interaction.

Hence, performing ePractice as a community demanded the production and consumption of specific accounts of membership performance, such as the development and circulation of templates. 'Good practice cases' were not alien to ePractice members but formed part of their professional work identity (of being a connected and networked European eGovernment practitioner). In making circulation possible, these accounts were what participants exchanged and what enabled 'exchange relations'. They were exchanged and—at the same time—constituted exchange relations. What may be found in the empirical material above is hence a different way of displaying membership than through the performance of a joint work practice. What can be found is rather the production and reconfiguration of accounts of its performance. What constitutes the exchange relations are practitioners, their diverse work practices and understandings, their ideas and motives, the materialities of ePractice as well as the interests of funding bodies such as the European Commission. The accounts in the form of the Measure Template display the work that was done in order to form and perform an eGovMeasureNet community. The display of membership was accomplished by contributing to these efforts of associating a diverse set of actors in a particular way. The exchange relations were not simply established between community members but rather enabled the circulation and association of a variety of heterogeneous entities.

9.5   Conclusion: Community by Template?

This chapter attended to how the communities of practice concept has been taken up by managers and policy makers in trans-local contexts. Although the concept was developed for co-located communities, it was transferred to distributed settings. In such settings, the sharing of practices is not necessarily active, and the performance of community not necessarily tied to their sharing. Some of the ambiguities of the original concept became problematic as it is not sufficient to passively and trans-locally share a practice—practitioners need to enter exchange relations in order to perform community membership.

The two vignettes demonstrate how community is understood by policy makers and managers as a form of organisation that needs to be cultivated and coordinated. Continuing on the success of the concept of communities of practice, a focus of such striving became the sharing of experiences (and 'good practices') in order to foster community-building. Hence, the sharing of practice came to be seen as co-terminus with enacting community. In a trans-local context, this meant—for the actors responsible for building community—a focus on how practices may be shared actively. One answer to this challenge was to describe local practices in standardised templates. However, different ways of organising the sharing of knowledge objects (e.g. who are the actors that define the structure of templates or how do they determine what counts as 'good practice') resulted in different forms of communality.

To conclude, digital communities of practice are not cultivated through the passive sharing of practices but rather through the performance of exchange relations. A crucial role for establishing and enacting exchange relations are knowledge objects such as templates. They are more than a 'common body of knowledge' (du Plessis 2008) or 'community knowledge pool' (Lesser and Storck 2001). Rather, they play a crucial role in the forming (and performing) of community by defining a communal practice and making membership performance visible. Hence, knowledge objects such as templates are not simply a means to actively share practices but rather a means for enacting membership. They enable exchange relations amongst individual practitioners, concepts such as eGovernment, practices, projects, and technologies that stand for a community's domain of interest, the diverse work practices and understandings of members, their ideas and motives as well as the interests of funding bodies such as the European Commission.

Acknowledgements   I would like to thank Karen Kastenhofer and Susan Molyneux-Hodgson for their constructive feedback and patience throughout the process. Furthermore, I thank Andreas Breiter, Andreas Hepp and his lab (in particular Matthias Berg, Anke Offerhaus and Susi Alpen) for providing feedback on an early version of this chapter. I am grateful to Theo Vurdubakis, Niall Hayes and Lucas Introna for their guidance during my PhD research and for encouraging me to expand it.

References

Ahn, M.J. 2012. Whither e-government? Web 2.0 and the future of e-government. In *Web 2.0 technologies and democratic governance. Political, policy and management implications*, ed. C.G. Reddick and S.K. Aikins, 169–182. New York: Springer.

Amin, A., and J. Roberts. 2008. Knowing in action: Beyond communities of practice. *Research Policy* 37 (2): 353–369.

Barad, K. 2007. *Meeting the universe halfway: Quantum physics and the entanglement of matter and meaning*. Durham: Duke University Press.

Bloomfield, B.P., and N. Hayes. 2009. Power and organizational transformation through technology: Hybrids of electronic government. *Organization Studies* 30 (5): 461–487.

Bobrow, D.G., and J. Whalen. 2002. Community knowledge sharing in practice. The Eureka story. *Reflections* 4 (2): 47–59.

Bowker, G.C., and S.L. Star. 1999. *Sorting things out: Classification and its consequences*. Cambridge, MA: MIT Press.

Brandi, U., and B. Elkjaer. 2011. Organizational learning viewed from a social learning perspective. In *Handbook of organizational learning and knowledge management*, ed. M. Easterby-Smith and M.A. Lyles, 23–42. Chichester: John Wiley.

Brown, J.S., and P. Duguid. 1991. Organizational learning and communities-of-practice: Toward a unified view of working, learning and innovation. *Organization Science* 2 (1): 40–57.

———. 2001. Knowledge and organization: A social-practice perspective. *Organization Science* 12 (2): 198–213.

Cooper, R., and J. Law. 1995. Organization: Distal and proximal views. *Research in the Sociology of Organizations* 13: 237–274.

Cox, A. 2005. What are communities of practice? A comparative review of four seminal works. *Journal of Information Science* 31 (6): 527–540.

du Plessis, M. 2008. The strategic drivers and objectives of communities of practice as vehicles for knowledge management in small and medium enterprises. *International Journal of Information Management* 28 (1): 61–67.

Easterby-Smith, M., and M.A. Lyles. 2011. The evolving field of organizational learning and knowledge management. In *Handbook of organizational learning and knowledge management*, ed. M. Easterby-Smith and M.A. Lyles, 1–22. Chichester: John Wiley.

European Commission. 2007. *High-level report -provide e-government good practice portability*. Brussels, available at: www.euforum.org/IMG/pdf/D1.6-Final_report-v1.pdf: European Commission. accessed 27 July 2010.

Fischer, G. 2001. Communities of interest: Learning through the interaction of multiple knowledge systems. *Proceedings of the 24th IRIS conference*, pp. 1–14.

Gherardi, S. 2006. *Organizational knowledge: The texture of workplace learning*. Malden, MA: Blackwell Pub.

Gherardi, S., and D. Nicolini. 2002. Learning in a constellation of interconnected practices: Canon or dissonance? *Journal of Management Studies* 39 (4): 419–436.

Haas, P.M. 1992. Epistemic communities and international policy coordination. *International Organization* 46 (1): 1–35.

Jarke, J. 2015. "Networking" a European Community: The case of a European Commission egovernment initiative. *ECIS 2015 Completed Research Papers*. Paper 86.

———. 2017. Community-based evaluation in online communities: On the making of 'best practice'. *Journal of Information Technology and People* 30 (2): 371–395.

Lave, J. 2008. Epilogue: Situated learning and changing practice. In *Community, economic creativity, and organization*, ed. A. Amin and J. Roberts, 283–296. Oxford: Oxford University Press.

———. 2011. *Apprenticeship in critical ethnographic practice*. Chicago: Univ. of Chicago Press.

Lave, J., and E. Wenger. 1991. *Situated learning: Legitimate peripheral participation, learning in doing*. Cambridge: Cambridge Univ. Press.

Latour, B. 1987. *Science in action: How to follow scientists and engineers through society*. Harvard University Press.

Law, J. 1996. Organizing accountabilities: Ontology and the mode of accounting. In *Accountability: Power, ethos and the technologies of management*, ed. R. Munro and J. Mouritsen, 283–306. London: Cengage Learning EMEA.

———. 2004. *After method: Mess in social science research*. London: Routledge.

Lesser, E.L., and J. Storck. 2001. Communities of practice and organizational performance. *IBM Systems Journal* 40 (4): 831–841.

Millard, J. 2010. Government 1.5—Is the bottle half full or half empty? *European Journal of ePractice* 9: 1–16.

Miller, D. 2010. *An extreme reading of Facebook*. Working papers series #5. OAC Press.

———. 2011. *Tales from Facebook*. Cambridge/Malden: Polity Press.

Mol, A. 2003. *The body multiple: Ontology in medical practice*. Durham: Duke University Press.

Mosko, M.S. 2000. Inalienable ethnography: Keeping-while-giving and the Trobriand case. *Journal of the Royal Anthropological Institute* 6 (3): 377–396.

Munro, R. 2001. Calling for accounts: Numbers, monsters and membership. *The Sociological Review* 49 (4): 473–493.

Orr, J.E. 1996. *Talking about machines: An ethnography of a modern job*. Ithaca: ILR Press.

———. 2006. Ten years of talking about machines. *Organization Studies* 27 (12): 1805–1820.

Østerlund, C., and P.R. Carlile. 2005. Relations in practice: Sorting through practice theories on knowledge sharing in complex organizations. *Information Society* 21 (2): 91–107.

Pollock, N. 2012. Ranking devices: The socio-materiality of ratings. In *Materiality and organizing: Social interaction in a technological world*, ed. P.M. Leonardi, B.A. Nardi, and J. Kallinikos, 91–112. Oxford: Open University Press.

Roberts, J. 2006. Limits to communities of practice. *Journal of Management Studies* 43 (3): 623–639.

Rowley, J. 2011. e-Government stakeholders—Who are they and what do they want? *International Journal of Information Management* 31 (1): 53–62.

Scott, S.V., and W.J. Orlikowski. 2012. Great expectations: The materiality of commensurability in social media. In *Materiality and organizing: Social interaction in a technological world*, ed. P.M. Leonardi, B.A. Nardi, and J. Kallinikos, 113–133. Oxford: Open University Press.

Star, S.L. 2010. This is not a boundary object: Reflections on the origin of a concept. *Science, Technology, and Human Values* 35 (5): 601–617.

Swan, J., H. Scarbrough, and M. Robertson. 2002. The construction of 'communities of practice' in the management of innovation. *Management Learning* 33 (4): 477–496.

Vaast, E., and G. Walsham. 2009. Trans-situated learning: Supporting a network of practice with an information infrastructure. *Information Systems Research* 20 (4): 547–564.

Vann, K., and G.C. Bowker. 2004. Instrumentalizing the truth of practice. In *The Blackwell cultural economy eader*, ed. A. Amin and N. Thrift, 40–57. Oxrford: Blackwell Publishing Ltd.

Wenger, E., R. McDermott, and W.M. Snyder. 2002. *Cultivating communities of practice: A guide to managing knowledge*. Boston: Harvard Business School Press.

Part II
Troubled Identities

Chapter 10
Performing Science in Public: Science Communication and Scientific Identity

Sarah R. Davies

This chapter asks: what kind of scientific identity work is going on within public communication of science? Or, to put it a different way: how do scientists use science communication to construct identities as scientists and members of scientific communities? These questions take for granted that public communication is, in fact, implicated in scientific identity-building, and that's something I'll return to. But—to jump straight to my conclusions—the chapter will not answer these questions in any tidy way. In line with the conceptual framework presented in the introduction to this volume, I will suggest that categories like 'identity' or 'scientific community' are unstable. Indeed, when we apply them to my data—interviews and participant observation of a large science festival—they blur, become unusable. They will not fit. This in itself is one of my findings.

Why am I confident that science communication[1] does scientific identity work? There are both historical and contemporary reasons for this. History of science has shown that science and its publics have always been co-produced (Shapin 1990). Early scientists needed reputable (meaning: white, male, upper class) public witnesses to vouch for the trustworthiness of their work (see, e.g. Shapin 1994, 2010). The public performance of experimentation, or at least the possibility of such public performance, was essential to portraying science and scientists as reliably authoritative. Scientific identity since then has trodden a delicate line between being the *same* as such public audiences—with scientific knowledge framed as

---

[1] Here I am concerned solely with science communication to public audiences, rather than intra-scientific communication (which is certainly also involved in identity construction). For more on this distinction see Horst et al. 2016; Palmer and Schibeci 2014; in the rest of this essay, 'science communication' should be understood as referring to public science communication.

S. R. Davies (✉)
Department of Science and Technology Studies, University of Vienna, Vienna, Austria
e-mail: sarah.davies@univie.ac.at

K. Kastenhofer, S. Molyneux-Hodgson (eds.), *Community and Identity in Contemporary Technosciences*, Sociology of the Sciences Yearbook 31,
https://doi.org/10.1007/978-3-030-61728-8_10

commonsensical, accessible, or mundane—and being exotically different and uniquely authoritative, and therefore worthy of special treatment (Broks 2006; Shapin 1990). For instance, Thomas Gieryn describes the way in which battles around the status of phrenology as (pseudo)scientific were played out in nineteenth century Edinburgh through cultural and political debates over who should be awarded a vacant university chair (Gieryn 1999, p. 115–182). The question of what made for good 'science'—and therefore of who deserved the chair—was not an internal debate; rather, it involved public canvassing, political machinations, and debate of scientists' characters. Nineteenth and twentieth century educational and popularisation activities (from Mechanics Institutes to educational pamphlets or museums) similarly answered a number of boundary work and identity-building purposes, from defining science's 'others' to reinforcing particular social orders or constructing a shared national identity (Bensaude-Vincent 2009; Shapin and Barnes 1977; Topham 2009). Scientists have always used public communication to craft shared understandings of what science is, and what it means to be a scientist.

Scholarship on contemporary science has also emphasised the entanglement of public communication and scientific identity. Felt and Fochler (2012, 2013), for instance, use the notion of 'epistemic living spaces' to describe how scientists' worlds may be shaped by public narratives about science. Epistemic living spaces are 'researchers' individual or collective perceptions and narrative re-constructions of the structures, contexts, rationales, actors and values which mould, guide and delimit their potential actions' (2012, p. 136). They represent what is thinkable and possible in a particular environment, including about one's role and community. These living spaces are flexible and require constant work and adaptation. Felt and Fochler describe how junior scientists, in particular, are affected by the 'excellence narratives' that circulate in public—stories within which science is clearly linear, progresses towards applications, and where one's career will ultimately lead to a stable academic position (2013). Held up against such stories, 'the individual lives of young scientists are deficient nearly by default' (ibid., p. 10). In a similar vein, Baudouin Jurdant has argued that popular science writing is in fact more for scientists than for publics. It can be understood as the 'autobiography' of science and thereby as a means of 'making real' its discourses and theories (1993). Popularisation, he writes, 'seems to have an epistemological role to play within science itself' (ibid., p. 371).

The shaping of epistemic living spaces through public communication is not stable. Maja Horst has shown that scientists may take on different roles within science communication, relating to their representation of different kinds of scientific collectives (2013). They might perform as an 'expert', representing their field of expertise, as a 'research manager', representing their organisation, or as a 'guardian of science', representing the institution of science as a whole. The part they play in public communication therefore relates to different scientific roles they may choose to take on (as a manager, for instance, or as a member of a particular discipline); importantly, individual scientists do not exclusively inhabit particular roles but may take one or the other depending on the immediate context. At a micro level, we can thus view identity as hybrid, flexible, and subject to shifting structural and contextual needs (Lam 2010). One such need may be the relation between individual and group identity. Are scientists speaking for themselves, and thereby performing an individual identity, or are they involved in constructing organisational or disciplinary identity? Brosnan and Michael (2014) have shown, for example, that a single group leader may come to represent the possibilities of a whole field of research, along with symbolising their group, while Horst (2013) also argues that scientists' performances of organisations have a particular character and tone. We might understand tensions between university communication departments and communicating scientists as arising from these dynamics. Scientists may see themselves and their public communication as their concern and their activities as reflecting on themselves alone; communication or press departments, however, view the organisation as inevitably implicated in scientists' public representations (Autzen 2014; Davies and Horst 2016).

These accounts demonstrate, in different ways, how scientific identity is co-produced with public communication. In this chapter I am concerned with how this happens in practice, and what we can learn about this process through close empirical examination of an example of science communication. In what follows I unpack findings from a study of a large science festival, Science in the City, held in Copenhagen in June 2014. First, I describe the conceptual framing I am mobilising; second, in a series of three empirical sections, I present a series of vignettes from the Science in the City festival, using these to explore how identity was performed; and third, I draw together my findings to offer some reflections on the wider question of the relation between scientific identity and the practice of public communication.

10.1 Identity as Performance

It will be clear from the discussion above that I am working with a generous definition of identity: I see it as relating to role, boundary work (or role differentiation), and community. Most importantly I view identity as processual. It is not something that is fixed or stable, but a performance that may be carried out differently in different contexts. (With regard to the becoming/performing dicotomy described in the introduction to this volume, this analysis is firmly oriented to performance.) This view can be traced back to the work of Erving Goffman and the notion that the self is not something that is expressed or experienced within interaction in any pure or authentic way, but rather is *performed* (Goffman 1955, 1959). Social interaction, Goffman argues, involves the performance of particular characters. As in the theatre, everyday life requires individuals to convincingly produce a certain version of the self. A 'character' is a 'figure, typically a fine one, whose spirit, strength and other sterling qualities the performance [is] designed to evoke' (1959, p. 244). Understanding social interaction in these terms has a number of implications. One is that performances are contextual: a certain setting (what Goffman calls a region) may require one kind of character, while another calls for one that is very different. We perform our selves in different ways with our lovers and our bosses;

analytically, those performances can reveal something about what qualities we believe are prized in each context. A further implication is that work is involved in the presentation of the self. Goffman calls the performer a 'harried fabricator of impressions' (p. 244) and writes that to '*be* a given kind of person … [is] to sustain the standards of conduct and appearance that one's social grouping attaches thereto' (p. 81, emphasis in original). Even if the maintenance of these standards feels and appears effortless—and may well be entirely unselfconscious—this should not disguise the fact that performances are not innate but learned and require the management of numerous sociocultural cues. The notion of this effortfulness is developed in particular in the idea of 'face-work' (Goffman 1955; Myers 2003) which emphasises that any encounter is liable to fail—that is, that one or more of those involved may 'lose face' as their self-presentation falters or falls down completely.

Goffman's work has been widely taken up in science studies. His notion of 'front stage' and 'back stage', for instance, has been used to discuss the formal and informal presentation of science (see discussion in Greiffenhagen and Sharrock 2011) and resonates with STS concepts such as the use of empiricist and contingent repertoires (Gilbert and Mulkay 1984) or Latour's two-faced Janus as a representation of science (Latour 1987). Greg Myers and other discourse analysts have used Goffman to examine the dynamics of science-based interactions, for instance by exploring how face can be maintained when discussing risk-based topics (Davies 2011; Myers 2003, 2007). Perhaps the most extended use of the performance metaphor, however, and the one that is most relevant to my discussion, is in Stephen Hilgartner's volume *Science on Stage* (2000), in which he argues that scientific policy advice can be understood as a kind of public drama. Hilgartner opens up the ways in which science advice—and scientists—are perceived as authoritative and relevant; science advisors, he writes, 'use a variety of techniques to create—or better, to *enact*—the basis of their authority as experts' (p. 6). Similarly to Goffman, he emphasises that understanding the production of authoritative characters (or any other convincing role) as a performance does not imply cynical or manipulative performers. Scientists are not 'con men', they simply have 'multiple identities' (p. 13):

> Performers typically seek to display themselves in ways that make them appear to conform to the "identity norms" befitting a person of their status given the nature of the occasion … In the world of science advice, one of the central identity norms concerns the objectivity (in the sense of disinterestedness) with which advisory bodies approach their charge. Thus … advisors work hard to enact objectivity (Hilgartner 2000, p. 14)

Again, this highlights the importance of context—the 'nature of the occasion'. Identities are performed in ways appropriate to the current setting: in science, and science advice, unity and objectivity might be fitting to one front stage context, while controversy and contingency are performed on the back stage. In this sense, scientific identity is always a production—requiring the identification of relevant 'identity norms'—and is always shifting.

How to apply this framework to public communication? Viewing scientific identity as processual and hybridic already starts to muddy the waters of my initial question (how do scientists use science communication to construct scientific identities?) by fragmenting 'scientific identity' into many different possibilities. A more relevant question might be *which* scientific identities are performed in public communication. It also raises the question of what front and back stages might be involved in science communication. Much analysis of scientific communication has assumed that this is a process by which science presents a cleaned-up public face to the world—that it is the front stage to the back stage of scientific uncertainty, debate, and hedging. Rae Goodell, for instance, in her seminal account of 'visible scientists' (1975), sees celebrity scientists as 'representatives of the scientific community as a whole' (p. 203) despite the fact that their views may diverge wildly from the consensus; more recently, Declan Fahy has argued that celebrities such as Stephen Hawkings or Richard Dawkins are less communicators than brands encapsulating particular public narratives about science (2015). But, as Hilgartner has suggested, the distinction between front and back stage is not always clear cut: 'each backstage region is the theater for additional layers of performance' (Hilgartner 2004, p. 447). While some public communication may present the glossy, finished version of front stage science, it seems possible that, at times, back stage identities may leak through or are even deliberately performed.

The rest of this chapter reflects on these dynamics through the example of a large public science festival, Science in the City, which was held in Copenhagen in June 2014 over the course of 6 days. The festival as a whole involved dozens of distinct exhibits, activities, and events—from the opportunity to visit a marine research vessel to a dialogue on IP issues in biohacking and an exhibit, based on veterinary research, featuring live pigs. As such, it—like similar festivals—was a distinctive form of science communication, one that offered particularly rich opportunities for multiple identity performances and for slippage between front and back stages. The variety of formats present, and the nature of a live science event in which performances cannot readily be fixed or rendered static, mean that the festival offers a kind of extreme case of identity work within science communication. While I interviewed the lead organiser of the festival and took fieldnotes from participant observation at the festival site, my research largely focused on six science communication projects in particular, seeking to produce a 'thick description' of their production and reception (Ponterotto 2006). I carried out interviews with the organisers of these and focused my participant observation on their physical instantiations (variously, a photo exhibition, two interactive installations, an art piece, a movement-based workshop, and a public lecture). I also collected public responses to them via questionnaires, though these data will not be discussed here. The chapter will not attempt to present results from this research as a whole.[2] Rather, I will use a number of key vignettes from the festival to illuminate some of the dynamics through which scientific identity was (or was not) negotiated within it.

---

[2]Further information about the festival, the research, and findings from it can be found in Davies and Horst (2016).

10.2 Science Communication Is Heterogeneous

I suggested at the start of the chapter that one finding from this research was of a complexity that overtook the research questions I began with. I was interested in scientific identity, and how it might be mediated within science communication, but my engagement with the festival, from my starting point of identifying project 'organisers' to interview, rapidly began to fracture the categories I had begun with. Contra classical accounts such as Goodell (1975) or guides to science communication such as Bennett and Jennings (2011), which position scientists as the key actors within public communication, Science in the City and its constituent communication projects were highly diverse assemblages. Academic scientists were certainly involved (sometimes as lead 'organisers', though sometimes not), but the projects also enrolled, for example, research administrators, artists, professional science communicators, philosophers, event managers, students (paid and as volunteers), university press officers, a local quarry, businesses, social researchers, freelance designers, international organisations, local government, and organisers' friends and family. Enquiring about *scientific* identity immediately becomes more complex in such a situation. Is this the only identity being performed? And who exactly is able to perform it?

As a brief example it is worth sketching out the organisation of one project, the *Seeing the Unseeable* installation/exhibition. *Seeing the Unseeable* was designed to explain and promote the science behind the European Spallation Source (ESS), a particle accelerator then under construction in Lund, in the nearby Skåne region of Sweden. The project was a collaboration between Region Hovedstadet (Copenhagen's city council), the ESS organisation, MAX IV (another local accelerator), the University of Copenhagen, and the Technical University of Denmark. The final installation involved an 'immersive experience' featuring strobe lights in a tunnel-like space, which led into a larger area containing table-top demonstrations of physics experiments, videos and computer interactives, and a lounge area with free coffee and sandwiches. The conceptual development of the space was contracted out to a local science centre, and it was staffed by volunteer students and scientists from the participating universities. These scientists were the 'face' of the exhibition, and their contact with visitors was key to how the project was experienced. But they played only a minor role in its production and development. Far more important were actors such as Mads, the science centre project manager who led the exhibition concept development and training for the volunteers, and Kell, the Region Hovedstadet manager who was seeking to ensure that the ESS would be a 'growth motor' for the region and who was behind the project's funding.[3] As the project came to fruition, scientific identities were certainly at stake (not least in the context of Mads' interest in promoting a particular version of science communication and his concerns that the scientist volunteers were not in line with this). But so were other

---

[3]The names of individuals have been changed.

identities and communities, from 'science communicator' to 'public-spirited volunteer'.

The point here is not to outline in detail how these identities were produced. Rather it is to illustrate some of the diversity of the roles and groups that were implicated within these different science communication projects. One of the reasons it becomes difficult to talk about the production of scientific identity within Science in the City, then, is that so many other personal and professional roles, and thereby identities, were also being performed within these public communication activities. As we will also see, at least some of these overlapped with scientific identities.

10.3   Scientific Identities Are Multiple

If the organisation of the different science communication projects indicates some of the complexity of pinpointing performances of scientific identity, so too does the consideration of their timelines. The literature I discussed in the first section often looks at scientific identity at the moment of its contact with public audiences, exploring how science is produced as Other to lay witnesses or public groups even as it interacts with them (see Gieryn 1999; Shapin 1990). There is a rich literature that has explored the versions of science and scientists that are represented within science communication (e.g. Chimba and Kitzinger 2010; Jacobi and Schiele 1989; Weingart et al. 2003), and, certainly, physical instantiations of Science in the City projects promoted some images of scientists over others—as accessible, capable, and humorous, for instance, rather than elite or incompetent. Such representations are not surprising in an environment that explicitly aimed to present science in a positive light. But as the history of *Seeing the Unseeable* indicates, public communication projects have a hinterland of preparation and development within which scientific identities may also be performed (in Mads' training programme for volunteer scientists, for example, and discussions of good communication behaviours therein). In engaging with the projects, it became clear that scientists framed their identities differently at different moments within them and that particular roles or versions of the self came to the fore at different points. Identities and community memberships were multiple and flexible.

I will illustrate this multiplicity with a couple of examples from two Science in the City projects. The first is taken from *Ocean of Resources*, which was a large format photography exhibition featuring eight images, each with a paragraph of explanatory text underneath (Fig. 10.1). This was a straightforward and unpretentious project, led in its practical aspects by a university research administrator (Nete) and in spirit by a marine scientist (Harro), who had taken the photographs and developed the text. The project was funded by the Norwegian Research Council and was in fact inspired by the Council's call for projects to participate in the festival; Science in the City was, in other words, an opportunity for Nete and Harro to further develop Harro's photography—which had previously featured in smaller scale, local exhibitions—as a means of public engagement. Both were clear that the project was largely about

Fig. 10.1 Ocean of resources. (Image by Alexander Dich Jensen (permission granted for use))

branding, gaining visibility, and 'showcasing' university research. This is Harro, for instance, talking about his aims for the project:

> . . .we are really trying to connect each picture to ongoing research activity at the university, and especially the ones that are financed by the research council. So we're trying to use it also as a showcase of what kind of research we are doing . . . I think the main goal for me is to really have people to see that, you know, marine resource is not only the fish we eat but it's a lot of other things.

Harro's interests are twofold. He is first concerned with publicising university research: each of the pictures would be explained using examples from his home university and would thereby highlight its activity in this area (and indeed its very existence: Nete and Harro were based at a small university in the far north of Norway and wanted, as Nete said, to 'show off' that it was there and active in research). In this respect, the Norwegian Research Council is not an incidental audience; *Ocean of Resources*, by seeking out funding from it and by showcasing research that it funds, is able to perform the university in general and Harro in particular as an active and engaged partner of the Council, and thus as a well-established research actor. Along with this performance of organisational identity, Harro also has more didactic aims. The title, *Ocean of Resources*, refers to marine environments as an important natural resource, and he wants audiences to understand that the oceans are important for 'a lot of other things' besides fishing (for tourism, for example, or preserving biodiversity). The project therefore had some quite traditional science communication goals: Harro sought to promote and publicise 'sound science' in the public domain.

At the same time, Harro also presented his involvement in the project as part of a quite different identity, one related to leisure and pleasure. One of the first things I asked him was about the relation of the images to his research. Given that he was a marine scientist, I had assumed that the photos were taken during the course of his work. But no, he said, this was 'personal interest. It's not something I use daily in my research'. This was exactly why he and Nete had had to invest energy in—as he says in the quotation above—connecting 'each picture to ongoing research activity at the university'. It similarly emerged that his commitment to the images went far beyond the need for the public to understand that the oceans are an important resource; they were, for him, an affective technology that could convey something of the experience of diving and of encountering underwater life. Though this was difficult for him to articulate, the sense was that images such as those of 'beluga whales who are looking straight at you' were intrinsically powerful and emotional such that 'people [feel] a connection to what they're seeing'. This version of himself had little to do with the dry promotion of research or even with public education. Rather, it was an identity that was about leisure and 'personal interests', but perhaps even more about having a connection to the natural world. Though there was an obvious link to his research activities as a marine scientist, he framed this way of being as rather different—an identity that was more personal than professional and thus a 'back stage' performance.

A second example comes from a public lecture-style event at Science in the City, the Ig Nobel Prize show. This was a local iteration of what has become an international franchise. The Ig Nobel Prizes (for science that 'first makes people laugh, and then makes them think'[4]) are awarded annually at Harvard and receive extensive international press and media coverage; the organiser, Marc Abrahams, also carries out other shows around the world and runs associated activities such as the Annals of Improbable Research. The local organiser, Niklas, was an Ig Nobel laureate and a scientist at a Danish university. The Science in the City festival shows (there were two, identical in format) built on other shows that Niklas had been involved in; as with these previous versions, Abrahams and a number of other Ig Nobel laureates travelled in to present their stories.

Though the Ig Nobels—including the two Science in the City shows—are humorous in tone, and aim to entertain, Niklas was clear that this was a *scientific* endeavour. All content is 'science based', and the shows involved occasionally 'distinguished' but always 'serious' scientists. Everything presented is 'solid science'. Niklas was also keen to point out the prize's origins at Harvard and the fact that, at the annual prize-giving ceremony, it is actual Nobel winners who give out the awards. Much of the discursive work Niklas did within our interview, then, focused on protecting an identity as a robust, credible scientist against an implied charge of dumbing down or producing mere entertainment. The Ig Nobel Show as authentically scientific and trustworthy was key to his involvement in it and, relatedly, he presented it as a means of educating the public, ensuring their support for basic

---

[4]See http://www.improbable.com/ig/

research (which he saw as under attack) and allowing them to appreciate the wonders of scientific thinking. His aim, he said, was for audiences 'to understand how fun and interesting and beautiful science is and that the most beautiful thing science can do to you is it can make you think. It can enrich your brain'.

The Ig Nobel Show can thus be understood as a means of propagating a particular kind of scientific identity and version of science. It was designed to present and protect science as entertaining and accessible while also being curiosity-driven, reliable, and important. Again, however, there were moments when Niklas took up other identities, in particular when the show was constructed less as a showcase for science and more as an opportunity for particular social identities to come to the fore. There was his conviction, for instance, that the stories the show featured—about different pieces of 'improbable' research—were the narrative equivalents of earworms, such that they would stay with audience members and inevitably spark further reflections and conversations. But it was also clear that his involvement in the show was a means for him to nurture and enjoy a particular community. Having won his Ig Nobel in 2004—and having greatly enjoyed the experience of participating in the Harvard prize-giving ceremony—Niklas kept in touch with Marc Abrahams and had several times discussed the possibility of participating in a tour. By 2009 or 2010

we started to invite some of the Ig Nobel prizewinners for a small tour in Denmark and we have also been in Sweden some years so we have been there a couple of times before and had shows. And it's always been really nice events. It's almost become like a small tradition, but it's never been as big as [the Science in the City show] is really. This is the biggest event we have been to, so we are very excited about that.

Science in the City was an opportunity not only to publicise the kind of science promoted through Ig Nobel but also to continue the 'small tradition' of a particular group of laureates coming together to perform the show. This was a social event as much as an educative one. Niklas took great pleasure in the sense of community and belonging that the show seemed to involve, continuing to be excited about the stories the laureates told and the chance to work together with them. The Ig Nobel prizes, he said, was a 'movement . . . to fight for science', but they also functioned as a way of building community and shared identity for those—like Niklas, Abrahams, and the other laureates involved in the show—who were committed to this aim.

For both Harro (in the *Ocean of Resources* project) and Niklas (in the Ig Nobel Prize Show), public communication functioned to disseminate particular versions of science and scientific identity. But it also involved other identities, versions of the self connected to leisure, to engagement with nature, or to a particular social group. It is worth noting that it is not easy to pin these different characters (to use Goffman's term) down to front or back stage performances. Harro-the-promoter-of-Norwegian-research or Niklas-the-serious-scientist might be foregrounded in the final realisation of their two different communication projects, but the other selves that were implicated in those projects were not understood as entirely inappropriate for public view (if they were, it is unlikely they would have been discussed with me in an interview). The Ig Nobel Show was very explicitly a practiced and glossy performance, but I watched Niklas before and after that performance as he chatted and joked with his fellow presenters; elements of those interactions—those friendships—found their way into the style of the 'front stage' show. What we might say, then, is that it is not straightforward to distinguish between the front and back stage, at least in this context. It seems possible to respond to multiple 'identity norms', both scientific and from other domains, at the same time.

10.4   Audiences Are Heterogeneous

If public communication can involve scientists inhabiting multiple roles at the same time, it seems likely that the same kind of complexity is true of audiences. In this section I look at a final example of how a shared sense of organisational identity was produced within the Science in the City festival, this time by exploring the results of a particular combination of material infrastructure, scientist-communicators, and audiences. Here I am interested in how a specific assemblage of objects and diverse human actors—including the audiences of public communication—allow particular scientific identities to be performed.

My example here moves beyond a single project—the *Breaking and Entering* interactive installation (Fig. 10.2)—to take in the subsection of the Science in the City site it was situated within. *Breaking and Entering* was organised by an interdisciplinary group of University of Copenhagen employees and external consultants (researchers, administrators, students, designers), and, as such, it was placed

Fig. 10.2   Breaking and entering. (Images by Alexander Dich Jensen (permission granted for use))

in what became known colloquially as the 'KU tent' (KU being the abbreviation for *Københavns Universitet*).[5] This was a large marquee which housed a number of University of Copenhagen communication projects. In addition to *Breaking and Entering*, these included a 'teddy bear hospital' run by medical students, a café, an installation developed by the University's International Office describing the experiences of international employees, and a small stage and seating area used for 5 min 'lightning' research talks and longer chemistry shows. Though the projects had been developed independently—to the extent that there was occasional competition over the use of space or the noise levels—the space had a common design aesthetic and was linked together through the use of wooden pallets as seating throughout (these also appeared in a 'spillover' seating and lounge area outside the marquee). The projects were also linked through the figure of Emilie, the University of Copenhagen project manager who had been tasked with organising the university's presence at Science in the City, and who had liaised with all the projects in advance of the festival and was present throughout.

Six days—the length of the festival—is a long time to spend running a science communication project. Though there were shifts in staffing the KU tent projects at any particular time (the teddy bear hospital, for instance, only ran over the weekend), there was also a core cadre of individuals who were present for the majority of the festival period. These people (as a University of Copenhagen employee, I count myself among them) were there to set up and install their projects before the festival opened; they ate lunch (provided by the university) at around the same time each day, and they shared in the ebbs and flows of visitor numbers and the dynamics of the space. Over the course of the festival the tent became lived in: things broke and were stuck back together; the marquee took on a distinctive smell; we became familiar with each other and with the tent's soundscape. Many of my fieldnotes relate to the KU tent and its dynamics, in part because it contained *Breaking and Entering* but also because it was simply an easy place for me—a University of Copenhagen staff member—to be. (Unlike some of the other projects I was observing, the space was also protected from the vagaries of the weather.) The community of the tent, both human and non-human, became deeply familiar. By the second half of the festival I could write in my fieldnotes that I had been

> reflecting on how mundane and bedded down our activities have become—how everyday the environment is now. At the start it was new and exciting and confusing, but now we have eaten the same lunch in the same place, and sat around the tent, and gauged the dynamics of the space, for several days in a row, and somehow it starts to feel like home.

'We' refers here to myself and other University of Copenhagen staff involved in the festival, in particular those connected to *Breaking and Entering* but also to the

---

[5]*Breaking and Entering* was led by my department, and I was one of the researchers involved in its development; on this project, then, my role as not just an observer but also a participant was particularly heightened. Further discussion can be found in chapter three of Davies and Horst (2016). The title refers to the idea of science as an ivory tower which may be infiltrated by society through an act of breaking and entering.

other projects in the tent. Our involvement had become rhythmic and routinised; more than that, it had become collectivised. There was a developing sense of camaraderie as people relaxed with one another and shared the experience of carrying out public communication. Emilie and others held business meetings during quiet moments; volunteer communicators drifted round to look at the other exhibits and chat. The 'KU tent' became an outpost of the university as a whole. Over the course of the festival it turned into a space where, together, we could perform an identity as dedicated university employees concerned with public outreach and collectively construct the university as a socially engaged, relevant organisation. We were concerned not just with personal scientific identity but with ensuring the reputation of a particular scientific organisation.

The work of co-constructing organisational identities was not limited to those involved in running the KU tent projects. Many of the visitors to the space were linked to the university in some way. Senior managers were given tours of the different projects while visiting the festival, project organisers' friends and family came to see what they had been working on, and university colleagues visited projects their departments had been involved in. At moments it was clear that the 'public' audience was largely comprised of those working in research or the academy: a physics-based lightning talk, for instance, that prompted high level technical questions; or another talk for which I sat wedged between my head of department and a senior manager at the faculty. The University of Copenhagen and its members, of course, had an interest in what was being presented at the festival. It was the university, and the nature of those who were part of it, that was being represented through the different KU tent projects as much as the research that those projects focused on.

We can thus suggest—as the title of this section implies—that the audiences of public communication are heterogeneous. In the case of the KU tent, an important (perhaps even primary) audience was not the lay public but university staff, managers, and others with connections to academic research.[6] In this sense, *Breaking and Entering* and the other projects located in the tent were doing what marketing scholarship has called 'auto-communication' (Christensen 1997): performing the identity of the university (and perhaps of institutional science more generally) to existing organisational members. More than this, though, I would argue that we can view the KU tent assemblage as doing significant new identity-building work. The activities, exhibits, social relationships, and atmosphere that developed over the course of Science in the City together helped construct a shared identity as university employees; similarly, they produced the university as a particular kind of research organisation (socially engaged, lively, open). Collegiality, collaboration, and openness were performed, and they thereby in some way became part of the 'identity norms' of University of Copenhagen researchers and staff. In this sense we might not view the KU tent as a 'front stage' performance at all. Rather, the activities it hosted

---

[6]We also saw this in my discussion of *Ocean of Resources*. The Norwegian Research Council was both an important partner in and audience for the project.

were doing the back stage work of crafting the public characters we would go on to perform at other sites. The real identity work was done as we fixed broken exhibits and fetched lunch together, with the public performance and communication of science a useful side-effect.

10.5 Discussion

I noted at the start of the chapter that this analysis served to destabilise some of the categories of my own research and of the literature on science communication and scientific identity more generally. How to look at scientific identity when it emerges that scientists are only a small part of the assemblage that brings science communication projects into being? Or how to capture what is produced in the interface between 'science' and 'publics' when those publics may themselves take up scientific identities? What my engagement with the Science in the City festival has done is not so much answer my initial question—what kind of scientific identity work is going on within public communication of science?—as fracture it into many different possibilities. Science communication is used by scientists for many different identity-building purposes, in many different ways. It further relates to different kinds of communities: as the guest editors write in their introduction, "entanglements between collectivity and identity are clearly visible, albeit not trivial or tangible in most respects" (p. 30). It thus appears impossible to be definitive—to say, for instance, that public communication is always about boundary work and the protection of distinctive scientific identities, or that 'objectivity' is a stable identity norm (cf. Hilgartner 2000). 'Science communication', to use a single term to represent what appears to be a highly heterogeneous activity, may involve the performance of characters from 'science communicator' to 'nature lover' or 'representative of the University of Copenhagen' rather than, or as well as, that of 'scientist'. In addition, all such characters offer scope for interpretation. The role of scientist is not highly scripted: one might play it as an addendum to a leisure identity, as Harro did, or, as Niklas did, as a social character, part of the production of enjoyable community.

My discussion has similarly muddied the waters of notions of front and back stage within identity construction.[7] It is not always clear where public performances begin and the back stage work of preparing them ends: the KU tent, for example, might be construed as being only secondarily for a public audience while primarily functioning for internal organisation-building; and even cleaned-up, glossy presentations of science such as the Ig Nobel Prize show may involve back stage-style moments of relaxation. This again encourages us to constantly interrogate the work

---

[7]I am by no means the first person to problematise a clear distinction between front stage and back stage performances. Most recently, social media has been viewed as one way in which such distinctions are systematically breaking down (Lee 2006).

that is going on in public representations of science. Celebrity scientists may symbolise coherent scientific 'brands' (Fahy 2015), and increasingly public presentations may depict increasing certainty (Greiffenhagen and Sharrock 2011), but, again, this may not always be the case. We cannot take the nature of science, as constructed in public, for granted.

An understanding of science communication as highly complex, and as capable of supporting multiple and flexible identity performances, is thus a key outcome of this analysis. Indeed, I believe that exploring these complexities and contingencies is a productive way forward for studying science communication generally and for understanding the work that such communication does to science and scientists (Davies and Horst 2016). Performances are contingent, identity is shifting, and any instance of public communication is under-determined, offering scope for different kinds of work on the self. In the midst of such complexity, is it possible to say something about the patterning of identity work or the ways in which performances might settle into familiar and repeated repertoires? Certainly, these data do not reveal many hints as to what such patterns and repertoires might involve. As I have argued, diversity is the most striking point. In closing we might look instead to the kinds of notions and resources that could help identify moments of stability—ways in which identity-formation becomes more settled or routinised—in future research. One such resource might be to explore what is clearly taboo or unsayable within science communication. I have noted, for instance, that Science in the City representations of science and scientists tended to portray them as accessible, humorous, and in a positive light; future research might thus interrogate how it is *not* possible to portray science in public settings (as boring, perhaps, or insignificant).

A further tool for conceptualising the space between actants' agency and the settled patterns through which we experience the world is the concept of 'narrative infrastructure' (Felt 2017). Narrative infrastructure forms an 'ambient discursive environment' (ibid., p. 56): it is the web of stories and taken-for-granted ideas and arguments that crystallise around a particular issue. Importantly, though, such infrastructure is dynamic. It is 'temporally stabilised' and therefore transient, and it contains 'actors who create, adapt, multiply, support and entangle ... dominant narratives' (ibid.). Narrative infrastructure, as a concept, could provide one way of looking for wider patterns in the public performance of scientific identity. Which narratives and assumptions frame events such as Science in the City and make particular kinds of performances more likely than others? We might, from the STS literature, start to suggest some possibilities, hinted at by the following terms: public understanding of science, promotion of basic research, democracy, deficit and dialogue, innovation (Davies and Horst 2016; Irwin 2006). But these can only be tentative. One key direction for future research should therefore be further exploration of the framing narratives of science communication. Such work will help articulate the relation between situated performances of scientific identity and the narrative infrastructures that make them possible.

References

Autzen, C. 2014. Press releases—The new trend in science communication. *Journal of Science Communication* 13 (3): 1–8.

Bennett, D.J., and R.C. Jennings. 2011. *Successful science communication: Telling it like it is.* Cambridge: Cambridge University Press.

Bensaude-Vincent, B. 2009. A Historical Perspective on Science and Its "Others". *Isis* 100 (2): 359–68. https://doi.org/10.1086/599547.

Broks, P. 2006. *Understanding popular science.* Maidenhead: McGraw-Hill Education (UK).

Brosnan, C., and M. Michael. 2014. Enacting the 'neuro' in practice: Translational research, adhesion and the promise of porosity. *Social Studies of Science* 44 (5): 680–700.

Chimba, M., and J. Kitzinger. 2010. Bimbo or boffin? Women in science: An analysis of media representations and how female scientists negotiate cultural contradictions. *Public Understanding of Science* 19 (5): 609–624.

Christensen, L.T. 1997. Marketing as auto-communication. *Consumption Markets & Culture* 1 (3): 197–227.

Davies, S.R. 2011. How we talk when we talk about nano: The future in laypeople's talk. *Futures* 43 (3): 317–326.

Davies, S., and M. Horst. 2016. *Science communication: Culture, identity and citizenship.* New York: Palgrave Macmillan.

Fahy, D. 2015. *The new celebrity scientists: Out of the lab and into the limelight.* Lanham: Rowman & Littlefield.

Felt, U. 2017. 'Response-able practices' or 'new bureaucracies of virtue': The challenges of making RRI work in academic environments. In *Responsible innovation 3: A European agenda?* ed. L. Asveld, M. van Dam-Mieras, T. Swierstra, et al., 49–68. Cham: Springer.

Felt, U., and M. Fochler. 2012. Re-ordering epistemic living spaces: On the tacit governance effects of the public communication of science. In *The sciences' media connection: Public communication and its repercussions,* ed. S. Rödder, M. Franzen, and P. Weingart, 133–154. Dordrecht: Springer.

———. 2013. What science stories do: Rethinking the multiple consequences of intensified science communication. In *Science communication today: International perspectives, issues and strategies,* ed. P. Baranger and B. Schiele, 75–90. Paris: CNRS Editions.

Gieryn, T.F. 1999. *Cultural boundaries of science: Credibility on the line.* Chicago: University of Chicago Press.

Gilbert, N., and M. Mulkay. 1984. *Opening pandora's box: A sociological analysis of scientists' discourse.* Cambridge: Cambridge University Press.

Goffman, E. 1955. On face-work: An analysis of ritual elements in social interaction. *Psychiatry: Interpersonal and Biological Processes* 18 (3): 213–231.

———. 1959. *The presentation of self in everyday life.* London: Penguin.

Goodell, R. 1975. *The visible scientists.* Boston: Little, Brown & Company.

Greiffenhagen, C., and W. Sharrock. 2011. Does mathematics look certain in the front, but fallible in the back? *Social Studies of Science* 41 (6): 839–866.

Hilgartner, S. 2000. *Science on stage: Expert advice as public drama.* Stanford: Stanford University Press.

———. 2004. The credibility of science on stage. *Social Studies of Science* 34 (3): 443–452.

Horst, M. 2013. A field of expertise, the organization, or science itself? Scientists' perception of representing research in public communication. *Science Communication* 35 (6): 758–779.

Horst, M., S.R. Davies, and A. Irwin. 2016. Reframing science communication. In *The handbook of science and technology studies,* ed. U. Felt, R. Fouché, C.A. Miller, et al., 4th ed., 881–907. Cambridge: MIT Press.

Irwin, A. 2006. The politics of talk: Coming to terms with the 'new' scientific governance. *Social Studies of Science* 36 (2): 299–320.

Jacobi, D., and B. Schiele. 1989. Scientific imagery and popularized imagery: Differences and similarities in the photographic portraits of scientists. *Social Studies of Science* 19 (4): 731–753.

Jurdant, B. 1993. Popularization of science as the autobiography of science. *Public Understanding of Science* 2 (4): 365–373.

Lam, A. 2010. From 'Ivory Tower Traditionalists' to 'Entrepreneurial Scientists'? Academic Scientists in Fuzzy University—Industry Boundaries. *Social Studies of Science* 40 (2): 307–340.

Latour, B. 1987. *Science in action: How to follow scientists and engineers through society.* Cambridge Mass: Harvard University Press.

Lee, H. 2006. Privacy, publicity, and accountability of self-presentation in an on-line discussion group. *Sociological Inquiry* 76 (1): 1–22.

Myers, G. 2003. Risk and face: A review of the six studies. *Health, Risk & Society* 5 (2): 215–220.

———. 2007. Commonplaces in risk talk: Face threats and forms of interaction. *Journal of Risk Research* 10 (3): 285–305.

Palmer, S.E., and R.A. Schibeci. 2014. What conceptions of science communication are espoused by science research funding bodies? *Public Understanding of Science* 23 (5): 511–527.

Ponterotto, J.G. 2006. Brief note on the origins, evolution, and meaning of the qualitative research concept thick description. *The Qualitative Report* 11 (3): 538–549.

Shapin, S., and B. Barnes. 1977. Science, Nature and Control: Interpreting Mechanics' Institutes. *Social Studies of Science* 7 (1): 31–74.

Shapin, S. 1990. Science and the public. In *Companion to the history of modern science*, ed. R.C. Olby, G.N. Cantor, J.R.R. Christie, et al., 990–1007. London: Routledge.

———. 1994. *A social history of truth: Civility and science in 17th century England.* Chicago: The University of Chicago Press.

———. 2010. *Never pure: Historical studies of science as if it was produced by people with bodies, situated in time, space, culture, and society, and struggling for credibility and authority.* Baltimore: JHU Press.

Topham, J. 2009. Rethinking the history of science popularization/popular science. In *Popularizing science and technology in the European periphery, 1800–2000*, ed. F. Papanelopoulou, A. Nieto-Galan, and E. Perdiguero, 1–10. Aldershot: Ashgate.

Weingart, P., C. Muhl, and P. Pansegrau. 2003. Of power maniacs and unethical geniuses: Science and scientists in fiction film. *Public Understanding of Science* 12 (3): 279–287.

Chapter 11
Being a 'Good Researcher' in Transdisciplinary Research: Choreographies of Identity Work Beyond Community

Andrea Schikowitz

11.1 Introduction—The Multiplication of Belongings in Research

Since the institutionalisation of modern science, disciplinary communities and sub-communities have been the basic structural, epistemic, and cultural units of academic knowledge production and education (Kuhn 1962; Knorr Cetina 1999; Stichweh 2008). Disciplinary communities have been described as providing shared understandings of what should count as relevant problems, suitable methodological and theoretical approaches, and of the kinds of findings that should be accomplished (ibid.). This includes understandings of how to be a proper scientist in terms of practices and skills as well as virtues and motivations (Ylijoki 2000; Daston and Galison 2007; Shapin 2008). Shared understandings of what makes a 'good' researcher[1] (comprising norms of thorough professional knowledge and skills as well as moral attitudes) have been conceptualised as 'role identities', 'social identities', or as 'subject positions' in different disciplines (for an overview see Cerulo 1997). In this chapter, I understand identity as a shared conception of how researchers should position themselves within different value repertoires (Fochler et al. 2016), which differs between disciplines (Becher 1989; Ylijoki 2000). These disciplinary identities are passed on through socialisation, and they are internalised,

---

[1] In this chapter, I will not address how being a 'good' researcher is related to research ethics and (avoiding) misconduct—issues that are currently debated in terms of research integrity and responsibility—because these issues hardly come up in the empirical material at hand. The researchers in my sample were first and foremost concerned with which methods, skills, and motivations were valued in different research fields.

A. Schikowitz (✉)
Munich Center for Technology in Society, Technical University Munich, Munich, Germany
e-mail: andrea.schikowitz@tum.de

© The Author(s) 2021
K. Kastenhofer, S. Molyneux-Hodgson (eds.), *Community and Identity in Contemporary Technosciences*, Sociology of the Sciences Yearbook 31,
https://doi.org/10.1007/978-3-030-61728-8_11

adapted and negotiated, or resisted by individual researchers through 'identity work' (Swidler 1986; Butler 1990).

Yet, different contemporary developments are challenging the notion of a rather exclusive and lasting belonging of individual researchers to the one disciplinary community into which they had been socialised, to which they subsequently contribute, and which they reproduce (Knorr Cetina 1999). On the level of research governance, a range of initiatives have been installed to facilitate researchers' mobility between disciplinary communities and between scientific and practical fields (such as networking initiatives, collaborative research programmes, and possibilities to combine different fields of study in the European higher education sector; see Gibbons et al. 1994; Kerr and Lorenz-Meyer 2009). Thus, the belonging of individual researchers is becoming more fluid and diverse (Galison 1996; Henkel 2005; Darbellay 2015; Hackett et al. 2017). This means that it is no longer the case that a researcher is necessarily educated, socialised, and employed in one disciplinary community that is also home to her collaborators, her reviewers, her publication media, and audiences. Rather, researchers maintain different and changing kinds of relations to different communities and collectives. Under these circumstances, researchers cannot draw on a singular, coherent imagination of how to be a good researcher. In turn, the very meaning of community is challenged when there is a perpetual exchange of community members.

Felt and Fochler (2012) argue that broader policy changes do not directly influence research practices but affect them indirectly via alterations of collective imaginations about what makes a good researcher. They argue that this results in

> new kinds of images of what doing science and being a scientist means, both in terms of the aims of scientists' epistemic pursuits, as well as in terms of the skills and virtues expected from the scientist as a person … *Whether they embrace or reject these new images, researchers have to position themselves with respect to them in their orientation work and when developing their own professional identity.* (Felt and Fochler 2012, p. 152, emphasis added)

For this reason, the loosening and multiplying of researchers' belonging can be expected to severely challenge both research communities and research identities. Thus, there is a need for new ways of conceptualising the dynamic and fluid relations between collectivity and individuality (see also Brew 2007; Kerr and Lorenz-Meyer 2009).

In this chapter,[2] I ask how researchers with diverse and dynamic relations to different collectives develop a self-understanding of what it means to be a good researcher, i.e. what the normative ideals are that they should strive for. In doing so, I empirically analyse how researchers who engage in transdisciplinary research

―――――――――――――――

[2]The research this chapter is based on was conducted at the Department of Science and Technology Studies (University of Vienna). This work was supported by the Austrian Federal Ministry for Science and Research under the funding scheme proVISION (funding of the project 'Transdisciplinarity as Culture and Practice') and by the Dr. Maria Schaumayer scholarship for returners of the Vienna University of Economics and Business (amongst others, funding for publication projects).

occasionally or regularly narrate, adopt, translate, resist, and combine the different imaginations of being a good researcher that they encounter. I focus on transdisciplinary research because it attempts to cross different kinds of boundaries (Klein 1996; Hirsch Hadorn et al. 2008)—including intra-scientific boundary transgressions between disciplines and boundary transgressions between science and other societal areas. How and which boundaries are addressed, transgressed, challenged, or maintained in researchers' identity work is an empirical question.

When I refer to transdisciplinarity in this paper, I mean research that aims to solve a societal problem by bringing together actors from different areas who are considered to contribute expertise, knowledge, or experiences that are relevant for solving that problem (compare Hirsch Hadorn et al. 2008; Stichweh 1994).[3] Here, it is important to note that transdisciplinarity, understood in this sense, aims not only to integrate diverse actors but also to abandon these collaborations again when the concrete problem that brought them together has been solved or re-defined (Gibbons et al. 1994). Actors are meant to then return to their respective domains or assemble in different constellations around other problems. This (by definition) permanently changing character of belonging is one aspect that distinguishes transdisciplinarity from emerging interdisciplinary fields which aim to build lasting communities—at least in theory (Molyneux-Hodgson and Meyer 2009). Thus, looking at transdisciplinary research will trigger new insights about 'identity work' beyond coherent and stable communities.

11.2   State of the Field—Identity beyond Scientific Disciplines

Literature that addresses inter- and transdisciplinary identities mostly highlights the inherent dilemmas, tensions, and paradoxes faced by researchers who move between disciplines and domains. This hints at the fact that identities between and beyond communities are not unproblematic and that they at least reflect the 'essential tensions' (Kuhn 1977; Hackett 2005) of research explicitly (Kerr and Lorenz-Meyer 2009; Andersen 2013; Felt et al. 2013; Woelert and Millar 2013; Darbellay 2015; Turner et al. 2015; Schikowitz 2017).

Empirical literature that analyses the identities of researchers who engage in inter- or transdisciplinary research practice focuses on how researchers position themselves in relation to and as demarcated from disciplines, and thus on the attachments and identities they develop (see Hacking 2004; Brew 2007; Granjou and Arpin 2015). In addition, they analyse challenges and tensions that go along with a position

---

[3]Earlier definitions of 'transdisciplinarity', as for example described by Nicolescu (2006), do not refer to joint problem solving but rather to a unitary scientific approach based on metaphysics and quantum physics that should go beyond disciplinary specifics (see also Klein 2014). In this chapter, I do not refer to this meaning of transdisciplinarity.

between or beyond disciplines (Turnhout et al. 2013). Studies that address how these attachments and identities develop beyond a single discipline, and how they are maintained or shift because of 'identity work', mostly focus on the process of role negotiation in inter- or transdisciplinary project teams (see Lingard et al. 2007) or on processes of socialisation in inter- and transdisciplinary education programmes (Hackett and Rhoten 2009; Felt et al. 2013). In a recent body of literature, social scientists have reflected upon their collaboration with natural scientists in the research area that considers ethical, legal, and social implications (ELSI), in post-ELSI research, and in responsible research and innovation (RRI). They have reflected on how this collaboration has challenged the identities and roles of these social science researchers (Fitzgerald et al. 2014; Balmer et al. 2015).

In this chapter, I contribute to the empirical literature on identities beyond disciplinary communities in two ways: first, I investigate not only the attachments and positioning of researchers who engage in transdisciplinary research but also the practices of identity work they do to establish, adapt, and maintain these positions. Second, I go beyond researching identity work and role allocation in one specific collaborative setting and analyse researchers' identity work across and beyond their engagement in different groups and collectives. I thus analyse not only what kinds of identities are emerging in transdisciplinary research but also what kinds of identity *work* emerge when researchers repeatedly move between different collectives with different understandings of being a good researcher. In order to analyse identity work, I develop the sensitising concept of 'choreographies'.

11.3   Approaching Identity Work beyond Disciplines as 'Choreography'

Starting from the insight that relations between collectives and individuals are becoming more fluid and dynamic, this chapter draws on concepts that do not presuppose a certain relation but make it possible to capture different kinds of relating that can be found empirically. Thus, the term 'togetherness' (based on Felt 2009; esp. Kerr and Lorenz-Meyer 2009) is used, understood, and further conceptualised as constellations of specific understandings and practices of collectivity and individuality in research (Schikowitz 2017). In this sense, collectivity and individuality are always regarded as related, mutually constitutive, and mutable.

In this chapter, the focus is on 'identity work' (Swidler 1986; Butler 1990) as one way of relating collectivity and individuality, understood as practices of interpreting, narrating, adapting, maintaining, and resisting collective imaginations of being a good researcher. The term emphasises that identity is not to be understood as something fixed. Instead, identity is continuously enacted and stabilised through practices of coping with tensions and articulating different ideals, as well as articulating these ideals with self-understandings and practical circumstances, and thus it represents an achievement. Yet, identity work is not arbitrary either. Instead, practices of identity work are embedded in and entangled with other practices and routines (methodological procedures, education, communication, dissemination habits, etc.) and institutional arrangements (research funding and evaluation, career imaginations, job descriptions, etc.) that stabilise them. Thus, the notion of identity work adopts a relational understanding of agency (Giddens 1984; Law and Mol 2008). It is about how individuals translate collective imaginations and how collective imaginations are reproduced through individuals' practices. Looking at identity work makes it possible to capture the mutual constitution of broader imaginations of what makes up a good researcher and individuals' struggles to build and maintain a satisfactory self-understanding within a certain context. Thus, identity work simultaneously constitutes collectivity and individuality in specific and related ways, which shapes togetherness in research.

Building on the insight that individual movement within and between different collectives reproduces and shapes togetherness, I propose using the sensitising concept of 'choreographies' to analyse the identity work done under conditions of multiple and flexible belongings. This understanding of choreographies is inspired by Cussins' (1998) notion of 'ontological choreography', which she introduced as a way to analyse how, in an infertility clinic, body parts that are objectified and treated separately, different technical procedures, legal and bureaucratic procedures, emotional moments, etc. retain their affiliation to a whole through coordinated spatiotemporal movements. Similarly, identity work as choreography consists of constant movement, a constant back and forth between contradictions. Thus, the notion of choreography includes performance, and it directs attention to how different belongings and ideals in different moments might be related through '*dance instead of design*' (Law 2003, p. 58). In this sense, a specific way of moving constitutes an identity in the first place by aligning otherwise separate belongings. Yet, there might be restrictions about which moves are attainable and worthwhile for whom in which situations (e.g. in different career stages, institutional positions, etc.). While some researchers might be able to dance creatively according to their own inspiration, others might feel forced to move in a specific way or avoid certain moves that appear to be risky. In this way, I understand choreography as a subtle relation of rules, routines, situated responses, and improvisation. A choreography enacts power relations as it distributes and orders possibilities of moving, foreground and background, and degrees of freedom. Simultaneously, choreography includes possibilities for resistance and variation.

Summing up, while the notion of choreography allows for an analysis of multiple aspects in relation to changing practices of togetherness in contemporary research (see Schikowitz 2017), for the purpose of the chapter at hand I focus on choreography as a way of working on and building an identity through moving within and between different belongings when engaging in transdisciplinary research that is held together by a certain style, rhythm, and pattern. Thus, choreographies might develop momentum and effects that were unintended when the individual steps and moves were designed. This makes it possible to reflect on how change in collective orders happens—often in subtle and non-linear ways.

11.4   Material and Methods

11.4.1   Case

As a case for analysing identity work beyond community, I use the major Austrian research programme on transdisciplinary sustainability research, 'proVISION'[4] (2004–2012), and the projects funded by this programme. A central funding requirement of the programme was to address problems at the interface of nature and society and to elaborate them in a transdisciplinary way, together with partners from different scientific disciplines and areas of practice, such as scientists, stakeholders, locally affected people, local administrations, NGOs, etc. However, the programme was clearly located in the scientific realm. It was operated by the Austrian Federal Ministry of Science and Research and only scientists could apply for funding and be project leaders.

In the programme description that was published on the website and in the two calls for proposals, the envisioned identity of 'transdisciplinary researchers' was discussed in detail. These researchers were ascribed the role of mediators who should achieve a balance between different perspectives and interests and integrate different ways of knowing and experiences based on a 'neutral' stance (see Schikowitz 2017). The programme description explicitly addressed the fact that there are not yet any clear norms or criteria for 'good interdisciplinary science'. That is why humility, i.e. the readiness to sacrifice one's own scientific prestige to work together on complex problems, was demanded. In contrast, the programme's evaluation criteria also included high-ranked peer-reviewed publications. Thus, the programme requirements and imaginations reflected different kinds of tensions that have been described as being challenging for identities beyond communities (see Turnhout et al. 2013)

11.4.2   Empirical Approach

The empirical material used for this paper was produced collectively in the project 'Transdisciplinarity as Culture and Practice'.[5] The data was collected within eleven

---

[4]The original website, which also published the two calls, was online until the end of the programme at http://www.provision-research.at (accessed 12 May 2013). Due to the discontinuation of the programme and a reorganisation of the responsibilities of the Austrian ministries, the website was shortened and relocated to https://bmbwf.gv.at/forschung/national/programme-schwerpunkte/provision (accessed 30 December 2018).

[5]The project was conducted at the Department of Science and Technology Studies (University of Vienna) from 2009 to 2013. It was funded by the Austrian Federal Ministry for Science and Research. The project leader was Ulrike Felt, and project collaborators were Judith Igelsböck, Andrea Schikowitz, and Thomas Völker. See https://sts.univie.ac.at/en/research/completed-research-projects/transdisciplinarity-as-culture-and-practice (accessed 30 December 2018).

of the projects that were funded by the programme proVISION and in one doctoral training programme that was co-funded by the research programme. The empirical material for this article mainly consists of 28 semi-structured interviews, two focus groups, and observation protocols of eight meetings held in three different projects. As contextual information, I looked at researchers' (self-)presentations on their websites, CVs, and publication lists.

The researchers had different positions in the projects (project leaders and collaborators), were at different career stages, had different institutional affiliations (specifically employed for the project, faculty positions at a university, employed in private research institutions, etc.), and came from different disciplinary backgrounds (mostly from natural sciences that identify themselves as interdisciplinary, such as ecology, but also from social sciences and humanities such as economics, sociology, history, or theology). For the analysis, I oriented on grounded theory in its constructivist version (Clarke 2005; Charmaz 2006). I used the notion of choreography as a sensitising concept to analyse how to align diverging belongings, but I adapted this notion during the analysis according to the further empirical insights.[6] Analysing the accounts of this diverse set of informants indicated that researchers with different backgrounds and in different positions move in different ways and that they experience different degrees of freedom in their choreographies.

11.4.3 Analytical Strategy and Presentation of Findings

My initial interest in identity work emerged out of researchers' utterances in the material. In the interviews, many of them explicitly addressed the question of 'how to be good' in transdisciplinary research because they experienced diverging ideals and expectations. Researchers talked about their self-understanding and if and how it is congruent with different identities in transdisciplinary research and beyond. I coded the interviews focussing on accounts of what it means to be a good researcher in transdisciplinary research and in other research areas, what tensions researchers encounter between these areas, and how they deal with these tensions.

From the analysis of the researchers' accounts, four different choreographies emerged that differ in terms of how researchers move within and between different belongings and commitments (see Sect. 11.5). I aggregated researchers' utterances in the material into these four distinct choreographies that cut across individuals' stories. Thus, an individual researcher potentially enacted parts of different choreographies. Yet, most of the interviewees drew predominantly on one of them. The notion of choreography also drew my attention to three aspects of identity work: the specific 'tone' or 'style' (the overall emotional expression, for example, if a dance appears rather cheerful or grave, etc.), the rhythm of identity work (the time-structure

---

[6]This chapter is based on the analysis conducted in the course of my dissertation project (Schikowitz 2017). Here I focus on a specific question and develop it further.

as a combination of speed and duration of moves and steps, for example, an emphasis on continuity or change of pace), and the pattern or scope of movements (if the dancer stays in the foreground or background, or on which parts of the stage s/he moves). In this way, analysing identity work as choreography also made it possible to reflect on the collective orderings that frame and shape which moves are attainable for whom (see Sect. 11.6).

11.5 Findings—Identity Work as Choreography

11.5.1 Being an 'Explorer'—Undertaking Temporary Trips into Unknown Territory

One way of being a 'good researcher' in transdisciplinary research that comes up in the empirical material is by undertaking temporary trips into unknown territories. Such unknown territories become accessible by engaging in transdisciplinary research projects and returning 'home' in-between these trips. The notion of 'explorer' captures how this choreography resembles the journeys of explorers from the fifteenth to the eighteenth century. What these explorers did in unknown territories was collect and bring home valuable specimens and goods to study them within their own thought system (see Pickstone 2009). In a similar way, a choreography of exploring involves moving into inter- and transdisciplinary engagements, gaining inspiration or new data resources there (which includes data produced by extra-scientific actors, such as NGOs or administrative bodies), bringing this back into a researcher's usual disciplinary thought system, and processing it there according to her usual working methods. Thus, 'explorers' move mostly within their disciplinary field but occasionally, driven by curiosity, they temporarily undertake trips into foreign areas.

In interviews, researchers (both senior and early stage) who enacted a choreography of exploring seemed very relaxed and satisfied with their work. They conveyed the impression of feeling comfortable in the interview situation—sitting casually, using colloquial language, and making jokes. They were generally very open towards us 'outsiders' and willingly shared all kinds of information—they would, for example, put us on the internal project mailing list, invite us to all their project meetings and other events, send us the slides of their presentations, and talk to us about their work practices, obstacles, and setbacks in detail. Researchers enacting an explorer choreography stemmed mostly from the natural sciences, covering different career stages and positions, from tenured professors to predoc project collaborators. They exhibited a stable disciplinary affiliation accompanied by a simultaneous attraction to working in very different areas, collaborating with proponents of different disciplines and professions, a supplement to their disciplinary careers they would refer to as *'the salt in the soup'*.[7] While often foregrounding the interdisciplinary nature of their home disciplines and their scientific approaches, they did not see themselves as transdisciplinary researchers but as *'not too far away'*, as researchers who occasionally engaged in transdisciplinary collaborations.

Thus, the relaxed attitude of 'explorers' might be related to the fact that they hardly experience tensions with their disciplinary identities when engaging in transdisciplinary research. For them, being a good researcher means to follow their scientific curiosity, even if that means engaging and exchanging data with extra-scientific actors. Cooperating with new actors in this way does not challenge their disciplinary research practices; rather it expands the scope of the empirical field into new areas and makes new bodies of data available to them. In this way, the choreography provides both security and freedom for researchers who enact it.

11.5.2 Being a 'Caring Broker'—Creating Ad-Hoc Social Bonds while Keeping Epistemic Independence

Another way of creating a researcher identity through moving between different belongings consists of engaging in ad-hoc social bonds that can later be turned into professional cooperation while leaving the epistemic independence of the respective partners intact. Here, the moves and relations are not framed as motivated by a strategic aim, a moral goal, or by a common epistemic interest. Instead, a 'caring broker' enacts a social motivation (*'The first contact came because [at this event] I knew no one, and then someone was nice and I talked to him for a while, and that is actually how that [cooperation] emerged'*). A 'caring broker' enlarges his or her scope through initial social contacts with single actors that later open up larger territories to move into. Thereby, 'caring brokers' do not exploit other scientific or societal actors, but they take care that each and every member of the project team yields outcomes that are respectively relevant for them. In this sense, they do not only broker between actors, but they do so in a caring way—trying to understand the individual needs and balancing them within a project. They often act as facilitators who organise knowledge exchanges between other actors (compare Meyer 2010). Accordingly, researchers who enact this choreography are often in a coordinating position—they are project coordinators, heads of departments, or they are in charge of research platforms or groups.

In the interviews, this choreography is often covered by small anecdotes that start with (sometimes incidentally) emerging social relations, such as getting into a nice

---

[7]If not marked otherwise, all quotations stem from interviews with researchers. The original language of the interviews is German. Quotations were translated by the author. To prevent endangering the anonymity of the informants when providing longer descriptions of identities and contextual information, it is not specified which quotations are from which interview.

conversation at an event, establishing an informal get-together within a project team, or getting to know the members of an organisation during a longer research stay on a professional as well as personal basis. 'Caring brokers' frame these relations as emerging, not as strategically planned. Further on in the interviews, 'caring brokers' explained how (sometimes even after the end of a project) these relations later had unexpected beneficial effects that came as a pleasant surprise. Examples of these effects included being invited to follow-up projects with former clients or partners or the promotion of project findings in different contexts. Thus, the choreography of caring does not produce short-term outputs but rather indirect and long-term effects. To describe these indirect effects, a researcher used the metaphor of a stone thrown into water, which causes waves that can be observed at a distant place, or of yeast resting under a cover, fermenting out of sight, and suddenly causing surprising effects. Thus, moving within and between engagements is explained as emerging from social relations of mutual care and as developing according to its own dynamics.

While the choreography of caring links different actors in common projects, the respective epistemic enterprises remain clearly separated. Each scientific and extra-scientific partner gets what is respectively valued in their worlds, and there is no aspiration to blur epistemic boundaries—relevant research questions, methods, and outcomes are defined by scientific disciplines or areas of practice; what gets exchanged is data and readymade findings. 'Caring brokers' argue that every partner is '*top-quality in his [sic] area*', and including others in genuine research would diminish this quality. In this way, a 'caring broker' builds a network of actors who trust him or her in the sense that if s/he asks them to collaborate in another project they would agree because they are convinced that s/he would guard their interests and that it would be beneficial for them epistemically, allowing them to '*still [do] hardcore research*'.

To create a common transdisciplinary project, 'caring brokers' relate these separated disciplinary areas through an overarching issue and story line ('*We put the mortar between the bricks ... but the bricks were disciplinarily focused*'). They take care that the single parts fit into this overall frame and they act as the link to the funding agency. The practices of 'caring brokers' involve coordination and emotional work. As a way of '*getting them on board*', 'caring brokers' create social relations within the project team by making it a comfortable and interesting space for social exchange—'*We looked forward to the meetings and we had a party at the end*'. Things like nice locations, food and drink, and a nice atmosphere are often mentioned. The social relations create loyalty and trust towards the project as a whole and towards the 'caring broker'.

11.5.3 Being a 'Moral Manager'—Heading for a 'Greater Good'

Yet another choreography that makes it possible to develop a researcher identity in transdisciplinary research is to align different activities and belongings by directing them to the 'greater good' of sustainable development. All the individual steps and moves in this choreography are heading for this 'greater good'.

The identity that emerges through this choreography can be labelled as being a 'moral manager' because the main motivation is an explicitly normative one (to foster sustainability rather than to produce knowledge as a goal in itself). As one researcher explains: '*Sustainability is based on values that drive us, or me, you likewise. So that . . . makes [the research] meaningful*'. Management techniques are applied in order to reach the goal of sustainability, coordinate different activities, and organise the projects efficiently. Management techniques—in the sense of outcome-oriented steering and detailed plans, or even contracts with project partners—are meant to ensure that no one loses sight of the project goal or spends too much time on activities that do not produce immediate outcomes. For example, as one researcher explains: '*I mean, it is rather a huge challenge or restriction that due to this openness you never know if the loops you spend your resources on actually lead to success . . . And we didn't have these resources—that causes stress*'. This means that the choreography of heading for a 'greater good' includes controlling and resisting all moves that might distract from this ultimate goal.

In the interviews, as well as at public events, researchers who enacted this choreography spoke in an earnest way and were aware of the importance of their message. The overall tone of this choreography was alarmed and somewhat defensive—different interests, scarce resources, and inefficient processes appeared as distractions from the goal of working towards sustainability. For example, when we asked one project leader if we could observe a project meeting, this was rejected because the project leader was afraid that we might disturb the research process. 'Moral managers' also modified their own behaviour, sacrificing individual benefit for the sake of the overall goal of sustainability. Personally, they abstained from unsustainable behaviour. In their research, they avoided or restricted activities that count only within science but seem to have no immediate practical effect on the ability to foster sustainability, such as producing specialised publications for a scientific audience. They described this as '*staying pragmatic*'.

While the overall scope and direction of this choreography is rather targeted and straight, its rhythm is unsettled and wavering. 'Moral managers' emphasised that to keep everything in line, they need specific qualities, skills, and competences, such as coordination, multitasking skills, and '*juggling*'. While never losing sight of the 'greater good', 'moral managers' also seemed to enjoy the more playful moves that allow them to introduce their own creativity and skilful performance. The requisite qualities were described as a certain readiness to get involved, risk-affinity, or courage. Thus, going beyond a discipline was framed not only as a question of cognitive capacity but primarily as a question of moral strength and courageousness.

Overall, moral qualities were emphasised as a crucial part of a *'transdisciplinary researcher personality'*, going hand-in-hand with the willingness and ability to acquire knowledge from several areas. This was also reflected in the professional career path of many 'moral managers'. They often did not spend their whole education and professional life in one discipline. Furthermore, in many cases they had had some additional training or previous experience in areas like project management or science communication. Here it is interesting that the willingness (or a *'serious want'*[8]) to abandon a disciplinary home, not only temporarily but permanently, was often explicitly mentioned as necessary for implementing trans-disciplinary research in a meaningful way. Such mental strength was needed because not belonging to a specific community means that the relevant communities need to be identified and assembled anew for each and every project. When asked about the relevant communities for their research in the interviews, 'moral managers' sounded rather frustrated: *'Yep, that's the question for every transdisciplinary project'*. This signifies extra effort vis-à-vis disciplinary researchers who stay within one community in which they move more intuitively.

11.5.4 Being a 'Polymath'—Integrating Encounters with Others into One's Own Life Story

When researchers who enact the choreography of a 'polymath' talk about what it means to be a good researcher, it feels a bit anachronistic and nostalgic. They evoke the idea of a researcher who is directed by her engagement with a certain phenomenon rather than steered by a strict project plan. Here, the researcher appears not as a representative of a certain discipline nor as 'broker' who facilitates encounters with or for other actors but as a 'polymath' who moves back and forth between different communities and belongings and who integrates the different experiences and encounters within her personality and her personal body of knowledge. This choreography puts the person of the researcher in the centre, and developing personally—intellectually, emotionally, and morally—is its main imperative. The main motivation is not, in the first place, the *production* of knowledge as a commodity or as a means to reach a 'greater good' but rather *understanding* in a more comprehensive, personal, and emotional sense—*'experienced through one's own brain and through one's own emotions'*. This is also what distinguishes the choreography of a life story from the explorer choreography that focuses on objective facts that would reflect knowledge about the world, independently of the person of the researcher.

To develop a deep and personal understanding of a certain phenomenon, researchers reported that it is necessary to engage with others on a personal basis, and that transdisciplinary research allows for such encounters. Encounters should be characterised by openness, mutual interest, and respect, and they trigger reflection,

---

[8]This quotation was from a focus group with researchers of different projects.

creativity, and the development of new perspectives on both sides. Thus, researchers enacting this choreography generally welcomed our interview requests as they see reflection as a crucial part of their research, and they also regarded the interviews we conducted with them as an opportunity for such reflection. For example, we were asked for the transcript of our interview for 'internal reflection'. The rhythm that goes along with this choreography is that of a biographical narrative, of working on different experiences and encounters through reflection and of creating a coherent and holistic story. Thereby, the continuous process of learning is what matters rather than single episodes or specific outcomes. Togetherness appears in the form of open reflection and mutual learning—however, it is restricted to the individual persons involved in such encounters and to specific moments.

The choreography of a life story has a rather epic scope; it does not restrict itself to specific kinds of problems or to a specific research area but includes universal and interconnected claims. In this way, transdisciplinarity is seen not only as an approach for handling specific problems but as a better way of doing science (epistemically and morally). The choreography follows a unitary understanding of science, also evoking the earlier meaning of transdisciplinarity as a unitary scientific approach (Nicolescu 2006; Klein 2014) that demands a holistic researcher identity (Giri 2002). It frames transdisciplinarity as an attainable standard for scientific research more generally and something that should be institutionalised. Thus, a 'polymath' does not feel committed to her discipline but rather follows an overarching view. She experiences institutional obstacles to such an overarching approach as illegitimate and seeks to overcome them by promoting transdisciplinarity. One researcher even stated that the disciplinary organisation of science would be '*an absurdity*' and '*naturally completely unproductive and utterly, utterly stupid*'.

11.6   Collective Ordering

Building on the analysis of four different choreographies, I reflect in this section on how identity work as choreography impinges on collective orders and how collective ordering shapes the conditions of possibility for individual moves. Overall, the analysis suggests two broader collective ways of ordering togetherness: first, researchers who do not question their disciplinary belonging use transdisciplinary research as a temporary extension that provides resources to inspire and enrich their disciplinary research ('explorers' and 'caring brokers'). They engage in practices of networking and 'trade' but do not blur epistemic boundaries. Second, researchers who try to position themselves and their research as genuinely transdisciplinary rely on improvised and more personal choreographies ('moral mangers' and 'polymaths'). They cannot draw on a collective set of stories about how to be a 'good researcher' but need to invent themselves permanently anew. Thus, they engage in community-building and institutionalisation. They try to establish and stabilise a transdisciplinary community that could provide legitimacy for their own way of being in research.

11.6.1 Transgressing but Maintaining Boundaries through Trade

The choreographies of undertaking trips into unknown territories and of creating social bonds are both firmly anchored in disciplinary belongings and their epistemic standards and practices of how to be a good researcher. They allow for togetherness as 'trade' (in the sense of Galison 1996). Thus, they can often be found within the same projects. Trade means that while established boundaries are maintained, single actors can temporarily cross them, exchange goods, and subsequently integrate these goods into their own value system. Thus, practices of trade work to build networks rather than communities. Engaging in trade allows researchers to break out of their densely structured and institutionalised research environments without abandoning the security that their disciplinary belongings provide. Trade makes it possible to keep disciplinary knowledge production within interdisciplinary settings separate, to a large extent. Relations are developed via clearly defined interfaces—encounters serve to transfer knowledge as 'input' and 'output'. The value of transdisciplinary research would be to link knowledge about single aspects of a given problem in a systematic way, or to transform them into a comparable and linkable scale.

Organising togetherness as 'trade' provides security and orientation for researchers who move between different collectives. In projects that enact togetherness as trade, most members (researchers as well as stakeholders) expressed in the interviews that they perceive the introduction of predefined procedures—rather than negotiating relations during the process—as relieving. As a result of practices of 'trading', existing boundaries are often reinforced or left untouched. Researchers who engage in 'trade' are mostly from the natural sciences or quantitative social sciences. They perceive a division of (epistemic) labour through the input and output of separately produced data to be unproblematic. Yet, those who do not share a positivist understanding of knowledge as reflecting parts of one common reality, but who see knowledge and knowing as situated within the context of its production, face problems engaging in the collective order of trade. This is the case for qualitative interpretive or constructivist knowledge as well as for embodied knowledge.

11.6.2 Establishing New Boundaries through Attempts at Building a Transdisciplinary Community

Analysing different choreographies shows that staying within established epistemic boundaries seems rather safe, while trying to blur them requires personal sacrifices or personal engagement without knowing if this will work out in the long run (see also Blättel-Mink and Kastenholz 2005). The latter is the case mostly for researchers who see themselves as transdisciplinary researchers, even when they are senior researchers in rather secure positions. It is even more pressing for early stage researchers who often describe the feeling of 'belonging nowhere' in a very personal way, as exemplified in the following quotation from a student in a transdisciplinary doctoral college:

> My disciplinary home has gotten a little lost now, because it is not X any more—and even before it was only with some reservation—and, so, I can use a lot of it, I can really use very, very much of it, but it is not the case that I would be at home in this community then.

Here, we encounter a distinction between drawing on knowledge and approaches from different disciplines and a feeling of belonging. Also, other researchers use a very emotional and personal language when narrating this impression, talking of '*not feeling at home*' or '*being an outsider*'. The metaphor of the home reflects the desire to belong and be accepted as an insider, a desire hardly fulfilled in transdisciplinary research. '*Belonging everywhere and belonging nowhere*' are described as two sides of the same coin. The identity work of 'transdisciplinary researchers' is thus not so much about negotiating their self-understanding vis-à-vis an ideal but also about *creating an ideal* that would prove stable across different belongings and engagements. In many cases, researchers locate this overarching ideal on a moral level, while they feel the need to vary epistemic standards and practices from project to project. Thus, while it seems to be quite clear for them how to be morally good in transdisciplinary sustainability research, on an epistemic level, what it means to be a good researcher must perpetually be established anew (compare also Felt et al. 2013).

Moreover, in the choreography of heading for a 'greater good', tensions with striving for good epistemic work in a traditional disciplinary sense can arise. This goes along with insecurity and precarity, since intra-scientific assessment criteria and the attribution of worth, as well as legitimacy and credibility in the policy domain, are often related to epistemic aspects (compare Turnhout et al. 2013). In addition, the choreography of a life story, where a personal development process provides the connective ideal, falls short of meeting standardised assessment criteria. 'Being a polymath' is thus only attainable for already-established researchers with a fixed institutional position, while their younger colleagues describe it as '*completely naïve*' within the current science system.

Several researchers who position themselves as transdisciplinary researchers express the desire for a common epistemic framework that would provide them with some common understandings about legitimate questions, methods, conceptual approaches, and procedures, so that these do not need to be negotiated and justified every time. In this way, they take into account or even wish to normalise in the sense of building a '*standard routine*', '*certain guidelines, where to look*', and '*the agreement on a structured process*' that would be '*basically . . . the core of such a thought collective that should develop*'. This leads to the slightly paradoxical situation where something that, by definition, should always be assembled anew ends up being established in a more permanent manner. Such tendencies of institutionalisation are hence seen by some researchers as actually contradicting the transdisciplinary claim for inclusiveness (this is also discussed in the literature; see Blättel-Mink and Kastenholz 2005; Jahn et al. 2012; Turner et al. 2015). An early stage researcher engaged in one of the projects describes this exclusive nature of community-building by stating that '*community actually already demarcates vis-à-vis others ... so that is more or less an in-group*'. In this way, practices of community-building are necessarily exclusive and 'disciplining'. Institutionalisation and flexibility are adjured simultaneously—the exclusiveness and rigidity of disciplines is criticised as being inadequate for overarching and complex problems, but at the same time researchers strive for the stability and orientation that such institutionalised structures would provide. As a result, institutionalisation and de-institutionalisation of transdisciplinarity, building and re-building communities, stays in a permanently provisional state.

11.7 Conclusions

My concern in this contribution has been with transdisciplinarity, which makes claims about identity work and belonging beyond disciplinary communities. Rather than building a new community, researchers are urged to abandon all long-term belongings. This goes along with specific kinds of tensions that differ from the tensions that occur in emerging interdisciplinary fields such as nanoscience and synthetic biology, where providing new identities and building new communities is an explicit goal (Molyneux-Hodgson and Meyer 2009). The findings presented here suggest that community *still* matters as a basis for providing an understanding of how to be a proper scientist in terms of practices, virtues, and motivations; thus identity work beyond community is a paradoxical undertaking.

I analysed four different choreographies that go beyond single projects or belongings and that allow for being a good researcher in transdisciplinary research. 'Explorers' follow their epistemic curiosity, which is firmly based in their home discipline but which leads them to undertake trips into unknown territories from time to time where they engage and exchange with different actors. 'Caring brokers' align their different encounters and belongings through social bonds. They cooperate in common projects where they work on the same topic or exchange data without intermingling their disciplinary epistemic practices. 'Moral managers' are heading for a 'greater good', a goal that unites all their different engagements and belongings, and they are ready to sacrifice personal epistemic interests and disciplinary success to that goal. 'Polymaths' pursue the path of a life story, translating all kinds of encounters and belongings into their personal development and embodied knowledge—which proves to be only attainable for researchers who are already in an institutionally secure position.

These choreographies of identity work go along with two broader collective ways of ordering togetherness. 'Explorers' and 'caring brokers' are engaging in networking and trade (Galison 1996), which maintains existing boundaries but allows for crossing them in predefined ways. 'Moral managers' and 'polymaths' engage in attempts of community-building and institutionalisation of transdisciplinarity, which stays an ongoing struggle as they simultaneously abandon all long-term belongings.

These findings suggest that togetherness or communality in transdisciplinary research does not necessarily mean membership in a community. Togetherness also appears as engagement in ever changing strategic and personal networks (that does not, however, challenge belonging to the home discipline) or as (ongoing) engagement in the paradoxical undertaking of building community while abandoning belonging to a community. Both kinds of togetherness include ongoing movement and oscillation between groupings beyond community but which simultaneously gravitates around community or the desire for a community. Identity work in situations of multiple belongings only works out for researchers when they come from a 'home base', from a stable community that they can return to and refer to (compare Hacking 2004). This holds true across career stages and institutional positions. Thus, there is a tension between stability and openness. Being open towards others seems to be satisfactory only when one is part of a stable and necessarily closed community.

Against this backdrop, I return back to the initial question whether it is possible to imagine transdisciplinary choreographies of identity work that would allow for a blurring of epistemic boundaries between different disciplines and between science and society in a more radical way, transcending practices of trade (that 'explorers' and 'caring brokers' enact) but without self-denial or precarious working circumstances (as is the case for 'moral managers' and 'polymaths' who abandon disciplinary belongings). The findings suggest that this would require collective coping strategies beyond single programmes or projects (compare Schikowitz 2017). In order to develop such collective coping strategies, policy would need to reflect upon what the intended forms and practices of togetherness actually are, what kinds of individuality specific forms of collectivity would imply, and vice versa.

Existing literature that develops concepts and imaginations on what 'good research' in both epistemic and moral terms could mean in a transdisciplinary frame—such as Funtowics and Ravetz (1993) on 'post-normal science' or Pohl (2005) and Hollaender et al. (2008) as well as Pohl and Hirsch Hadorn (2008) on 'transdisciplinarity'—does not, however, provide imaginations of how 'good transdisciplinary research' could be aligned with 'good disciplinary research' when researchers move between different belongings. This literature focuses on transdisciplinary research projects and mostly identifies methodological issues or insufficient personal or institutional commitment as hindering transdisciplinarity from fulfilling its promises. Yet, the analysis at hand and its focus on choreographies that go across and beyond different projects and work contexts suggests that beyond single projects, individual researchers (even if they are committed to the idea of transdisciplinarity) are not able to both satisfy the normative claims of transdiscipinarity and the epistemic requirements of disciplinary research. Thus, without cultural change within science and science policy, transdisciplinary researchers still risk falling short in terms of disciplinary assessment systems (compare Turnhout et al. 2013). A broader debate is needed on how incremental change can be valued beyond measurable and countable output (compare Fochler et al. 2016)—a debate that needs to include not only transdisciplinary research but also disciplinary research. This could involve understanding transdisciplinarity as an opportunity to imagine and probe desirable and attainable research futures that go beyond the dominant research regime based on quantification (compare, for example, Klenk and Meehan 2015).

The analytical utility of the notion of choreography beyond the specific case of transdisciplinary research seems promising. While the specific choreographies and the identities that go along with them might be different in different contexts, dealing with contradictions between different belongings can be analysed through the lens of choreographies on a general basis. The notion may well provide a sensitising concept for research in all fields that feature change, tensions, and contradictions, such as emerging interdisciplinary fields.

Acknowledgements First of all, I owe thanks to the project team—Ulrike Felt (lead), Judith Igelsböck and Thomas Völker, with whom I collaboratively produced the data this article is based on—and to Ulrike Felt as the supervisor of my PhD project that is the basis for the analysis conducted in this paper. I also want to thank Maximilian Fochler, Claudia Schwarz-Plaschg, and Lisa Sigl for providing feedback on earlier versions of this paper. I am very grateful to the guest-editors of this volume and to the two anonymous reviewers who engaged thoroughly and productively with my text and thus helped to enhance it considerably. Not least, I thank all the researchers, practitioners and programme representatives of the proVISION programme, who took the time to share their experiences and reflections.

References

Andersen, H. 2013. The second essential tension: On tradition and innovation in interdisciplinary research. *Topoi* 32 (1): 3–8.

Balmer, A.S., J. Calvert, C. Marris, S. Molyneux-Hodgson, E. Frow, M. Kearnes, K. Bulpin, P. Schyfter, A. MacKenzie, and P. Martin. 2015. Taking roles in interdisciplinary collaborations. *Science and Technology Studies* 28 (3): 3–25.

Becher, T. 1989. *Academic tribes and territories: Intellectual enquiry and the cultures of disciplines*. Milton Keynes: Open University Press.

Blättel-Mink, B., and H. Kastenholz. 2005. Transdisciplinarity in sustainability research: Diffusion conditions of an institutional innovation. *International Journal of Sustainable Development & World Ecology* 12 (1): 1–12.

Brew, A. 2007. Disciplinary and interdisciplinary affiliations of experienced researchers. *Higher Education* 56 (4): 423–438.

Butler, J. 1990. *Gender trouble: Feminism and the subversion of identity*. New York: Routledge.

Cerulo, K.A. 1997. Identity construction: New issues, new directions. *Annual Review of Sociology* 23: 385–409.

Charmaz, K. 2006. *Constructing grounded theory. A practical guide through qualitative analysis*. London/Thousand Oaks/New Delhi: Sage.

Clarke, A.E. 2005. *Situational analysis. Grounded theory after the postmodern turn*. Thousand Oaks: Sage.

Cussins, C.M. 1998. Ontological choreography: Agency for women patients in an infertility clinic. In *Differences in medicine. Untraveling practices, techniques and bodies*, ed. M. Berg and A. Mol, 166–201. Durham/London: Duke University Press.

Darbellay, F. 2015. Rethinking inter- and transdisciplinarity: Undisciplined knowledge and the emergence of a new thought style. *Futures* 65: 163–174.

Daston, L., and P. Galison. 2007. *Objectivity*. New York: Zone Books.

Felt, U., ed. 2009. *Knowing and living in academic research. Convergence and heterogeneity in research cultures in the European context*. Prague: Institute of Sociology of the Academy of Sciences of the Czech Republic.

Felt, U., and M. Fochler. 2012. Re-ordering epistemic living spaces: On the tacit governance effects of the public communication of science. In *The sciences' media connection—Public communication and its repercussions*, Sociology of the sciences yearbook, ed. S. Rödder, M. Franzen, and P. Weingart, vol. 28, 133–154. Bielefeld: Springer.

Felt, U., J. Igelsböck, A. Schikowitz, and T. Völker. 2013. Growing into what?' The (un-) disciplined socialisation of early stage researchers in transdisciplinary research. *Higher Education* 65 (4): 511–524.

Fitzgerald, D., M.M. Littlefield, K.J. Knudsen, J. Tonks, and M.J. Dietz. 2014. Ambivalence, equivocation and the politics of experimental knowledge: A transdisciplinary neuroscience encounter. *Social Studies of Science* 44 (5): 701–721.

Fochler, M., U. Felt, and R. Müller. 2016. Unsustainable growth, hypercompetition, and worth in life science research. Narrowing evaluative repertoires in doctoral and postdoctoral scientists' work and lives. *Minerva* 54 (2): 175–200.

Funtowics, S.O., and J. Ravetz. 1993. Science for the post-normal age. *Futures* 25 (7): 739–757.

Galison, P. 1996. Computer simulations and the trading zone. In *The disunity of science: Boundaries, contexts, and power*, ed. P. Galison and D.J. Stump, 119–157. Stanford: Stanford University Press.

Gibbons, M., C. Limoges, H. Nowotny, S. Schwartzman, P. Scott, and M. Trow. 1994. *New production of knowledge: Dynamics of science and research in contemporary societies*. London/Thousand Oaks/New Delhi: SAGE Publications.

Giddens, A. 1984. *The constitution of society. Outline of the theory of structuration*. London: Polity Press.

Giri, A.K. 2002. The calling of a creative transdisciplinarity. *Futures* 34 (1): 103–115.

Granjou, C., and I. Arpin. 2015. Epistemic commitments: Making relevant science in biodiversity studies. *Science, Technology & Human Values* 40 (6): 1022–1046.

Hackett, E.J. 2005. Essential tensions: Identity, control, and risk in research. *Social Studies of Science* 35 (5): 787–826.

Hackett, E., and D. Rhoten. 2009. The Snowbird charrette: Integrative interdisciplinary collaboration in environmental research design. *Minerva* 47 (4): 407–440.

Hackett, E.J., J.N. Parker, N. Vermeulen, and B. Penders. 2017. The social and epistemic organization of scientific work. In *The handbook of science and technology studies*, ed. U. Felt, R. Fouché, C.A. Miller, and L. Smith-Doerr, 4th ed., 733–764. Cambridge: MIT Press.

Hacking, I. 2004. *The complacent disciplinarian*. Retrieved from https://sciencestudiesintro.files. wordpress.com/2015/02/hacking-complacent-disciplinarian.pdf.

Henkel, M. 2005. Academic identity and autonomy in a changing policy environment. *Higher Education* 49 (1): 155–176.

Hirsch Hadorn, G., H. Hoffmann-Riem, S. Biber-Klemm, W. Grossenbacher-Mansuy, D. Joye, C. Pohl, U. Wiesmann, and E. Zemp, eds. 2008. *Handbook of transdisciplinary research*. Bern: Springer.

Hollaender, K., M.C. Loibl, and A. Wilts. 2008. Management. In *Handbook of transdisciplinary research*, ed. G. Hirsch Hadorn et al. Bern: Springer.

Jahn, T., M. Bergmann, and F. Keil. 2012. Transdisciplinarity: Between mainstreaming and marginalization. *Ecological Economics* 79 (0): 1–10.

Kerr, A., and D. Lorenz-Meyer. 2009. Working together apart. In *Knowing and living in academic research. Convergence and heterogeneity in research cultures in the European context*, ed. U. Felt, 127–167. Prague: Institute of Sociology of the Academy of Sciences of the Czech Republic.

Klein, J.T. 1996. *Crossing boundaries. Knowledge, disciplinarities and interdisciplinarities*. Virginia: University Press of Virginia.

———. 2014. Discourses of transdisciplinarity: Looking back to the future. *Futures* 63: 68–74.

Klenk, N.L., and K. Meehan. 2015. Climate change and transdisciplinary science: Problematizing the integration imperative. *Environmental Science & Policy* 54: 160–167.

Knorr Cetina, K. 1999. *Epistemic cultures. How the sciences make knowledge.* Cambridge, MA/-London: Harvard University Press.

Kuhn, T.S. 1962. *The structure of scientific revolutions.* Chicago: University of Chicago Press.

———. 1977. *The essential tension: Selected studies in scientific tradition and change.* Chicago: University of Chicago Press.

Law, J. 2003. *Traduction/Trahison: Notes on ANT.* Lancaster: Lancaster University. Retrieved from http://www.lancaster.ac.uk/fass/resources/sociology-online-papers/papers/law-traduction-trahison.pdf.

Law, J., and A. Mol. 2008. The actor-enacted: Cumbrian sheep in 2001. In *Material agency*, ed. C. Knappett and L. Malafouris, 57–77. Lancaster: Springer.

Lingard, L., C.F. Schryer, M.M. Spafford, and S.L. Campbell. 2007. Negotiating the politics of identity in an interdisciplinary research team. *Qualitative Research* 7 (4): 501.

Meyer, M. 2010. The rise of the knowledge broker. *Science Communication* 32 (1): 118–127.

Molyneux-Hodgson, S., and M. Meyer. 2009. Tales of emergence—Synthetic biology as a scientific community in the making. *BioSocieties* 4 (1): 129–145.

Nicolescu, B. 2006. Transdisciplinarity: Past, present and future. In *Moving worldviews: Reshaping sciences, policies and practices for endogenous sustainable development*, ed. B. Haverkort and C. Reijntjes, 142–166. Leusden: ETC/COMPAS.

Pickstone, J.V. 2009. Ways of knowing: Towards a historical sociology of science, technology and medicine. *The British Journal for the History of Science* 26 (4): 433–458.

Pohl, C. 2005. Transdisciplinary collaboration in environmental research. *Futures* 37 (10): 1159–1178.

Pohl, C., and G. Hirsch Hadorn. 2008. Methodological challenges of transdisciplinary research. *Natures Sciences Sociétés* 16: 111–121.

Schikowitz, A. 2017. *Choreographies of togetherness. Re-ordering collectivity and individuality in transdisciplinary sustainability research in Austria (Doctoral thesis).* Vienna: Department of Science and Technology Studies/University of Vienna.

Shapin, S. 2008. *The scientific life. A moral history of a late modern vocation.* Chicago: University of Chicago Press.

Stichweh, R. 1994. *Wissenschaft, Universität, Professionen. Soziologische Analysen.* Frankfurt am Main: Suhrkamp.

———. 2008. The sociology of scientific disciplines: On the genesis and stability of the disciplinary structure of modern science. *Science in Context* 5 (1): 3–15.

Swidler, A. 1986. Culture in action: Symbols and strategies. *American Sociological Review* 51 (2): 273–286.

Turner, V.K., K. Benessaiah, S. Warren, and D. Iwaniec. 2015. Essential tensions in interdisciplinary scholarship: Navigating challenges in affect, epistemologies, and structure in environment–society research centers. *Higher Education* 70 (4): 649–665.

Turnhout, E., M. Stuiver, J. Klostermann, B. Harms, and C. Leeuwis. 2013. New roles of science in society: Different repertoires of knowledge brokering. *Science and Public Policy* 40 (3): 354–365.

Woelert, P., and V. Millar. 2013. The 'paradox of interdisciplinarity' in Australian research governance. *Higher Education* 66 (6): 755–767.

Ylijoki, O.H. 2000. Disciplinary cultures and the moral order of studying—A case-study of four Finnish university departments. *Higher Education* 39 (3): 339–362.

Chapter 12
Constructing (Inter)Disciplinary Identities: Biographical Narrative and the Reproduction of Academic Selves and Communities

Carlos Cuevas-Garcia

12.1 Introduction

Interdisciplinarity has become so prominent in academia and science policy that scholars have started paying attention to its implications for the development of academic careers and individual and collective identities (OECD 1972; RCUK 2015; European Commission 2011; National Academy of Sciences 2005; Holley 2015; Calvert 2011; Bartlett et al. 2016; Leahey et al. 2017; Klein and Falk-Krzesinski 2017). There are studies that point to a number of concerning findings—for example, that interdisciplinary work is perceived as work of low quality; that it involves personal, emotional, and intellectual challenges because interdisciplinary researchers may not be perceived as experts, and because they may not fulfil the expectations of their disciplinary peers (Buanes and Jentoft 2009; Pfirman and Martin 2010). Interdisciplinarity has also been described as a professional challenge because it does not offer a 'clear career track from an undergraduate major to professorial appointment' as traditional disciplines do (Turner 2000, p. 60; Abbott 2001). Although interdisciplinarity has gained popularity, Henkel (2005) argues that having a disciplinary identity has remained relevant because in a competitive environment such as academia, having a track record in a single discipline increases individuals' chances of attracting funding, accumulating rewards, and obtaining a permanent position. Despite the existence of inter- or trans- disciplinary doctoral programmes, Felt et al. (2013) argue that these are 'not necessarily perceived as being particularly supportive when aiming for an academic career' (p. 520). In their introduction to this volume, Kastenhofer and Molyneux-Hodgson summarise

C. Cuevas-Garcia (✉)
Munich Center for Technology and Society, Technical University of Munich, Munich, Germany
e-mail: carlos.cuevas@tum.de

© The Author(s) 2021
K. Kastenhofer, S. Molyneux-Hodgson (eds.), *Community and Identity in Contemporary Technosciences*, Sociology of the Sciences Yearbook 31,
https://doi.org/10.1007/978-3-030-61728-8_12

coherently these concerns observing that 'attachments to older labels remain strong even when a scientist ventures into new waters (Molyneux-Hodgson and Meyer 2009), often due to concerns about the transitory nature of funding and research policies' (p. 4).

Other studies, however, provide a more nuanced picture. There are studies that identify variations in how interdisciplinary research is perceived by different disciplinary communities (Calvert 2011; Klein 2010; Huutoniemi et al. 2010; Barry et al. 2008). In a survey that asked 600 social scientists from different disciplines to agree or disagree with the statement '*In general, interdisciplinary knowledge is better than knowledge obtained by a single discipline*', 57.3% of economists disagreed or strongly disagreed with the statement, compared to 25.3% of sociologists and 9.4% of psychologists (Gross and Solon 2007). In addition to these variations, Garforth and Kerr (2011) note that interdisciplinarity is regarded differently in different institutional contexts; for example, in traditional university departments and in applied contract research units. Accordingly, scholars can accumulate different forms of scientific, academic, and symbolic capital (Bourdieu 1988) depending on the type of work (disciplinary or interdisciplinary) that they do and the type of institution in which they work. Based on these observations, they suggest that discussions about the personal and professional implications of interdisciplinarity should pay more attention to institutional contexts and to individual academic trajectories.

This complex scenario makes it challenging to understand the meaning of increasing calls for interdisciplinarity for the articulation of individual and collective academic identities. How do scholars involved in interdisciplinary research conceive of their place within academia and how do they negotiate their academic identities? Do they focus on developing and performing clearly distinguished disciplinary identities, or rather on negotiating an identity combining elements from different disciplines? Besides disciplinary labels, what other details are relevant in the construction of an academic identity? And how do negotiations of identity differ across disciplinary communities? The literature provides initial insights for addressing these questions, but the complexity of the topic calls for more refined studies and methodological approaches. Intending to contribute to addressing these questions, this chapter examines how scholars involved in interdisciplinary research construct and negotiate their academic identities through biographical narratives, paying particular attention to how the local contexts of the individuals' main disciplines shape their identity constructions.

Academics' biographical narratives are an interesting but underexplored type of data. They provide access to individuals' unique trajectories, but they are more than neutral descriptions of life events. Through biographical narrative, individuals can negotiate their identities as specific types of people and as members of particular communities, they can attribute desirable personal characteristics to themselves and to others, and they can justify past and future actions and decisions. Furthermore, as it will be argued in this chapter, academics' biographical narratives might contribute to the establishment, maintenance, and reproduction of academic disciplines. The chapter will focus on the accounts provided by individuals from four different disciplines, some of which are characterised by their interdisciplinary nature and others by their insularity, including mathematics, economics, computer science, and archaeology. These data belong to a broader sample of 27 interviews with researchers and administrators of all ranks and faculties from a large, research-oriented British university, which were collected between December 2012 and September 2013 for a project that looked at the discursive construction of the interdisciplinary self.

The analysis presented in this chapter draws on Taylor and Littleton's (2006, 2008, 2012) narrative-discursive analytical approach, a novel approach that 'offers a way of investigating the social nature of biographical talk' (2006, p. 23). According to Taylor and Littleton (2006), biography is 'a situated construction' in which the 'wider discursive environment is implicated' (p. 23). Rooted in Foucauldian theory and ethnomethodology, the theoretical assumption is that meanings of objects and subjects are constructed through talk and social interaction. It follows from this assumption that individuals are, *to some extent*, active in claiming, constructing, and negotiating their identities. They are not *entirely free* because they are constrained by their unique life trajectories and by meanings that are socially and historically available within the communities they inhabit, such as values attached to specific categories and to the ways in which lives are expected to develop (Taylor 2007). For example, how individuals are meant to become members of a disciplinary community, and how interdisciplinarity is valued within specific academic contexts. This approach provides a number of analytical concepts (namely *canonical narrative*, *breach*, *identity trouble*, and *repair*) that facilitate investigating how academics construct their identities within the interactional context of an interview while dealing with these widely established meanings.

The intention here is not to suggest that biographical narrative is the only or the most important resource for the construction of an academic identity. Scholars' engagement with particular communities, their research outputs, their grant submissions, and their engagement with other nonhumans (i.e. laboratories and research settings, materials, and instrumentation) are also relevant (Michael 1996). Yet, by suggesting an in-depth study of biographical narrative, the intention is to contribute to the accumulation of diverse and detailed methodologies for the study of academic identities. Biographical narratives are interesting because of their relative flexibility, variability, and adaptability to the interactional and the institutional context in which they are produced. In the next sections, the chapter will examine the interrelated histories of the concepts 'identity' and 'discipline' (12.2). It will describe the methodological approach, the analytical categories, and the data in more detail (12.3), and then the analysis will be presented (12.4). The chapter will conclude by suggesting that individuals' biographical narratives are valuable research objects that deserve detailed analytical attention since they may contribute to rendering the narratives that constitute the 'core' of a scientific discipline, little by little, more heterogeneous (12.5).

12.2   How Disciplines and Identity Interrelate

The study of disciplinary and interdisciplinary identities requires a reflexive and cautious approach because the notions of 'academic discipline' and 'identity' interrelate in two distinct and peculiar ways (Gleason 1983; Wetherell 2010; Forman 2012; Sugimoto and Weingart 2015). The first one has to do with the fact that constructing disciplinary boundaries and creating identities are closely related activities; the second one is concerned with how uses of the term 'identity' reinforce disciplinary boundaries. This section begins with the first form of interrelation. Reviewing accounts about the origin of disciplines should contribute to understanding not only how these acquired their current appearance and their outwardly independent identities but also the relevance of narrative for the achievement of these endeavours.

Debates about how knowledge should be produced, documented, and transmitted, whether by a specialised or by a holistic approach, have existed since the eighteenth and nineteenth centuries (Klein 1990). According to Forman (2012), in the nineteenth century the notion of disciplinarity was only an ideal, but by the middle decades of the twentieth century the ideal became taken for granted, as if it had eventually been achieved. Forman contends that historians of science contributed extensively to establishing the idea of disciplinarity as a given, and that the compilation of disciplinary histories during the eighteenth and nineteenth centuries was motivated 'in part by the aim of creating clear and distinct disciplinary identities' (2012, p. 62). He argues that in writing the history of individual disciplines, historians of science took for granted the postulate of social or institutional differentiation, a postulate that explains that societies follow 'a natural and inevitable process of increasing functional differentiation of social, cultural, and occupational roles, resulting in increasingly autonomous institutions operating to ever-higher functional standards' (Forman 2012, p. 69). Scientific disciplines represented the highest level of institutional differentiation, and it is likely that historians produced disciplinary histories that intended to validate this postulate.

In a similar way, Schaffer (2013) argues that disciplines seem to be different and independent from each other because they tell stories about where they come from, where they are, and where they are going. To Schaffer, these stories are performative in the sense that they 'provide a rationale and means for the pursuit of the disciplinary enterprise' (2013, p. 57), and they make disciplines look like 'well-institutionalized homogeneous systems of formal behaviour' (p. 58). However, sceptical of these stories, he argues that all disciplines are hybrid and internally fragmented, and that all stories of disciplinary unity are, in fact, misleading. In a more recent study, Sugimoto and Weingart (2015) suggest that different disciplinary histories provide different and even contradictory criteria of what counts as a discipline, and that these texts, in telling the history of a discipline, establish the criteria of disciplinarisation.

These studies suggest that the boundaries that demarcate one discipline from another, and thus where disciplinarity ends and interdisciplinarity begins, are not as clearly defined as it might be assumed. From this assumption it follows that disciplinary and interdisciplinary identities can be permanently open to negotiation. Yet, this is not to deny that disciplinary boundaries are real in their consequences: they define how funding is distributed, who has authority over specific problems, and, many times, who gets hired and who doesn't (Henkel 2005).

The second way in which disciplines and identity interrelate stems from the fact that different uses and understandings of the term 'identity' allow for the construction and maintenance of disciplinary boundaries. According to Gleason (1983), since the term 'identity' became popular in academia during the 1950s and 1960s, it quickly started to be used in two forms: one that referred to it as a personal and subjective achievement, and other that referred to it as a social category. Since then, the study of identity has been marked by a bifurcation that distinguishes personal identity from social identity, as if the distinction between the internal, personal and psychological, and the external and social were clearly demarcated. However, as Wetherell (2010) argues, such binaries are to some extent a product of 'priority disputes between sociology and psychology and [of] accusations of blindness on both sides' (p. 10). She explains that research on identity has both built on and criticised the division between the personal and the social sides of identity. Moreover, she asserts that the debate will not cease any time soon because 'it is not possible to investigate any aspect of identity without being struck by the extent to which the social is personally owned, and by the myriad ways in which people make social locations psychological' (Wetherell 2010, p. 12).

Studies of disciplinary and interdisciplinary people and identities tend to overlook that the division between the personal and the social dimensions of identity is to some extent a product of disciplinary disputes, and they tend to focus on either one side or the other. Amongst the studies that look at the social side of academic identity, one feature that has been explored is the use of disciplinary labels. Pinch (1990) argues that scientists use disciplinary labels in flexible ways, shifting from one to another in order to claim certainty over specific problems and uncertainty over others. Brew (2007) illustrates this flexibility in great detail. In a study that included interviews with 71 researchers from several fields in the UK and Australia, less than one third identified themselves with single disciplinary labels. By contrast, most of the interviewees provided tentative 'nested' or 'confluent' identities between or across different disciplines and specialist areas, using clauses such as 'I suppose', 'I think you could say', 'I guess', and 'it could also be'. She concludes that studies of disciplinary and interdisciplinary identity should take into account more fluid metaphors that could 'capture the ... rhetorical and reflexive nature of academics' disciplinary affiliations' (Brew 2007, p. 423). Moreover, she suggests that interdisciplinarity should be seen as a common practice of academics rather than a deviant one.

Pinch's and Brew's studies challenge the view that disciplines are fixed entities, and they point to the flexible and rhetorical character of disciplinary identities. With that, they avoid taking the dividing line between disciplinary and interdisciplinary identity for granted, just as the historical studies described earlier suggest. However, that flexibility and the search for more fluid metaphors say little about the individuals' unique trajectories and about the institutional contexts in which they work. It is as if Pinch and Brew looked at the most obviously 'social' side of identity at the expense of the 'personal' and individual side.

Analyses that are limited to features of identity that can be categorised as 'social' overlook other important elements of identity construction that would count as 'personal' or even 'psychological'. By contrast, other studies suggest that interdisciplinary researchers have very peculiar personal characteristics. Klein (1990), for example, suggests that interdisciplinary researchers are divergent thinkers 'who may not be too narrow to deal with cross-cutting issues', who have 'a high degree of ego strength, a tolerance for ambiguity, considerable initiative and assertiveness, a broad education, and a sense of dissatisfaction with monodisciplinary constraints' (p. 183). Similarly, Castán Broto et al. (2009) claim that 'if people combine knowledge and have a certain quality of mind and personality they will enjoy conducting interdisciplinary research despite, and because of, its challenges' (p. 928). What is striking in these studies is that, paradoxically, by describing these personal characteristics, they contribute to the construction of the 'interdisciplinary researcher' as a distinct social category.

The arguments above underline that the study of how academic, disciplinary, and interdisciplinary identities are constructed should not be restricted to analyses of the use of disciplinary labels alone. Moreover, they suggest that the study of interdisciplinary identities has to be critical of fixed categories such as 'disciplinary', as opposed to 'interdisciplinary', and 'personal', as opposed to 'social'. For instance, scholars who identify themselves as critical social psychologists, such as Gergen (1973) and Billig (1991, 2008), argue that the boundary between the study of psychology and the study of history is blurry, and they suggest that the content of our thoughts has a social history and depends on the ideologies of our time. Applying this same argument to the content of psychology itself, they contend that the way we understand and talk about psychological phenomena (attitudes, identity, rationality, remembering, forgetting, etc.) is historically contingent. Thus, whatever we understand now as social and as psychological should not be so easily taken for granted.

Gergen and Billig's thinking has been influential in the development of discursive psychology (Potter and Wetherell 1987; Wetherell 2008; Billig 2012), the field that provides the background for the analysis presented in this chapter. The purpose of this field is to explore the ways in which individuals use psychological language and categories in conversation and social interaction to construct particular versions of the self and the world. The purpose is not to reduce psychological phenomena to social phenomena, but to illustrate how effective and flexible psychological language is in the production of justifications and accountability.

12.3 Methodology

12.3.1 Analytical Approach and Analytical Categories

The analysis of the data presented in this chapter draws on the 'synthetic narrative-discursive approach' that Taylor and Littleton (2006, 2008, 2012) developed in a long-term study of creative identities. They call it a synthetic approach because it pays attention both to how subjects are positioned by wider discourses as well as to how they can challenge and re-negotiate the meaning of those positions.

The analytical categories used in this approach are *canonical narrative, breach, trouble*, and *repair*. The notion of the *canonical narrative* (Bruner 1990) refers to the ways in which lives are expected to unfold or to the 'expected connections of sequence and consequence which create narrative structure and trajectories' (Taylor and Littleton 2006, p. 26). A *canonical narrative* 'can provide a logic for talking about personal circumstances, life stories and decisions' (Taylor 2007, p. 6). The relevance of this category does not rely on its accuracy but on its strength as a discursive resource.

A rupture from the *canonical narrative* is called *breach*, and, as Taylor and Littleton (2006) suggest, 'unexpected associations' or connections become a source of *trouble* which requires *repair* (p. 27). The notion of *identity trouble* refers to identities that are inconsistent with conventional expectations and also to those that are difficult to adopt because they are inconvenient, undesirable, or incompatible with other identities that speakers may have previously adopted. In most occasions, *breaching* from the *canonical narrative* leads to *trouble*, and in those circumstances *repair* is expected.

12.3.2 Empirical Material

The empirical material examined in this chapter was collected as part of a PhD project focused on the discursive construction of the interdisciplinary self. The purpose was to identify how interdisciplinary researchers construct narratives of who they are and how they became interdisciplinary, also asking questions such as what are their professional challenges and expectations, how are they perceived and described by others, and what is expected of them. Semi-structured interviews were conducted with 14 male and 13 female researchers and administrators whose research descriptions suggested their involvement in interdisciplinary work. Potential interviewees were selected after searching the university website, and others were recommended by the interviewees. The interviews lasted between 40 and 70 minutes, and all of them were fully recorded and transcribed. The research obtained ethical approval from the Department of Sociology of the same university.

In order to secure anonymity, all names of individuals were changed to pseudonyms, and some details of the fields in which or with which they work were also changed.

The analysis will focus particularly on the narrative constructions of interviewees working on mathematics, computer science/bioinformatics, economics, and archaeology. These interview extracts were selected because they illustrate some of the different possibilities in which identities can be negotiated across disciplines. The four interviews used in this chapter are representative of the whole sample in the sense that they demonstrate patterns of how the interviewees drew on personal details of their lives, on conventional narratives and wider social meanings, and on articulations of repair provided whenever they distanced from the conventional narratives. These four interviewees worked in traditional disciplinary departments at the moment of the interview.

In order to inform the analysis, it is valuable to provide some information about the particular fields of the interviewees examined in this chapter. To begin with, mathematics and mathematicians' identity have a distinct place in discussions of interdisciplinarity and of academic identity because, on the one hand, 'mathematics enables work in many different disciplines, from the natural and physical sciences to the social sciences and fine arts', and because it 'can also support knowledge integration across disciplines' (Fisher and Beltran-del-Rio 2010, p. 88). However, despite the contact of mathematics with other disciplines, the identity of the mathematician is a particularly strong one. For instance, mathematicians, and possibly physicists—who draw strongly on mathematics—have received more attention than other scholars in popular culture (Haynes 2016), where they are often portrayed as clever individuals with a great passion for knowledge, and at times embodying the prototype of the lone genius (Mialet 1999). One can think of Albert Einstein, Richard Feynman, John Nash, and Stephen Hawking. Other disciplines lack this sort of well-known personalities.

Bioinformatics, which draws on biology, computer science, and mathematics, is a clear example. The issue of bioinformatics is that it 'is often seen as being neither good biology, nor good computer science, but, rather, as a service provider to biology' (Bartlett et al. 2016, p. 188). With such a connotation, it is interesting to examine how researchers involved in bioinformatics negotiate their identities.

The negotiation of disciplinary and interdisciplinary identities in economics is also an interesting subject. Fourcade et al. (2015) note that economists have a high regard for scientific purity and that they rely strongly on the use of formal methods. Moreover, economics has strong connections to wealth and power, and these conditions provide the discipline with a sense of superiority. A consequence is that economics is more insular than other disciplines, and there is distrust on the interpretivist methods of other social sciences. According to one of Fourcade's (2009) interviewees, economists condemn highly those who do not follow the epistemological preferences of the discipline and who may present data that sounds 'anecdotal'. In this context, engaging with other disciplines might be risky.

At the opposite extreme, archaeology is a discipline that draws on a huge variety of methods and theories from other disciplines, such as the natural sciences, the arts, and the humanities. Kristiansen (2009) argues that the discipline's ambitious aim of studying and preserving the past would not be possible without the contributions of zoology, environmental sciences, physics, medicine, history, and anthropology, amongst other disciplines. Furthermore, he notes that studies that are perceived as the first modern archaeological studies 'were based upon interdisciplinary work, which was to remain a dominant feature in archaeological practice until this day' (p. 28).

12.4   Analysis

The analysis presented below focuses mainly on the interviewees' responses to the first question I asked, 'can you please tell me about your background?' The first extract comes from Lawrence, a professor of applied mathematics who spent most of his career in a publicly funded laboratory before moving to academia. Curiously, he never did a PhD, and his first academic post was a professorship. In the years he has worked at the university he has collaborated with people in engineering, and he often participates in interdisciplinary workshops.

12.4.1   Lawrence – Mathematics

I think it was always my ambition to become a scientist of some kind or another. From as long ago as I can really remember—probably goes back to primary school. But then there was a period when I went to a grammar school and I guess some of the way science was taught didn't captivate me, and the mathematics I was taught wasn't that exciting. And in fact I come from a farming background; I decided after I've done my O levels that I was gonna basically go and work at my father's farm because that was kind of work he'd done and his father before him. But he did me the biggest favour he could, he gave me all the dirty nasty jobs, and I decided there must be an easier way making a living than this. And I went back and I started doing A levels and I did very traditional hard science—A levels, maths, further maths, physics and chemistry. And I've been an average student, or probably better than average student, I would say. But, I found when I started doing A level and further maths I really enjoyed it, and I was pretty good at it, actually. And I enjoyed the physics; chemistry not so much, but I basically decided that I really wanted to go to university and do science. So I went to Cambridge and I did mathematics there. I got a first *[class degree]* and I was pretty good. I was actually, to my surprise, better than most people, not everybody, but most people.

In this extract, it is interesting to note that even though the interview was about interdisciplinarity, Lawrence gave priority to presenting himself as a gifted mathematician. Secondly, one can note that the extract contains references of personal traits, social categories, and even a description of Lawrence's family environment. In that way, one can see that even though the narrative is uniquely personal, it is also shaped by widely established social meanings. To some extent, the details of the family environment can be interpreted as a confirmation of the skills that Lawrence attributes to his young self.

As mentioned earlier, mathematicians are often portrayed as clever individuals and as the prototype of the lone genius. In line with this image, Lawrence describes his early ambition, passion, and proficiency in science and mathematics. One crucial part in the story is that he refers to a time when he lost his usual motivation and thought he would work on a farm just as his ancestors did. Yet, Lawrence narrates that his father gave him only dirty and nasty work, which to him was like a favour because this encouraged him to reconsider his plans for the future, to take his fate back into his own hands, so to speak, and return to the academic track. The story can be read as if Lawrence's father knew he had a gifted son whose place was in academia, not on a farm, and this story seems to corroborate that Lawrence did indeed have the skills he claims to have. Since the different elements included in the narrative are well integrated with each other, it is hard to challenge the identity that Lawrence articulated during the interview, and thus he is out of 'identity trouble', and 'repair' is not required. The case is different for the following interviewee, a computer scientist working in bioinformatics.

First, however, it is relevant to mention identity troubles that two other computer scientists referred to during the interviews. Jane, who works in systems and synthetic biology, argued that the 'scientific' status of computer science is often played down by specialists from other disciplines when it is called 'information technology' rather than computer *science*. Similarly, Robert, who works in computer vision and systems biology, argued that in his interdisciplinary collaborations he has to avoid being taken as somebody's *'IT guy'* or *'data monkey'* by doing work that is interesting *'from a computer science point of view'*. These arguments suggest that for computer scientists involved in interdisciplinary research, it is important to be recognised as scientific specialists rather than as service providers. These arguments also serve as further background to interpret Blanc's biographical narrative.

12.4.2   Blanc – Computer Science and Bioinformatics

> My background is computer science; I studied straight computer sciences from my under-graduate up to the PhD, which I did in the area of machine learning. And then I was looking for postdocs in my research area; I found one which was an application of machine learning to bioinformatics here at the university. And that's when I started to open to other, started doing a bit of interdisciplinary research. Was still core computer science but of course we had chemist collaborators, so it was the time I started understanding new languages and seeing different cultures. And now most of the work I do is applied and in collaboration with experimentalists, mostly in biology but across different schools.

Blanc describes a sequence of steps through which he has become a disciplinary specialist: he studied computer science from his undergraduate to his doctoral degree, and his postdoctoral career has led him to even more specialised research areas.

In this extract, it is important to pay attention to the rhetorical work that the little words do. First, Blanc describes that he studied *'straight'* computer science from the undergraduate degree to the PhD. He then emphasises that he was looking for postdocs in '*his*' research area and found one which was '*an application*' of this area to a particular field. Even when he describes the moment when he '*opened*' up to interdisciplinary research, he emphasises that it was only '*a bit*' of interdisciplinary research and that it was '*still core computer science*'. These little words help Blanc to adopt an identity that is hard to challenge: he is a specialist, not a bioinformatics technician working *for* the biologists or *for* any other specialists.

The following interviewee, Lindsay, also accounted carefully for her interdisciplinary engagement. During the interview, she said '*I have been interested in development since I was sixteen, I studied economics because I was interested in development*'. She also expressed critical views of interdisciplinarity and of interdisciplinary researchers, who, to her, '*may not be as rigorous as their disciplines may require*'.

12.4.3   Lindsay – Economics

> I'm an economist. I did my PhD at Cambridge. My PhD thesis was on family business networks in Peru. As a consequence of that research—and I suppose you can say that research was part of a process by which I became interested in kinship. In my early postdoc work I was seeking ideas about methods that could be used to do empirical work for the subject I was interested in and I didn't restrict myself to talk to economists. So basically I became interdisciplinarily interested in that time, I guess.

Similar to the case of Lawrence in a previous extract, Lindsay provides a clear disciplinary identity and the name of the prestigious university where she studied. It is common knowledge that prestigious universities have high admission requirements; therefore this detail reveals that she had a high quality academic performance before enrolling in her PhD programme. With this, she presents herself not as a common economist but as an economist with a very strong academic background and as a disciplined person with a clear ambition. In that way, details from Lindsay's unique trajectory and wider social meanings are combined so that she can portray herself as a hard-working academic specialist. In this extract and in Lindsay's argument that she was interested in development by the time she was sixteen, it is possible to identify a narrative that contains features also found in the accounts that Lawrence and Blanc provided. These include early interests and passion for a specific field, premature skills in this field, clear career goals, and a trajectory within that same field. Taking these features together, this narrative can be suggested as a canonical narrative that allows individuals to portray themselves as single discipline specialists.

In the second half of the extract, Lindsay specifies that her interests on the topic that led her to engage with other disciplines were an outcome of the research she undertook during the PhD. Her use of clauses such as '*as a consequence*' and '*part of a process*' point to a breach from the canonical narrative, but they also minimise the extent of this breach. Thus, these clauses can be read as rhetorically oriented to 'repair' the breaches from the canonical narrative and avoid giving the impression that she has driven her career in an undisciplined way or followed academic fashions. Moreover, the clauses suggest that Lindsay's interdisciplinary engagement was the product of a carefully calculated move. Finally, her claim that she *'didn't restrict [herself] to talk to economists'* suggests that being closed down to one's own discipline is, in fact, a restriction and not the most desirable way to conduct academic work. In that way, she 'repairs' or justifies her engagement with the 'outside' of the discipline.

The three interview extracts analysed so far have in common that the individuals present themselves with a clear disciplinary identity and as specialists. However, the following extract provides a different picture.

12.4.4   Julia – Archaeology

> Throughout my school years I was told that I was very, very stupid indeed and, you know, that I shouldn't go to college, I shouldn't go to university; there were all these things that I just wasn't clever enough to go. And they were probably right, but part of it was because I don't think I've found the thing that I was interested in. And then I went to university to study archaeology, and suddenly I didn't feel so stupid anymore, because archaeology is about every aspect of human life, society, culture, so it's kind of relevant to everything. So through one discipline I was able to study all the things that I'd done badly before but through a different kind of lens that for me was sort of engaging and fascinating. So yes, perhaps I have my poor performance at school and college to thank to sort of taking an interdisciplinary approach, because I still maintain that part. The way that I work is I know very little about lots of stuff, and it's the way I kind of bring them together that turns into something new that people perceive to be slightly more interesting.

As with the other extracts, this one represents an account that is personal and unique but at the same time embedded in wider social debates: who should go to university, what are the downsides of not being considered clever, and what is the path that has to be followed to become a specialist? It is crucial to note that the story Julia provides about herself and the characteristics of her discipline are presented as if they were closely related and highly compatible. To her, it was the interdisciplinary nature of archaeology and the fact that this discipline is *'about everything'* that made her interested in it, and once she found this discipline, she *'didn't feel so stupid anymore'*. Through her biographical account, she thus portrays herself as being naturally predisposed to work in an interdisciplinary field such as archaeology, just as Lawrence suggested he was somehow predisposed to become a mathematician.

For two of the previous interviewees, Lawrence and Lindsay, describing themselves as academically oriented and skilful since an early age was essential in their explanations of the professional path that they chose to take. By contrast, Julia's account is substantially distant from such a narrative. Rather than prematurely skilful and academically oriented, she was called stupid; instead of passionate about a single discipline since an early age, she found her interests only later on in life; and, instead of presenting herself as a specialist in a specific area, she argues that she now knows *'very little about lots of stuff '*. It might be that in a discipline as broad as archaeology this is a desirable personal characteristic rather than one to avoid. Curiously, while other interviewees intended to repair their breaches from the canonical narrative by playing down the extent of such breaches, Julia instead embraced the interdisciplinarity of archaeology to repair the troubled identity of being perceived as stupid. In that way, while to other interviewees being interdisciplinary was a personal feature to keep low and discrete, to Julia this was a personal feature to emphasise. This might suggest that the way interdisciplinary self-narratives are articulated depends on particular disciplinary contexts and also on the particular interactional and institutional contexts in which they are articulated. For example, in her current university webpage, Julia mentions her interdisciplinary engagement but not her 'stupid' past.

12.5 Discussion and Conclusion: Narrating Identities and Reproducing Disciplines

Drawing on Taylor and Littleton's approach and examining the biographical narratives of four scholars involved in interdisciplinary research, this chapter has contributed to the study of the construction of academic, disciplinary, and interdisciplinary identities. This final section summarises the findings of the analysis and discusses them against the backdrop of other existing studies.

The analysis illustrated that despite their involvement in interdisciplinary research, the interviewees provided narrative accounts that pointed to an 'expected' trajectory, namely to one that allowed them to identify themselves as members of specific disciplinary communities. To clarify, the features of this 'expected' trajectory include: (1) descriptions of early interests, passion, and skills in a specific discipline; (2) descriptions of early ambition and a clear career goal; and (3) training and experience within the same discipline from undergraduate degree to PhD, and even beyond. This expected life trajectory can be called the *canonical narrative of the single discipline specialist*. Consistent with Taylor and Littleton, when the interviewees' accounts deviated from or breached this canonical narrative—for example, because of their engagement with other disciplines—they provided some sort of justification, which could be interpreted as 'repair'.

The features of this narrative, including breaches and the subsequent repair, were more visible in the cases of Blanc, the computer scientist, and of Lindsay, the economist. Blanc argued that even though his work has involved chemists and biologists, this still counts as *'core computer science'*; and Lindsay described her engagement with other disciplines *'as a consequence'* of her research within economics. A contrasting case was that of Julia, the archaeologist, whose described life trajectory was at a significant distance from the canonical narrative described above. Curiously, and perhaps paradoxically, she drew on her deviation from the expected trajectory to justify her interests and her success in a discipline that is characterised, as Kristiansen (2009) argues, by its interdisciplinary nature and its possibility to make connections across disciplines in the arts, the humanities, and the social and the natural sciences.

The latter point reveals one of the most intriguing findings of the analysis, which is that the interviewees described themselves in ways that resembled their main disciplines. Julia's celebration of her broad interests mirrors the breadth of archaeology; by contrast, Lindsay's careful way of describing her engagement with interdisciplinarity resonates with the scientific purity and the insularity that Fourcade et al. (2015) attribute to the discipline of economics. Furthermore, Lawrence's described trajectory is similar to the way in which mathematicians are presented in popular culture. Yet, these 'disciplined' forms of describing themselves are also connected to details of the speakers' lives, such as Julia's and Lawrence's vivid descriptions of their childhood.

Based on these findings, a bold argument can be made. As more than one author has observed (Forman 2012; Schaffer 2013; Sugimoto and Weingart 2015), disciplines seem to be real, homogeneous, and different from each other because of the stories told about them. In a similar way, it would be reasonable to suggest that disciplines are also in part established, maintained, and reproduced through the narratives that individuals tell about themselves. Put differently, as individuals construct their identities, they also construct the identity of their disciplines. This is not an easy task, since individuals have to be able to articulate explanations that make their interdisciplinary engagements look like being part of the 'core' of the discipline and not alien to it. This skill requires an understanding of other wider discourses and narratives that constitute a particular discipline. At times, the chances of success might be low. And yet, the articulation of personal (inter)disciplinary identities may be a way through which the narratives that constitute the 'core' of a discipline become, little by little, more heterogeneous. This suggestion contributes to illustrating Kastenhofer and Monyneux-Hogdson's observation that "communities, their representatives and practitioners... are co-produced" (p. 10).

Pinch (1990) and Brew (2007) have contributed substantially to our understanding of the negotiation of academic identities, emphasising that disciplinary labels are not fixed but flexible. However, the analysis presented here demonstrates that the construction of an academic identity depends on more than the use of disciplinary labels. The successful (i.e. convincing and persuasive) articulation of personal and professional stories through a choreography of expected trajectories, breaches, and innovative repairs might contribute to the acquisition of epistemic power within and beyond an established discipline. In that way, this skill might represent a type of academic or scientific capital as Bourdieu and Garforth and Kerr (2011) understand the term. Garforth and Kerr have suggested that to understand the professional implications of interdisciplinarity, at a time in which disciplinary identity is still relevant (Henkel 2005), studies have to pay attention to specific institutional contexts and to individual academic trajectories. Intending to contribute to this research agenda, this chapter has demonstrated that these individual academic trajectories, as portrayed in biographical narrative, are carefully articulated interactional devices, and, as such, they represent valuable research objects that deserve detailed analytical attention.

References

Abbott, A. 2001. *Chaos of disciplines*. Chicago: University of Chicago Press.

Barry, A., G. Born, and G. Weszkalnys. 2008. Logics of interdisciplinarity. *Economy and Society* 37 (1): 20–49.

Bartlett, A., J. Lewis, and M.L. Williams. 2016. Generations of interdisciplinarity in bioinformatics. *New Genetics and Society* 35 (2): 186–209.

Billig, M. 1991. *Ideology and opinions: Studies in rhetorical psychology*. London: Sage.

———. 2008. *The hidden roots of critical psychology*. London: Sage.

———. 2012. Undisciplined beginnings, academic success, and discursive psychology. *The British Journal of Social Psychology* 51 (3): 413–424.

Bourdieu, P. 1988. *Homo academicus*. London: Polity Press.

Brew, A. 2007. Disciplinary and interdisciplinary affiliations of experienced researchers. *Higher Education* 56 (4): 423–438.

Bruner, J. 1990. *Acts of meaning*. Cambridge: Harvard University Press.

Buanes, A., and S. Jentoft. 2009. Building bridges: Institutional perspectives on interdisciplinarity. *Futures* 41 (7): 446–454.

Calvert, J. (2011). Systems biology, interdisciplinarity and disciplinary identity INNOGEN Working Paper No. 90.

Castán Broto, V., M. Gislason, and M.-H. Ehlers. 2009. Practising interdisciplinarity in the interplay between disciplines: Experiences of established researchers. *Environmental Science & Policy* 12 (7): 922–933.

European Commission. 2011. *Proposal for a regulation of the European Parliament and of the Council establishing Horizon 2020: The Framework Programme for Research and Innovation (2014–2020)*.

Felt, U., J. Igelsböck, A. Schikowitz, and T. Völker. 2013. Growing into what? The (un-)disciplined socialization of early stage researchers in transdisciplinary research. *Higher Education* 65: 511–524.

Fisher, E., and D. Beltran-del-Rio. 2010. Box: Mathematics and root interdisciplinarity. In *The Oxford handbook of interdisciplinarity*, ed. R. Frodeman, J.T. Klein, and C. Mitcham, 88–91. Oxford: Oxford University Press.

Forman, P. 2012. On the historical forms of knowledge production and curation: Modernity entailed disciplinarity, postmodernity entails antidisciplinarity. *Osiris* 27: 56–97.

Fourcade, M. 2009. *Economies and societies: Discipline and profession in the United States, Great Britain, and France, 1890s to 1990s*. Princeton: Princeton University Press.

Fourcade, M., E. Ollion, and Y. Algan. 2015. The superiority of economists. *Journal of Economic Perspectives* 29 (1): 89–114.

Garforth, L., and A. Kerr. 2011. Interdisciplinarity and the social sciences: Capital, institutions and autonomy. *The British Journal of Sociology* 62 (4): 657–676.

Gergen, K. 1973. Social psychology as history. *Journal of Personality and Social Psychology* 26 (2): 309–320.

Gleason, P. 1983. Identifying identity: A semantic history. *The Journal of Americal History* 69 (4): 910–931.

Gross, N., & Solon, S. (2007). The social and political views of American professors Working Paper.

Haynes, R.D. 2016. Whatever happened to the 'mad, bad' scientist? Overturning the stereotype. *Public Understanding of Science* 25 (1): 31–44.

Henkel, M. 2005. Academic identity and autonomy in a changing policy environment. *Higher Education* 49 (1–2): 155–176.

Holley, K.A. 2015. Doctoral education and the development of an interdisciplinary identity. *Innovations in Education and Teaching International* 52 (6): 642–652.

Huutoniemi, K., J.T. Klein, H. Bruun, and J. Hukkinen. 2010. Analyzing interdisciplinarity: Typology and indicators. *Research Policy* 39 (1): 79–88.

Klein, J.T. 1990. *Interdisciplinarity. History, theory & practice*. Detroit: Wayne State University Press.

———. 2010. A taxonomy of interdisciplinarity. In *The Oxford handbook of interdisciplinarity*, ed. R. Frodeman, J.T. Klein, and C. Mitcham, 15–30. Oxford: Oxford University Press.

Klein, J.T., and H.J. Falk-Krzesinski. 2017. Interdisciplinary and collaborative work: Framing promotion and tenure practices and policies. *Research Policy* 46 (6): 1055–1061.

Kristiansen, K. 2009. The discipline of archaeology. In *The Oxford handbook of archaeology*, ed. B. Cunliffe, C. Gosden, and R.A. Joyce, 3–46. Oxford: Oxford University Press.

Leahey, E., C.M. Beckman, and T.L. Stanko. 2017. Prominent but less productive: The impact of interdisciplinarity on scientists' research. *Administrative Science Quarterly* 62 (1): 105–139.

Mialet, H. 1999. Do angels have bodies? Two stories about subjectivity in science: The cases of William X and Mister H. *Social Studies of Science* 29 (4): 551–581.

Michael, M. 1996. *Constructing identities. The social, the nonhuman and change*. London: Sage.

National Academy of Sciences. 2005. *Facilitating interdisciplinary research*. Washington, DC.

OECD. 1972. *Interdisciplinarity, problems of teaching and research in universities*. Paris: Organisation for Economic Co-operation and Development.

Pfirman, S., and P.J.S. Martin. 2010. Facilitating interdisciplinary scholars. In *The Oxford handbook of interdisciplinarity*, ed. R. Frodeman, J.T. Klein, and C. Mitcham, 387–403. Oxford: Oxford University Press.

Pinch, T. 1990. The culture of scientists and disciplinary rhetoric. *European Journal of Education* 25 (3): 295–304.

Potter, J., and M. Wetherell. 1987. *Discourse and social psychology*. London: Sage.

RCUK. 2015. Institutional website. https://www.epsrc.ac.uk. Accessed 22 June 2018.

Schaffer, S. 2013. How disciplines look. In *Interdisciplinarity. Reconfigurations of the social and natural sciences*, ed. A. Barry and G. Born, 57–81. Abingdon: Routledge.

Sugimoto, C.R., and S. Weingart. 2015. The kaleidoscope of disciplinarity. *Journal of Documentation* 71 (4): 775–794.

Taylor, S. 2007. Narrative as construction and discursive resource. In *Narrative—State of the Art*, ed. M. Bamberg, 113–122. Amsterdam: John Benjamins Publishing Company.

Taylor, S., and K. Littleton. 2006. Biographies in talk: A narrative-discursive research approach. *Qualitative Sociology Review* 2 (1): 22–38.

———. 2008. Art work or money: Conflicts in the construction of a creative identity. *The Sociological Review* 56 (2): 275–292.

———. 2012. *Contemporary identities of creativity and creative work*. Surrey: Ashgate.

Turner, S. 2000. What are disciplines? And how is interdisciplinarity different? In *Practising interdisciplinarity*, ed. P. Weingart and N. Stehr, 46–65. Toronto: University of Toronto Press.

Wetherell, M. 2008. Subjectivity or psycho-discursive practices? Investigating complex intersectional identities. *Subjectivity* 22 (1): 73–81.

———. 2010. The field of identity studies. In *The SAGE handbook of identities*, ed. M. Wetherell and C.T. Mohalty, 3–26. London: Sage.

Chapter 13
'Big Interdisciplinarity': Unsettling and Resettling Excellence

Bettina Bock von Wülfingen

13.1 Introduction

This case study focuses on patterns of unsettling and resettling within an exemplary site of 'big interdisciplinarity'—the Germany-based *Interdisciplinary Laboratory Image Knowledge Gestaltung,* a large research cluster funded from 2013 to 2018 by the German Research Foundation within the *Clusters of Excellence* programme. This Cluster represented an international and multidisciplinary research group involving around 300 researchers embracing over 30 disciplines—from the natural and social sciences to the humanities, design, and arts. The explicit aim of the Cluster was to bring the natural sciences and humanities together in joint experiments (Bild Wissen Gestaltung 2015), while the Cluster saw itself as an experiment involving self-reflective feedback structures. Its self-proclaimed vision was not to dissolve the boundaries between disciplines, but rather to strengthen the disciplines through the ability to widen the range of possible-to-work-on research topics in interdisciplinary collaborations. Disciplinary differences were seen as engines of innovation, and hence explicit reflection on these differences was regarded as an essential element of the Cluster's work (Humboldt-Universität zu Berlin 2011).[1]

This cluster was the object of this study, which itself was part of the above self-reflective structures. This study focuses on the forms of knowledge, practices, and behaviours that intersect with differences of status, culture, disciplines, and *Erfahrenheit* (adeptness or acquired intuition, Rheinberger 2001, Fleck 1980; see

---

[1]The success of this interdisciplinary way of working was acknowledged by the German Research Foundation by granting a subsequent project on similarly broad interdisciplinary grounds, though with different research tasks.

B. Bock von Wülfingen (✉)
Department of Cultural History and Theory, Humboldt-Universität zu Berlin, Berlin, Germany
e-mail: bockvwub@hu-berlin.de

K. Kastenhofer, S. Molyneux-Hodgson (eds.), *Community and Identity in Contemporary Technosciences*, Sociology of the Sciences Yearbook 31,
https://doi.org/10.1007/978-3-030-61728-8_13

below) in the work and private life of this community. It applies a concept of a scientific community that, aside from epistemic phenomena, also accounts for those aspects from the sociopolitical world beyond the influence of the Cluster that impacted on the researchers' day-to-day work life, such as the funding conditions. The study uses ethnographic methods to identify whether and how researchers related differently to interdisciplinarity.

The study presented here takes the Cluster as a prominent example of the current scheme for fostering innovation in Europe. Previously, research in fields from cultural anthropology to the history of science and science and technology studies has investigated the 'projectification' (Torka 2018; Felt 2017; Vermeulen 2015; Midler 1995) of academia with regard to how it changes researchers' infrastructure and how they cope with organisational aspects. Since around the turn of the century, we have experienced a shift in research organisation in Europe toward more external and hence short-term funding for projects and the emergence of centres of excellence (see Sect. 13.3). While this shift is generally recognised, the claim that there is also a move toward interdisciplinarity or transdisciplinarity is contested (see e.g. Marcovich and Shinn 2011, Hubert and Louvel 2012, and the current discussion of convergence). Generally the term transdisciplinarity involves interaction between researchers from different disciplines combined with explicit collaboration with fields considered external to the sciences, such as industry, the hard-to-define public sphere, or politics (Huutoniemi 2010; Klein 2010). By contrast, interdisciplinarity— and this is the meaning that was applied in the *Interdisciplinary Laboratory*—refers to collaboration between different disciplines within the academic domain. Interdisciplinary collaboration usually takes place between fields of research that at least share some foundational education, which various authors call 'weak' or 'narrow' interdisciplinarity—such as literary studies and Romance philology, or biochemistry and molecular biology—in contrast to 'strong', 'wide', or 'broad' interdisciplinarity (Albert et al. 2017; Repko and Szostak 2016; Kastenhofer 2010; Klein 2010; Kelly 1996).

For this specific type of broad interdisciplinarity, I introduce the term 'big interdisciplinarity'. This interdisciplinarity is 'big' in the sense denoted above with the unusual number of disciplines spanning the entire academic spectrum. It is also and even more importantly 'big' in the sense of the concept of 'big science' (at least in some aspects). This concept was initially applied to science collaborations gathering around extraordinarily expansive instruments, involving big groups of scientists in large scale physics research, such as space research (Capshew and Rader 1992; de Solla Price 1963). Due to the general upscaling of science, the term became applicable to many more fields, such as large collaborations in biology with correspondingly big shared lab spaces (Vermeulen 2009a, b). I hence speak of 'big interdisciplinarity' with regard to the large number of researchers, the large scale of its funding, and the organisational format it necessitated.

Previous studies on broad interdisciplinarity suggest that collaboration between the natural sciences and humanities is difficult due to incommensurable epistemic differences, which hinder the emergence of collaborative communities (Albert et al. 2017; Repko and Szostak 2016; Klein 2010). This study asks: is something like a joint community and collaborative identity emerging? How do researchers stabilise their academic self-concepts when they cannot apply disciplinary frameworks? It shows that collaborative identities between the sciences and humanities can indeed be formed—with much extra effort and when a time span of several years of collaboration is allowed, which enables the development of interactional expertise (Evans and Collins 2010; Collins and Evans 2002; see below).

The unexpected results show that the new structure of 'big interdisciplinarity' in the Cluster, although it was perceived as a challenge in different ways by different members, provided an opportunity for those involved to form new (collaborative) identities. Crucial to the analysis of what occurs in this interdisciplinary structure is the awareness that disciplinary members maintain other—and sometimes competing—memberships in diverse sociocultural groups and subgroups aside from their professional training or experience. Interdisciplinarity costs time and produces different sorts of affect. For some it is an effort in scientific emancipation for which they sacrifice energy and even career opportunities; for others it is just another challenge in academic life, where they already experience difficulties in passing. In the context of the Cluster, it was not disciplinary identity that offered a sense of intellectual belonging to its members—researchers were thus 'unsettled.' I introduce this term to mark the productive and sometimes disorienting effect of detaching researchers from their known boundaries and disciplinary repertoires in interdisciplinary collaborations that involve not only epistemic and practical aspects in research but also affect. These researchers found other ways to identify, to 'resettle', be it in categories at the margins of or external to science, in the narrow interdisciplinary categories provided by the Cluster's structure itself, or in spaces of interactional expertise (Evans and Collins 2010, Collins and Evans 2002; see below).

The following section introduces the relevant concepts employed in this study—such as the specific notions of community, (intersectional) identity, interdisciplinarity, and expertise—and how they relate to each other. It also describes the empirical setting, in particular the setting for the interview study that generated the majority of the results discussed here. The third section is devoted to the funding context that shapes the Cluster's infrastructure and—in part—its epistemic and empirical practice. The fourth section, the analytical section, describes in greater depth the specificities of the Excellence Cluster *Image Knowledge Gestaltung* in relation to interdisciplinarity. Exploring the results of the interview study in detail, it unfolds the unsettling effects of interdisciplinarity, which simultaneously lead to the short-term stabilisation of other categories of identity and to new interdisciplinary identities. The fifth section summarises and discusses the results.

13.2 Communities, Identities, Interdisciplinarity, and Expertise

It is impossible to take the notion of community for granted in the case of an interdisciplinary community if we assume that disciplinarity provides that which is needed to be part of a scientific community: at a minimum, an overlap in identity and a common understanding of that community. A clearer understanding of this concept is therefore needed in order to discuss the type of community or types of communities we find (or do not find) in a large, temporary, and interdisciplinary research group such as a research cluster within the framework of the Excellence Initiative.

The notion of community has changed since its introduction into academic study (more specifically into sociology) by Ferdinand Tönnies, who distinguished between a romanticised concept of a good, traditional 'Gemeinschaft' [community] and society (Tönnies 1827). Weber criticised this idealised notion of a community of common destiny (groups whose members are mostly born into it, live and work together, and that offer their members social security), stressing the aspects of inclusion *and* exclusion as well as the (implicit or explicit) rules of behaviour that the group as a community necessarily imposes on its members (Weber 2002 [1922]). Thus for Weber a community does not necessarily only produce togetherness (belonging), but in doing so it shapes the acts of its members and provides a normative collective understanding of the shared community. Applying Weber to the concept of community distinguishes it from the very broad use of the term since the 1990s in empirical studies, such as studies of youth culture, in which a commonality of interest among the group's members is sufficient to constitute a community. A further distinction from Tönnies is useful for the description of a scientific community that is predominantly based on text—in Anderson's concept of an 'imagined community' (Anderson 1983), the community's members may not know each other personally but share more or less simultaneously experiences via print media, for example in a nation-state.

The publication-based togetherness of scientific communities is explicit in Ludwik Fleck's and Thomas Kuhn's description of scientific work. Building upon a similar thought tradition, community theorist Joachim Gläser applies the notion of narratives, prevalent in not only publications but also oral communication—both formal and informal, such as corridor chats—and oral and written stories about the community's history (Gläser 2015). When Kuhn introduced the notion of scientific community (Kuhn 1989 [1962]), he envisioned a community based on rational relationships and a structure in which science is separate from, and uninfluenced by, society. By contrast, earlier work by Ludwik Fleck, which Kuhn's ideas were originally based on, is more helpful and, again, more realistic as it does not depict science as an endeavour distinct and separate from society. Instead, the particular scientific thought collective—Fleck's term for specific communities in the sciences—is a part of society with specific rules, rites, and learned ways of seeing influenced by tacit sociocultural beliefs (Fleck 1980 [1925], 1983 [1929]). Scientific communities are in most cases—very similarly to Fleck's thought collectives—

depicted as constituted by members of a specific discipline or as gathering around a research topic. This community's rules, rites, and ways of seeing shape the specific experiences of the discipline's members and their adeptness ('*Erfahrenheit*'), acquired intuition, and thinking with tools and hands (Fleck 1980; Rheinberger 2001). And they ultimately shape disciplines and disciplinary cultures. Belonging to a discipline then provides for cultural identity as well as for identity based on group membership. This helps explain why generally it is assumed that, in order to be able to develop an identity on the basis of a scientific community, this community will be a *disciplinary* community. We should, therefore, as a presumption of the study presented here—of a community based on broad disciplinarity—expect that it is *difficult* to build communities on broad interdisciplinary grounds.

Moreover, we may challenge the assumption that the 'natural sciences' and 'humanities' are fundamentally distinct and incommensurable realms; indeed, between different disciplines of these fields, we can find similarities in epistemes, practices, and even in research objects. And even if we appreciate the differences in approaches and values between the more distal disciplines in these larger fields (Albert et al. 2017; Huutoniemi 2010), we can acknowledge that shared expertise can grow and bridge the gap to some extent: experience and being-experienced are not solid states but can change, and with the experience the expertise thereby acquired changes. Expertise, described by Collins and Evans (2010, p. 54) as a result of 'successful socialization', can be disciplinary, but it can also be generated among members of distant disciplines: Collins and Evans introduced the term 'interactional expertise' (Evans and Collins 2010; Collins and Evans 2002) to explain and justify how social scientists conducting fieldwork in areas of the natural sciences gain competences in those fields *without* sharing expertise in the respective *practices*. They distinguish between expertise as novice, contributory expertise of those within the field, which affords learned practices, and interactional expertise acquired by a 'deep sharing of discourse' (Evans and Collins 2010, p. 53).

As stated earlier, the Cluster under investigation in this study was subject to a specific funding structure. We will see further below that this funding structure had an influence on the community-building in the Cluster. This is not surprising in a non-internalist concept of epistemic models of scientific communities: critiquing internalist and functionalist concepts of such models, Karin Knorr-Cetina contributed the finding that much scientific work by a research group is achieved and guided by factors external to the respective community. Such non-epistemic factors shape day-to-day scientific work life and comprise factors that are restrictive and productive in equal measure, like technical infrastructure, career schemes, and funding opportunities (Knorr-Cetina 1981, 1982, 2003). She concludes that scientific communities understood as 'specialty' communities are 'largely irrelevant to scientific work'. Rather, the scientific collectives that shape research on a day-to-day basis are *transepistemic*. They 'include scientists and non-scientists, and encompass arguments and concerns of a 'technical' as well as a 'non-technical' nature' (Knorr-Cetina 1982, p. 101; see also Knorr-Cetina 2003).

Identity so far as depicted above could be understood as monolithic. We didn't discuss yet how biographic aspects of identity alien to the narrower scientific context shape the relationship between an individual's identity and that of the community. Gläser (2015) proposes a sharp definition of community as an 'identity-based collective'—the narratives of the community functioning as the medium of this identity construction. With narratives providing for identity construction and the narrated identity providing for community, identity is easily pictured as rather solid and solidifying. For this paper, it is helpful to depart from such an identity conception as solid, coherent, and consistent (even in scientific communities) and to focus on a more amalgamated or—to use a term introduced in gender and identity studies—a more intersectional identity (Crenshaw 1991; Choo and Ferree 2010). It is also relevant for the findings discussed further below that the community's identity is not just the sum of its member's individual identities. Gläser suggests that 'the collective self-perception, relying on the intersection of individual self-perceptions, constitutes a community'[2] (Gläser 2007, p. 86). In science (and probably in most other fields as well), this reading contrasts with the finding that not all individual self-perceptions or aspects of fragmented self-perceptions are equally welcome and that not all members of a community are equally privileged in the shaping of its image. Specifically, the image of the scientific persona (Daston 2003) may restrict certain self-conceptions from entering the scientific community's identity.

As stated earlier, this study draws on an intersectional conception of identity, viewing differences of disciplines, culture, language, social background, and gender, as well as other forms of categorical differentiation between individuals, as co-constitutive in identity construction as well as in the research practices and forms of knowing that go with it. Some of these co-constitutive traits are explicit (formal and can be taught); others are implicit (informal and learned by experience); some are conceived rather as individual traits, others as acquired throughout enculturation and socialisation processes as a distinct, field-related habitus (Bourdieu 1977). Such differences made by and between researchers can be analysed in combination, while acknowledging that 'disciplinary members maintain other—and sometimes competing—memberships in other cultural groups and subgroups, which include but are not limited to ethnicity, gender, sexuality, class, region, age, marital status, or even additional professional training and experience' (Reich and Reich 2006, p. 54). In the empirical study, specific differences between Cluster members based on usual categories were not taken as given from the outset but had to be identified from the material produced in the study. Not all aspects of difference are effective simultaneously but rather emerge as empirical factors in specific situations. In order not to reify categories, the study therefore centred on the processes in which 'people [are] recruited into categories' and yet still 'have choices in their subject positions' (Choo and Ferree 2010, p. 134).

In addition to regularly published mathematical data and statistical analyses of the diverse membership composition and public appearances within the Cluster, I used ethnographic methods in one empirical setting in which we held a one-day workshop

---

[2]Translation B.B.v.W.

with about 40 participants, and I conducted 20 semi-structured interviews with three graduate researchers, seven PhD candidates, five postdocs, and seven professors. The interviews took a minimum of an hour, with most of them lasting between 90 and 100 min. In addition, we conducted a study with the working title 'Diversity Moves' on the bodily use of space during oral presentations. The full results of this additional experiment will be expanded upon elsewhere (Bock von Wülfingen 2021) as this article focuses on the interview study.

As we have seen in the discussion of the concept of scientific community above, non-epistemic factors that are beyond the immediate influence of a research community contribute significantly to the shaping of day-to-day work in a scientific community. Funding structures and governance will therefore be discussed in the following section.

13.3   Context: The Excellence Initiative Funding Scheme

The Excellence Cluster *Image Knowledge Gestaltung* is part of a specific funding scheme that reflects a shift in research organisation and practice that can be observed across Europe. This shift marks a new direction in EU research policy since the early 2000s aimed at fostering what was then called 'excellence in science' (European Commission 2016). Excellence in this context is a rather evaluative aspect, a policy-tailored adaptation of the contested idea of quality (Hallonsten and Silander 2012). The fundamental characteristics shared by all the centres of excellence and programmes that have sprung up since then in Europe, from Gibraltar to Estonia, and shared with this Cluster are as follows: they are funded under a directive that requires them to be innovative, competitive, temporary, collaborative (interdisciplinary), and theme oriented.

This shift toward competitive funding schemes has been particularly significant in Europe, where universities have traditionally been more reliant on public funding (Bennetot Pruvot and Estermann 2014). German universities experienced major cuts in public spending on higher education in the early 2000s. These cuts were intended to mobilise universities to act in a more business-oriented way. They were closely followed by an intense problematisation of the resulting lack of university teachers and gaps in infrastructure. One reason for the general acceptance of the Clusters of Excellence national funding scheme when established in 2005 was that it promised to alleviate the tense situation, enabling the federal government to invest in research while at the same time contributing to a competitive funding environment administered by the German Research Foundation (DFG). A total of €4.6 billion was invested over the whole 10-year programme until 2018 (DFG 2013). The programme was split into two application phases. It was intended to create incentives for specialised and prominent universities and provided for a total of 46 Clusters of Excellence lasting five or—in most cases—10 years in total. After this decade of project-oriented, short-term funding, criticism of the continually required applications for funding (a topic that also emerged as an issue in the interviews of this case study) was such that politicians reacted: after the international evaluation of the programme (Internationale Expertenkommission zur Evaluation der Exzellenzinitiative 2016), the Ministry of Education and the government jointly decided to extend the Clusters of Excellence programme and to render the funding provided more reliable. Another call for applications was issued and the funding period extended to 7 years. This time 57 clusters were selected for funding with a total of €385 million (DFG 2018). For 2025 an evaluation is scheduled to choose from these a smaller pool of clusters. The resulting clusters and the respective universities that initiated them shall after another funding period be subjected to another selection that results in the creation of so-called 'federal universities' (Tagesspiegel 2016), receiving ongoing federal funding.

The past scheme, ending in 2018, was heavily invested in the lowest pay level of research—the PhD level. Only a minor fraction of funding went into full professorships as the universities were meant to finance these contracts after the termination of the cluster funding. However, in contrast to the usual funding of short projects addressing individual researchers, teams, or networks, excellence schemes of this scope with a duration of 5 years and more are aimed at the institutional level and involve strategic choices by, and require commitment from, the institutional leadership (Bennetot Pruvot and Estermann 2014), for instance when it comes to promises to consolidate the cluster's structures and research programme.

The instable funding situation (competition and terminability) was reflected in the structure and epistemology of the Excellence Cluster *Image Knowledge Gestaltung* as well as in the interviewees' accounts in explicit or tacit terms.

13.4 Communities and Identities within the Interdisciplinary Cluster *Image Knowledge Gestaltung*

The concept of this cluster involves strategic components that distinguish it from other clusters and specifically address the objective of successful interdisciplinary collaboration. These components will be described in Sect. 13.4.1. The remaining parts of this section are devoted to the results of the interviews, elaborating on how the members' academic identities were challenged by the aims and structure of the Cluster (see Sects. 13.4.2 and 13.4.3); on different ways they made sense of it, stabilised their identities, and resettled at the margins (Sect. 13.4.4); in narrower interdisciplinary contexts (Sect. 13.4.5); or in interdisciplinary interaction more generally (Sect. 13.4.6).

Fig. 13.1 Interdisciplinary Laboratory. Cluster of Excellence *Image, Knowledge Gestaltung* (Claudia Lamas, Image Knowledge Gestaltung 2016)

13.4.1   The Interdisciplinary Composition of Image Knowledge Gestaltung

The Cluster of Excellence *Image Knowledge Gestaltung* involved about 300 members, covering about 36 disciplines. The Cluster was part of a group of six clusters that the German Research Foundation grouped under the humanities and social sciences, while all the other 37 clusters of this funding period were grouped under the natural sciences and medicine. It was the only cluster to explicitly involve the natural sciences to a similar degree as the humanities, social sciences, arts, and design.

The directors until 2016 were a professor of cultural history and theory with expertise in architecture and a professor of art history with publications relating to the history of science. Since 2016, a professor of materials science has joined these two professors on the board of directors.

The Cluster consisted of a number of research areas, all of which were required to represent a good mix of these fields. All members worked together in experimental research groups, which ideally followed an empirical methodology. They worked on issues such as the evolution of form in nature and culture, historically neglected machines for the transmission of sound, or the development of an application known as a CarePad for use by patients in clinics (Fig. 13.1).

As is typically observed in interdisciplinary research units (Klonk 2016; Vermeulen 2009a, b), an important architectonic feature is a large shared space called the *Interdisciplinary Laboratory* for individual or group work and plenaries. This fosters face-to-face communication, a feature the Cluster emphasised. Members were encouraged to work on-site on the two floors of the old workshop building hosting the Cluster. Weekly face-to-face meetings were promoted. There was a fixed weekly talk, the LunchTalk, where members ate their lunch while listening to presentations. The interior design expressed a mobile and flexible mode of being and included sofas and desks on wheels. The furniture could be easily adapted to the needs of individuals or groups (comparable to descriptions of the James H. Clark Center at Stanford by Hall 2003).

Another distinctive feature was the self-reflective component in the Cluster's structure, epistemology, and practice mentioned above: the Cluster was seen as an experiment in conducting interdisciplinary research. This research applies both quantitative and qualitative research methods, involving statistics, the computational tracking of work objects and subjects, and ethnography. My own research on effects on identities and diversity within the Cluster was also subsumed here. The self-reflective component was not extant in other German clusters of that and earlier funding periods (for a discussion of a French cluster, cf. Cointe, this volume).

In general, interviewees from all status groups described the Cluster as a great opportunity for themselves and for the research community in general. They welcomed the fact that the creation of the Cluster *Image Knowledge Gestaltung* had opened up a space for interdisciplinary collaboration within a community that is usually strictly divided into disciplines, thereby enabling them and others to pursue research interests that would otherwise not be seizable. Much enthusiasm was expressed in the interviews about the sheer audacity and effort required to establish such a cluster.

The shared, but never explicitly mentioned, feature of the biographies of nearly all members of the Cluster was that they did not adhere exclusively to one particular academic field. Instead, they had either switched from one or more disciplines to their current one, or they had gathered a couple of years' professional experience outside academia; in some cases they still worked outside the university environment. The interviews discussed in more depth in the following section showed that members were unaware of this shared trait. Instead, a large number of the interviewees expressed unease about their mode of belonging in respect to the Cluster. For many, not sharing disciplinary identities with their collaborators was 'unsettling'. As mentioned in the introduction, I use this term in the following discussion to denote the productive delocalising effect that arises when an individual cannot rely on his or her habitualised vocabulary and conduct to bond with others in an academic environment. The 'unsettling' lack of (shared) disciplinary points of reference also posed an emotional strain and required members to create and adopt new schemes, positions, and interaction rituals to resettle.

13.4.2 Unsettling (Inter)Disciplinarity

The interviews show that, across all status and gender groups, inter- and transdisciplinarity was perceived as a great challenge and one that can only be overcome with much additional effort.

Many statements referred to unnamed *others* who seemed intellectually unsettled by the different demands; this was especially evident in reactions to the presentations given in the Cluster context (e.g. at the LunchTalks) that drew on various disciplinary backgrounds. It was also expressed in answers to the question about the necessary conditions for successful interdisciplinary cooperation. One might expect interviewees to refer to inter-disciplinary overlaps as a necessary condition; however, the interviewees' own discipline was never even mentioned. Instead, nine statements described having the self-confidence to distance oneself from one's own discipline as a necessary condition for collaborations, while others in a similar vein described the inability to let go of the standards of one's own discipline as a major obstacle to collaboration. A lack of ability to distance oneself from or let go of one's habitualised disciplinary standards was reported in early career members who had just finished their master's degree.

This added insecurity and extra labour could suggest that those groups who already struggle to find their place in the academic workspace might be expected to exhibit greater interdisciplinary discomfort. Comments in the interviews indicate that it is in fact the degree of academic acculturation that is decisive in determining whether the interdisciplinary work environment is perceived as a challenge: the more stable the academic culture is ingrained, be it by the provenance of a family with academic parents or long-term work in academia, the easier is the interdisciplinary work.[3] There was also agreement by all status groups that a self-assured personality was required for the ability to work in interdisciplinary teams.

Here we come to the crucial issue of disciplinary or simply academic identity intersecting with other self-concepts. This is where apparent differences in the identity stabilisation in the interdisciplinary context were observable, which I call the (re)settling. They appeared to be related to the specificities of the Excellence Cluster *Image Knowledge Gestaltung* as an explicitly interdisciplinary endeavour, and these findings will be expanded upon in the following section.

13.4.3 Resettling and Status

Besides disciplinary differentiation, one of the most obvious factors of difference in academic research communities is the status of researchers. Universities have instruments for democratic representation, the largest and most powerful of which is the academic senate. This body includes a whole range of status groups: professors,

---

[3]Gender was not mentioned as explanatory factor, however.

non-professorial faculty, management, postdocs, and students. Temporary research institutions, such as the Clusters of Excellence, have a different governance structure: one to three directors, together with the management and a steering committee, shaped day-to-day life in the cluster.[4] The German Research Foundation as the funder of the Clusters of Excellence does not require PhD candidates and postdocs to be involved in their governance. This was the case in the first years of *Image Knowledge Gestaltung*. In some interviews conducted in 2014, this structural phenomenon was viewed critically, especially in relation to the age gap between the steering committee and Cluster members. After 2 years, the *Interdisciplinary Laboratory Image Knowledge Gestaltung* amended its rules so that the steering committee subsequently included two postdocs with full voting rights.

Nearly all of the researchers shared the impression that a lack of time, which was in fact coproduced by the needs of the interdisciplinary Cluster as such, hampered interdisciplinary collaboration. This time-related issue intersected with academic status, as it was perceived differently in different status groups. Assistants and postdocs stressed that members with a permanent contract (in nearly all cases these were professors) were not existentially dependent on the Cluster in the same way as early-stage researchers. They could therefore not readily understand the impact of temporary contracts. One interviewee pointed out that creative freedom also required social security. At the same time, there was an apparent comprehension that the Cluster was subject to structural conditions beyond the control of the board of directors, which created some mutual understanding and bonding across status groups. Some of the professors raised the issue that the extra work involved in conducting interdisciplinary research was not taken into account in the amount of time granted by the funding initiative.

13.4.4 (Re)Settling at the Margins of Academia

Beyond *academic* status, several researchers marked their identity as not being academic at all. Five out of the thirteen non-professorial interviewees described themselves as not being representative of the Cluster. They offered different reasons for this, such as cultural differences due to atypical (non-academic) career paths, age, or having parents without high school qualifications. These members stemmed from all fields and they all shared an obvious feeling of unease. They described their difference as a personal defect in expressions such as 'I don't have the others' middle-class background'. They expressed unease when asked about their biography or referred to a 'class problem'. Several of them stated that moving in academic circles did not feel normal, which resulted in them never feeling confident. Even

---

[4]This is similar to other temporarily funded research institutions in Germany, such as the Max Planck Institutes (cp. Schikowitz, this volume, for a case study of a temporary research funding initiative in Austria).

among the full professors, one interviewee described having been selected for a professorial position as 'mere luck'.

13.4.5 Resettling in the Interdisciplinary Neighbourhood

With the absence of a disciplinary context, some members closed ranks within the broader, next-of-kin interdisciplinary field. Regardless of the interviewees' field of research and across the full spectrum of academic disciplines, more than half of all participants, aside from the professors, expressed in the interviews that the respective other field of research was represented more strongly in the Cluster and gaining more support from its governing boards. This perception did not reflect the absolute numbers of people involved in the different fields. It is also relevant that there was no incident described in which the interviewees' own group was perceived as larger or better represented than one of the others. For instance, one interviewee stated: 'The natural sciences and humanities go together well; design disciplines are rather at the margins'. Another said that the design disciplines were constantly flattered.

13.4.6 Resettling in the Interactional Space

Even though they sometimes shared some of the modes of identity stabilisation[5] mentioned in Sects. 13.4.2, 13.4.3, 13.4.4 and 13.4.5, many members of the Cluster resettled in the broader space of interdisciplinary interaction. Those participants who had already gained much experience in interdisciplinary work welcomed the realisation of interdisciplinarity within the Cluster or even the complete dissolution of disciplinary demarcations as a liberation from disciplinary restrictions. Indeed, some named interdisciplinarity as a key criterion for good research.[6] As their status and experience increased, members were more likely to express enthusiasm and a sense of relief at being freed from the restrictions of disciplinary boundaries, with this feeling being strongest among the professors.

In more general terms,[7] the work atmosphere between Cluster members changed between the founding of the Cluster in 2013 and mid-2015 in such a way that, when difficulties in understanding the other's approach emerged, reactions became more interested and respectful, resulting in the more frequent articulation of clarifying questions about the meaning, sense, and comparability of different terms and techniques. Another shift was observable in the following one-and-a-half years:

---

[5]Cp. identity breach, identity trouble, identity repair mentioned in the chapter by Cuevas-Garcia, this volume.

[6]Cp. the figure and choreography of the 'polymath' in Schikowitz, this volume.

[7]This insight draws on participant observation, field notes, and members' reports.

members from the humanities, design, and natural sciences began to actively apply 'alien' terminology and to cite prominent authors or studies that their co-researchers from the respective other fields rely on, or they would swop roles and read the other's part in a talk on a joint project, thereby reflecting the active development of new interdisciplinary techniques (Jany and Razghandi 2017). The changes between 2014 and 2016 described in the interviews indicate that more members became aware of their ignorance of others' research fields and methods, and began to listen, instead of promoting their own method(s) in opposition to others'.

13.5   Summary and Conclusions

What types of collective identities emerge in broad interdisciplinary constellations such as the Cluster of Excellence *Image Knowledge Gestaltung* empirically studied here? The formation of collaborative identities in 'big interdisciplinarity', i.e. between people in a large group formed by members of very distant disciplines at opposite ends of the academic spectrum, is usually deemed difficult, if not impossible, due to seemingly incommensurable epistemic differences. The results of this study show that collaborative identities can emerge in big interdisciplinarity under certain circumstances. This is based on sometimes distressing, individual efforts,[8] entailing detachment from one's own disciplinary framework, which I call 'unsettlement', and the resettling of the researchers, i.e. their relocation to other stabilising identity types and modes.

In comparison to typical project funding, the Clusters of Excellence scheme consists of medium-term funding for a large number of researchers, often involving strategic decisions and commitment by the respective universities. In addition to this, the Cluster studied here was also shaped in such a way as to force researchers to meet (physically) in interactional spaces.

On a *short-term* basis, alternative categories relating to academic status, family background, and self-esteem—intersecting with age and work experience—seem to step in when researchers are unable to rely on their disciplinary background in interactions with others or in self-perception. The interviews show that those members of the Cluster who cannot call upon an extensive academic background (from their family of origin or work life) positioned themselves at the margin of the Cluster. In other interviews or parts of the interviews, the idea was expressed that one's own group (whether this happens to be the humanities, natural sciences, or design) was in the minority in the Cluster. At the same time, by adhering to the distinctions between the narrow inter-disciplinary categories provided by the cluster (humanities, natural and technical sciences, design disciplines) cluster members located and identified themselves in the respective disciplinary areas. In this identity

---

[8]These efforts have an emotional component. Compare Schönbauer, this volume, for the emotional dimension of identity work.

they acted as an under-represented community on behalf of the respective community. These parts of the interviews differed from the sections in which issues shared by all Cluster members were discussed, such as loyalty toward the Cluster as a project that attempts to go beyond disciplinary knowledge production, and the conflict between the lack of time and the time needed for the development of interdisciplinary skills and collaborations.

New identity constructions emerged in the Cluster after changes in its governance structure had been implemented and members had become versed in interdisciplinary communication. Apparently, the lack of a disciplinary community was subsequently compensated by belonging to the broader interdisciplinary Cluster community.

Within this community a shared self-concept was developed, an identity as a good researcher devoted to knowledge production and assuming a great deal of unrewarded work for the sake of innovative research. An additional temporal dimension and often emotional labour were consistently apparent in the interviews, especially with the postdocs. Time was required to learn an interdisciplinary culture and to grow into a research community. It also became evident that, if interdisciplinarity is not institutionalised and structurally embedded, projects on a temporal basis do not allow researchers to firmly inhabit these new identities.

Analysing these results more closely, we find aspects specific to this Cluster and others related to the funding scheme or to academia in general. Clusters of Excellence provide a time frame (funding for several years) and an infrastructure (funding for buildings, architectonic design, and equipment) that are better suited to meeting interdisciplinary needs and coping with unsettledness than usual small scale projects.

The group minority/majority perception is specific to the Cluster *Image Knowledge Gestaltung,* as the interviewees explicitly referred to the different categories of disciplinary groupings (humanities, natural sciences, and technology and design disciplines) created by the Cluster. The statement that the other group (the out-group) was larger than one's own simultaneously characterises one's own group (the in-group). It clearly signals the minority's perception that there is competition for scarce resources, as the comparison of group sizes only makes sense in this context (Schlueter and Scheepers 2010; Seyranian et al. 2008),[9] and at the same time it indicates a feeling of being at a disadvantage within the Cluster with regard to status and room to manoeuvre.

---

[9]This is especially true if the existence of an in-group and an out-group has been postulated first— i.e. a clear demarcation is taken for granted by various subjects who share common characteristics, resulting in other subjects who do not belong to the in-group. Therefore, *all* '*others*' belong to *one* out-group *collectively,* which may achieve an impressive size as a consequence. Studies relating to migrant status, including experimental studies undertaken in the 1970s and 1980s and summarised by Messick and Mackie (1989), provide additional insights. Arbitrary allocation to a specific group (for instance, in an experiment in which pupils were randomly split into teams) results in favouritism of the in-group in contrast to the out-group, and this favouritism increases markedly if one of the teams is identified by an identity marker, such as an orange vest.

This phenomenon is not necessarily due to a lack of contact with the other group and resulting unrealistic prejudices. As has been experimentally demonstrated, the mere fact of being categorised in the first place may lead to lower self-esteem within the groups (Lemyre and Smith 1985). According to social identity theory, part of an individual's *self-evaluation* is formed by affiliation to a specific group. As the individual member's identity is based on the group's community identity, as is the case for many members in the cluster, it is important for self-esteem to view one's own group in a positive light (Messick and Mackie 1989).

We can conclude that, as a result of the intrinsically necessary grouping of Cluster members according to their disciplines, the Cluster inadvertently coproduced minority and majority perceptions. Continually marking Cluster members as belonging to a particular group in order to achieve and maintain a good mix of disciplines, which was an essential characteristic of the Cluster and one supported and desired by its members, further contributed to this effect. At the same time, these group perceptions allowed the members to resettle in narrower interdisciplinary communities.

There were other aspects shared with other institutions in academia in general which shaped the ability of the clusters member to develop more or less successful interdisciplinary collaborative identities: at first glance, the interview results described in Sects. 13.4.2 and 13.4.4 (unsettledness arising from interdisciplinarity as such) may suggest that, in contrast to early-stage researchers, who feel the interdisciplinary challenge more intensely, it is only increasing age and work experience that help more advanced researchers to feel more at ease when they are freed from their disciplinary framework and thus of all trained notions, habits, and ways of thinking, as described independently by Ludwik Fleck and Pierre Bourdieu (Fleck 1983; Bourdieu 1977). However, the results described in Sect. 13.4.4 revolve around the question of origin: 'Do I have the right background to fit into academia and to pass as an academic?' This question of academic pedigree, which is related to (non-)shared origins, is one that the Cluster has in common with other academic institutions. It relates to what is currently the subject of increasing discussion as the impostor phenomenon (Clance and Imes 1978): the impression of not belonging to, or not fitting well into, the academic (Cluster) community and the impression that one has become a member by accident and not by virtue of one's own achievements.

Literature in social anthropology, education research, and science studies (Bourdieu 2011; Kerr and Lorenz-Meyer 2009; Alheit 2009; Bourdieu and Wacquant 1992) shows that such uncertainty factors—being removed from the framework that one has only just learned to grow into—usually favour those who already belong to the majority group. In interdisciplinary settings, withdrawing the disciplinary framework as joint point of reference, seems to favour those who already feature a stable academic identity—irrespective of its disciplinary direction, be it by pedigree or long-standing experience. What is then shared with all the others is the general academic (and possibly middle-class) value system. To prove that academic disciplines are not in the least solid demystifies the academic community. This can also counteract the disruptive effects of interdisciplinary work.

The results show that the development of interactional expertise in big interdisciplinarity takes several years. This is true even when this is the explicit aim of the research context and corresponding infrastructure, for instance where a physical collaborative space and incentives for self-reflection and evaluation are provided, as in the case of this cluster. The notion of the 'deep sharing' (Collins and Evans 2010, p. 53) necessary for the development of interactional expertise explains the additional amount of time required.

In the case of *Image Knowledge Gestaltung*, the first skill in interactional expertise developed is a willingness to accept disciplinary differences and to learn from the other discipline's culture(s). The new research communities then gather around the interdisciplinary research objects of their interest instead of their disciplinary identities. It appears that the effort required to leave behind disciplinary security is then compensated by the new interdisciplinary 'identity-based collective' (Gläser 2015) that shares values about how to be a good researcher.

As was to be expected, we find links between the structural conditions of the Cluster described in Sect. 13.3 and the interview results: as the interviewees remarked, essential structural conditions that were beyond the control of the board of directors and the steering committee were strong factors in shaping day-to-day work life (compare Knorr-Cetina 1981, 1982, 2003). As indicated by earlier studies (Hallonsten and Silander 2012), this interferes with the need for PhD candidates to continue to think within the terms of their respective discipline so as not to lose sight of their next career steps, as well as with temporalities such as the limited-term contracts of those not in permanent professorial positions. Conversely, looking at higher status groups, the limited proportion of time available for research is a challenge for those steering the Cluster. As these results show, the time required to develop and work with interactional expertise is a crucial factor still underestimated in funding structures.

A major conclusion of the study is that even when researchers cannot apply disciplinary frameworks, a joint community and collaborative identity can still emerge. Researchers stabilise their academic self-concepts either by reverting to their already engrained general academic identity or by biographic aspects that are not part of their institutional academic experience. The study shows that collaborative identities between the sciences and humanities can indeed be formed—with much extra effort and when a time span of several years of collaboration is allowed, which enables the development of interactional expertise.

References

Albert, M., E. Paradis, and A. Kuper. 2017. Interdisciplinary fantasy. Social scientists and humanities scholars working in faculties of medicine. In *Investigating interdisciplinary collaboration: Theory and practice across disciplines*, ed. S. Frickel et al., 84–103. New Brunswick: Rutgers University Press.

Alheit, P. 2009. Exklusionsmechanismen des universitären Habitus: Unsichtbare Barrieren für Studierende auf dem „zweiten Bildungsweg". *Hessische Blätter für Volksbildung: Zeitschrift für Erwachsenenbildung in Deutschland,* 59(3), 215–226.

Anderson, B. 1983. *Imagined communities. Reflections on the origin and spread of nationalism.* London/New York: Verso.

Bennetot Pruvot, E., and T. Estermann. 2014. *DEFINE thematic report: Funding for excellence.* Brussels: European University Association.

Bild Wissen Gestaltung, Exzellenzcluster. 2015. Online profile. https://www.interdisciplinary-laboratory.hu-berlin.de/de/bwg/ueber-uns/. Accessed 26 Nov 2020.

Bock von Wülfingen, B. 2021. Universitäre Glaubwürdigkeit und Habitus – Vortragsgestik als ungleiches Familienerbe. In *Politiken der Artefakte und des Wissens, ed. P. Lucht (forthcoming).* Berlin/New York: De Gruyter.

Bourdieu, P. 1977. *Outline of a theory of practice.* Cambridge, MA: Cambridge University Press.

———. 2011. The forms of capital [1986]. In *Cultural theory: An anthology*, ed. I. Szeman and T. Kaposy, 81–93. Malden: Wiley-Blackwell.

Bourdieu, P., and L.J.D. Wacquant. 1992. *An invitation to reflexive sociology.* Chicago: University of Chicago Press.

Capshew, J.H., and K.A. Rader. 1992. Big science: Price to the present. *Osiris* 7 (1): 2–25.

Choo, H.Y., and M.M. Ferree. 2010. Practicing intersectionality in sociological research: A critical analysis of inclusions, interactions, and institutions in the study of inequalities. *Sociological Theory* 28 (2): 129–149.

Clance, P.R., and S.A. Imes. 1978. The imposter phenomenon in high achieving women: Dynamics and therapeutic intervention. *Psychotherapy: Theory, Research and Practice* 15 (3): 241–247.

Collins, H., and R. Evans. 2002. The third wave of science studies: Studies of expertise and experience. *Social Studies of Science* 32 (2): 235–296.

Crenshaw, K. 1991. Mapping the margins: Intersectionality, identity politics, and violence against women of color. *Stanford Law Review* 43 (6): 1241–1299.

Daston, L. 2003. Die wissenschaftliche Persona. Arbeit und Berufung. In *Zwischen Vorderbühne und Hinterbühne. Beiträge zum Wandel der Geschlechterbeziehungen in der Wissenschaft vom 17. Jahrhundert bis zur Gegenwart*, ed. T. Wobbe, 109–136. Bielefeld: transcript.

de Solla Price, D.J. 1963. *Little science, big science.* New York: Columbia University Press.

DFG. 2013. *Exzellenzinitiative auf einen Blick.* Bonn.

———. 2018. *Entscheidungen in der Exzellenzstrategie: Exzellenzkommission wählt 57 Exzellenzcluster aus.* Pressemitteilung Nr. 43, 27 September 2018. https://www.dfg.de/service/presse/pressemitteilungen/2018/pressemitteilung_nr_43/index.html. Accessed 26 Nov 2020.

European Commission. 2016. *Excellent science.* https://ec.europa.eu/programmes/horizon2020/en/h2020-section/excellent-science. Accessed 26 Nov 2020.

Evans, R., and H. Collins. 2010. Interactional expertise and the imitation game. In *Trading zones and interactional expertise*, ed. M.E. Gorman, 53–70. Cambridge, MA: The MIT Press.

Felt, U. 2017. Under the shadow of time: Where indicators and academic values meet. *Engaging Science, Technology, and Society* 3: 53–63.

Fleck, L. 1980. *Entstehung und Entwicklung einer wissenschaftlichen Tatsache* [1935]. Frankfurt am Main: Suhrkamp.

———. 1983. Zur Krise der Wirklichkeit. Erfahrung und Tatsache [1929]. In *Erfahrung und Tatsache. Gesammelte Aufsätze*, ed. L. Schäfer and T. Schnelle, 15–27. Frankfurt am Main: Suhrkamp.

Gläser, J. 2007. Gemeinschaft. In *Handbuch Governance: Theoretische Grundlagen und empirische Anwendungsfelder*, ed. A. Benz, S. Lütz, U. Schimank, and G. Simonis, 82–92. Berlin: Springer.

———. 2015. Stones, mortar, building: Knowledge production and community building in narratives in science. In *Narrated communities—Narrated realities*, ed. H. Blume, C. Leitgeb, and M. Rössner, 17–30. Amsterdam: Rodopi.

Hall, C.T. 2003, October 20. Formula for scientific innovation: Omit walls: Design of Stanford's Clark Centre fosters interdisciplinary research.*San FranciscoChronicle*, https://www.sfgate.com/bayarea/article/Formula-for-scientific-innovation-Omit-walls-2581751.php. Accessed 26 Nov 2020.

Hallonsten, O., and C. Silander. 2012. Commissioning the university of excellence: Swedish research policy and new public research funding programmes. *Quality in Higher Education* 18 (3): 367–381.

Hubert, M., and S. Louvel. 2012. Project-based funding: What are the effects on the work of researchers? *Mouvements* 3: 13–24.

Humboldt-Universität zu Berlin. 2011. *Initial proposal. Cluster of Excellence Image Knowledge Gestaltung. An interdisciplinary laboratory. Unpublished proposal.* Berlin: Humboldt-Universität zu Berlin.

Huutoniemi, K. 2010. Evaluating interdisciplinary research. In *The Oxford handbook of interdisciplinarity*, ed. R. Frodeman et al., 309–320. Oxford: Oxford University Press.

Internationale Expertenkommission zur Evaluation der Exzellenzinitiative. 2016. *Endbericht.* Internationale Expertenkommission zur Evaluation der Exzellenzinitiative betreut vom Institut für Innovation und Technik (iit), Berlin. http://www.gwk-bonn.de/fileadmin/Papers/Imboden-Bericht-2016.pdf. Accessed 24 Apr 2018.

Jany, S., and K. Razghandi. 2017, July 7. *Modelling 'Filter'* [Presentation at the anniversary celebration of *Image Knowledge Gestaltung*]. Berlin.

Kastenhofer, K. 2010. Zwischen 'schwacher' und 'starker' Interdisziplinarität: Sicherheitsforschung zu neuen Technologien. In *Inter- und Transdisziplinarität im Wandel?* ed. A. Bogner, K. Kastenhofer, and H. Torgensen, 87–122. Baden-Baden: Nomos.

Kelly, J.S. 1996. Wide and narrow interdisciplinarity. *The Journal on General Education* 45 (2): 95–113.

Kerr, A., and D. Lorenz-Meyer. 2009. Working together apart. In *Knowing and living in academic research*, ed. U. Felt, 127–168. Prague: Soú.

Klein, J.T. 2010. A taxonomy of interdisciplinarity. In *The Oxford handbook of interdisciplinarity*, ed. R. Frodeman et al., 309–320. Oxford: Oxford University Press.

Klonk, C., ed. 2016. *New laboratories. Historical and critical perspectives on contemporary developments.* Berlin/New York: De Gruyter.

Knorr-Cetina, K. 1981. *The manufacture of knowledge. An essay on the constructivist and contextual nature of science.* Oxford: Pergamon Press.

Knorr-Cetina, K.D. 1982. Scientific communities or transepistemic arenas of research? A critique of quasi-economic models of science. *Social Studies of Science* 12 (1): 101–130.

Knorr-Cetina, K. 2003. *Epistemic cultures. How the sciences make knowledge.* Cambridge: Harvard University Press.

Kuhn, T. 1989. *The structure of scientific revolutions* [1962]. Chicago: Chicago University Press.

Lemyre, L., and P.M. Smith. 1985. Intergroup discrimination and self-esteem in the minimal group paradigm. *Journal of Personality and Social Psychology* 49: 660–670.

Marcovich, A., and T. Shinn. 2011. Where is disciplinarity going? Meeting on the borderland. *Social Science Information* 50 (3–4): 582–606.

Messick, D.M., and D.M. Mackie. 1989. Intergroup relations. *Annual Review of Psychology* 40: 45–81.

Midler, C. 1995. 'Projectification' of the firm: The Renault case. *Scandinavian Journal of Management* 11 (4): 363–375.

Reich, S.M., and J.A. Reich. 2006. Cultural competence in interdisciplinary collaborations: A method for respecting diversity in research partnerships. *The American Journal of Community Psychology* 38: 51–62.

Repko, A.F., and R. Szostak. 2016. *Interdisciplinary research: Process and theory.* London: Sage.

Rheinberger, H.-J. 2001. *Experimentalsysteme und epistemische Dinge: eine Geschichte der Proteinsynthese im Reagenzglas.* Berlin: Wallstein Verlag.

Schlueter, E., and P. Scheepers. 2010. The relationship between outgroup size and anti-outgroup attitudes: A theoretical synthesis and empirical test of group threat- and intergroup contact theory. *Social Science Research* 39 (2): 285–295.

Seyranian, V., H. Atuel, and W.D. Crano. 2008. Dimensions of majority and minority groups. *Group Processes & Intergroup Relations* 11 (1): 21–37.

Tagesspiegel. 2016. *Im Elitewettbewerb soll es „Bundesunis" geben*. http://www.tagesspiegel.de/wissen/zukunft-der-exzellenzinitiative-im-elitewettbewerb-soll-es-bundesunis-geben/13038206.html. Accessed 26 Nov 2020.

Tönnies, F. 1827. *Gemeinschaft und Gesellschaft. Abhandlung des Communismus und des Socialismus*. Leipzig: Fues's Verlag.

Torka, M. 2018. Projectification of doctoral training? How research fields respond to a new funding regime. *Minerva* 56 (1): 59–83.

Vermeulen, N. 2009a. *Supersizing science: On the building of large-scale research projects in biology*. Maastricht: Maastricht University Press.

———. 2009b. *Supersizing science*. Maastricht: University of Maastricht.

———. 2015. From virus to vaccine: Projectification of science in the VIRGO consortium. In *Collaboration across health research and medical care: Healthy collaboration*, ed. B. Penders, N. Vermeulen, and J.N. Parker, 31–58. Farnham: Ashgate.

Weber, M. 2002. *Wirtschaft und Gesellschaft: Grundriss der verstehenden Soziologie* [1922]. Mohr Siebeck.

Chapter 14
A Passion for Science: Addressing the Role of Emotions in Identities of Biologists

Sarah M. Schönbauer

14.1 Scientists and Their Profession: An Emotional Relationship

An emotional charge of science as a profession is and has been a characteristic of living and working in the academe. Reflections on the role of the scientist have focused on emotions for a long time. Weber (1918/1946) proposed that the scientist is necessarily passionate and deeply involved in scientific life and work and must not be distracted by other more worldly matters. He argued that scientists need their passion in order to concentrate on their work as 'if the fate of their soul' would depend on it. Without passion, the scientist would lack a calling in science: being a scientist does not resemble a job but a vocation. In his famous book *The Protestant Ethic and the Spirit of Capitalism*, Weber (1905/1930) further depicts 'calling' as historically evolved and productive for a capitalist rationality in which some succeed and some do not. Thereby, success would actually show whether one is following ones' appropriate vocation. In Weber's conception, success, commitment, and dedication appear to be indispensable for passionate scientists that have found their vocation. An 'ideal' passionate scientist is envisioned as embracing love and affection for science while being committed to hard and enduring work. This imagination is, however, built along the tension between a scientist who is expected to produce 'objective' knowledge while being emotionally committed to the profession. Taking up this tension, Pickersgill (2012) has argued, for example, that scientific work and emotions are being co-produced. In a similar vein, Des Fitzgerald (2013) has stated that rather than seeing scientific work, affect, and emotions as

S. M. Schönbauer (✉)
Munich Center for Technology in Society, Technical University of Munich, Munich, Germany
e-mail: sarah.schoenbauer@tum.de

© The Author(s) 2021
K. Kastenhofer, S. Molyneux-Hodgson (eds.), *Community and Identity in Contemporary Technosciences*, Sociology of the Sciences Yearbook 31,
https://doi.org/10.1007/978-3-030-61728-8_14

distinct from scientific subjectivity, it is more important to re-think these terms as entangled.

Scientists not only acquire the skills and know-how of the job but also become part of 'an emotionally toned complex of rules, prescriptions, mores, beliefs, values, and presuppositions' (Merton 1973, p. 258). Thus, further studies have declared emotions as an 'unavoidable and essential aspect of science in all of its phases, and underlying the social basis of science itself' (Barbalet 2002, p. 134). Hence, aside from explicit training, one learns how to *be* a scientist while becoming part of a scientific community. In this vein, the study by Koppman et al. (2015) stated that a deep emotional commitment as counter-norm to a distanced standpoint would serve to 'orient and guide the work' of scientists (p. 33). In their work, the authors find emotional expressions of scientists to be context-dependent and specific to 'feeling styles' of research cultures and disciplines. Moreover, Mitroff (1974) and Collins (1998) showed that researchers' commitments and emotional involvements are essential for how they produce knowledge. Parker and Hackett (2012) described how researchers bonded together and stated that this involvement contributed to an increase in scientific productivity and cohesiveness. These and other empirical studies present salient examples of how emotions pervade every aspects of academic life.

Yet emotional relationships—specifically, the concept of passionate work—have also been shown to serve specific purposes. Passionate work might be regarded, for example, as a disciplinary mechanism that follows a logic of distinction, such as when the scientists who correspond to an ideal of a passionate worker are provided with symbolic capital that in turn helps to build a career in science. Such an ideal simultaneously excludes those who fail to meet the respective demands. This process has been described as a predominantly gendered one (e.g. Beaufaÿs 2015; Cech and Blair-Loy 2014), producing an academic ideal mostly as a worker 'whose life centres on his full-time, life-long job', leaving aside care responsibilities for a family (Acker 1990, p. 149). Furthermore, passionate work might also provide 'a compelling status justification (and also a disciplinary mechanism) for tolerating not just uncertainty and self-exploitation but also for staying (unprofitably)' in specific sectors (McRobbie 2004, p. 132).

This resonates with recent developments in academia, e.g. in the biosciences, making it increasingly important to analyse the emotional relationship that scientists have with their profession and work. Working in the biosciences nowadays means to be dependent on short-term contracts and funding possibilities based on competitive allocation (Slaughter and Leslie 1999; Hackett 1990). These circumstances create precarious environments for scientists (Laufenberg et al. 2018) who must then deploy coping strategies in order to manoeuvre insecurities and become a part of the exceedingly small number of scientists who *make it* to a stable job (Sigl 2016). At the same time, formal demands increasingly frame these work environments. This is expressed by the key currencies that scientists have to reach for, such as highly ranked publications, third-party funding, and international mobility. In a regime of uncertainty, scientists must constantly adapt their identity because they are on the move from one research group to another while engaging in research based on temporary funding and anticipating the need for future resources. Striving for stability and constructing a secure identity thus constitutes a major challenge for a life in science and is difficult to achieve. Ylijoki and Ursin (2015), for example, depict identities of scientists in today's transforming university environments within eight different narratives, spanning the diverse positions that scientists hold, such as rebel, loser, winner, and mobile careerist. In this work, they state that all these narratives co-exist, although with fundamentally different diagnoses of the academic environment, either regarding it as an exploitative or as an ideal institution.

Against the background of identities that 'are constructed in the midst of change and permanence' and 'linked to structural changes' (Hakala 2009, p. 177), I ask: what role does passion have in scientists' narratives? And how does this play out in researchers' identities? Or in other words, I analyse how scientists narrate passionate tales and how these tales create reference frames for their identity as scientists. I argue that this is ever more important considering the need for stability and coherence when living and working as a scientist. Accordingly, I explore how biologists recount their past and present in relation to their identity: for example, referencing their early career stage as they recall their fascination with science and striving for a career or imagining a future generation of scientists.

In so doing, I analyse the role of these narratives for the creation of self-understanding by group leaders and senior postdocs in the field of biology and how this self-understanding is used to create a reference frame for a future generation. I mostly draw on 11 biographical interviews with group leaders and senior postdocs as well as numerous informal conversations that were held as part of two ethnographic field studies conducted in the United States and Austria in 2013. The interviews were structured to follow the scientists' biographical stories and included questions concerning their past, present, and future in science. After extensive coding and memo writing (Charmaz 2006), I decided to explicitly focus on the scientists' emotional engagement with their profession since relating to their community by 'associating emotional states with certain activities' (Traweek 1988, p. 76) turned out to be one of the core concerns of the researchers for their lives in science.

Specifically, I analyse the scientists' stories of the past as productive for a particular rationale and storyline of their identity bringing forth qualities of a 'passionate' scientist. By recounting their past in science, the group leaders and senior postdocs I interviewed wistfully looked back at their time of initiation and early career stages while formulating concerns about the younger generation. I am not, however, inclined to analyse how scientists narrate their passionate past as potentially better or as 'golden' times in which they had more time and freedom to follow their interests (Holden 2015). Instead, I understand the scientists' tales of the past as indicating the norms, morals, and shared beliefs of a scientific community.

Accordingly, analysing these stories provides understanding of how members of a community build an identity and adhere to its culture and of how the past is formative for the scientist's identity (Traweek 1988). On the one hand, I show how the scientists create passionate tales that stabilise their identities. This is not to say that their identities become fixed but rather that emotional involvement provides a steadiness for how the group leaders and postdocs build their identity. On the other hand, I argue that tales of a passionate past are important for the ordering of scientific communities; their individual biographical tales not only follow a trajectory but also relate to institutional and cultural orders (Garforth and Červinková 2009). In this regard, I see scientists and their passionate tales as embedded in the social and cultural prerequisites that shape and sustain the science field (Beaufaÿs 2015). Accordingly, these tales not only represent individual experiences but show how an emotional relationship to the science profession provides an exclusive belonging. Thus, passionate tales serve to create and maintain a powerful disciplinary ideal, which is especially salient today as scientists are living and working in instable employment relationships.

Consequently, I provide a jigsaw piece for understanding further how scientists build their identities and how these identities mirror today's science cultures and their demanding environments. In the following section, I will first sketch my analytical understanding of the scientists' narratives of the past and the role of 'passion' therein.

14.1.1 Passionate Tales

In the stages of a scientific career, scientists learn how to craft their careers and develop culturally accepted ways of expressing emotions and feelings. A common 'emotional culture' with a specific 'feeling style' is built (Parker and Hackett 2012) that creates a collective cultural reference frame for the scientists' beliefs and feelings.

In this section, I concentrate on 'passion' as an emotional relationship between scientists and their profession that serves as a vantage point for understanding identity creation. Etymologically, passion ranges in meaning from the Latin word *passio* and its translation as physical suffering related to Christ's tale of woe to passion as an affect, an emotional relation with and to something. Hence, passion has a wide range—from its dual meaning of sacrifice and suffering versus love and affect and beyond to passion as an uncontrollable emotion.[1] In this paper, I take passion as an emotional state that creates a basis for the scientists' fascination with science, for their spatio-temporal commitment, and for an imagination of future scientists. Moreover, using narratives of biologists, I show how a specific 'scientific passion' is made up and how this can be characterised as the scientists' emotional relationship with the science profession and with their work.

In science, passion is expressed in, for example, stories about the scientists 'burning' interest, which depict how researchers immerse themselves into their

---

[1]Passion. (n.d.). In *Merriam-Webster*. Retrieved from https://www.merriam-webster.com/dictionary/passion; Passion. (n.d.). In *Oxford Living Dictionaries*. Retrieved from https://en.oxforddictionaries.com/definition/passion

professional lives while losing track of time and having fun. 'Burning' is also part of Daston's work (2003) on the scientific persona in which she argues that many stories—literary or real—portray scientists as being obsessive. Following Daston, the scientific persona represents a collective identity that does not necessarily have to account for every scientist's individual identity; rather, it is exemplary of the particularities, modes of life, and the bodily dispositions of a group that claims this identity.

Others have portrayed scientists as being passionate in the 'flow' of doing research, which refers to a 'heightened consciousness, sharpened attention, and total immersion in the task at hand' (Parker and Hackett 2012, p. 24, referring to Csikszentmihalyi 1996). The 'flow' is best demonstrated by musicians who lose track of time and space while playing music and become fully immersed in their play. Similarly, Neumann (2006) draws on the conception of 'flow' to define her understanding of 'passionate thought'. Her concept draws attention to the creative and emotional work of researchers, occurring in their present life as well as within memories of the past. She argues that a passionate experience does not only occur in adulthood or professional instances but might also date back, for example, to childhood experiences. Additionally, she characterises a passionate experience as an interactive moment that is shared with others, drawing attention to social interaction and relationships between people. In a similar fashion, I focus on passion as individual memory of the past and as interactive experience between biologists.

Instead of focussing on passion as 'flow', I concentrate on passion as a 'drive' that leads to the scientists' self-expression as scientists showing their commitment vis-à-vis the insecurities of today's science cultures. This passionate drive appears in recollections of their initial fascination with and motivations to pursue science as well as their enthusiasm for and commitment to work. In line with this, my conception of passion is tied to emotional commitment. Koppman et al. (2015) suspect that emotional commitment is especially prominent in the early career stage. This is consistent with the passionate tales that refer to the group leaders' and postdocs' early engagements with biology. They recall those in a reminiscent mood as they account for fun, commitment, and devotion. The mode of commitment also changes with the respective career stages of the scientists, such as when initial enthusiasm and fascination develops into an affective relation with the job, which allows the scientists to cope with temporal and spatial sacrifices. Commitment also plays an important role in managing 'insecurity over one's research', 'isolation', and 'negative evaluations' (Parker and Hackett 2014, p. 555). Consequently, passion also manages instabilities, and Gill (2010) has argued that being passionately attached to work provides a disciplinary mechanism for coping with prevalent insecurities in academia. On another note, the scientific 'drive' has also been shown to provide an exclusive mechanism, discriminating between those who work along such a 'drive' and those who fail to do so (Beaufaÿs 2003).

As the scientists narrate their past and present while imagining a future generation, they create a passion narrative that provides them with meaning and self-understanding. This narrative allows them to master their identities and accomplish reflexive control (Giddens 1991). Moreover, I regard the passion narrative as organised by a particular 'plot' that serves as an overarching theme, providing an 'intelligible whole that governs a succession of events in any story' (Ricoeur 1980, p. 171). Accordingly, the successive events are transformed into a story that makes them concordant elements of the stories' plots. The scientists move back and forth in time within these plots and make them congruent in their narratives. Hence, I argue that passionate tales provide identities with self-consistency and introduce a temporal relationship in the scientists' lives. The scientists then construct identities based on a past that follows a coherent storyline while also connecting to their profession and professional values.

At the same time, the plot represents the story of a culture that is filled with tales about success and heroism of the past, providing ground for the socialisation of future scientists (Traweek 1988; Law 1994). The scientists create an identity that resembles the cultural values and myths of a specific community and culture, which also defines what is good, bad, or worthless (Ylijoki and Ursin 2015; Schönbauer 2018). Accordingly, I regard passionate stories as evolving 'as a product of certain power structures' and at times also as functioning 'to produce, maintain and reproduce those power structures' (Mumby 1987, p. 113). Yet narratives also help the scientists to orient themselves in changing environments (Ylijoki and Ursin 2015). I thus argue, that the scientists gain stability in and for their identities as they connect past and present accounts of their lives while at the same time accommodating disciplinary ideals.

For the subsequent empirical section of this chapter, I have divided my analyses into three parts, focussing on the scientists' fascination with scientific research, their early career phase, and how they imagine future scientists.

14.2 Narrating Passionate Tales

14.2.1 A Fascination with Science

When describing their biographies, the group leaders and postdocs I interviewed narrated how their paths in science unfolded. The scientists recount a continuous fascination that bridges their present and past experiences (Neumann 2006). This fascination drives the love and appreciation for a profession that fosters their exploring and discovering selves. Following Neumann, I show how stories about the past often comprise a desire to do research on a particular subject or to do research as a practice. I also depict how this fascination creates a basis for scientists' self-understanding when reflecting back on earlier times.

Remembering certain times in their lives, the scientists recall an initial interest in scientific practice that makes them able to relate parts of their lives. By drawing on childhood experiences or remembering their graduate selves, the interviewees narrated the developing trajectory of their career and their fascination with scientific practice leading them to where they are now:[2]

> So I don't know quite where science came from for me [laughs]. But I have always been kind of very interested in very, like, kind of observational [tasks] and wanted to know a lot about just, like, the stuff around me, you know. (Daniel, postdoc)[3]

Having a working-class background, Daniel did not see the professional affiliation as something inherited from his parents. Yet he assumed that his interest in science had a specific location and origin, such as an inspirational teacher. Others tell similar stories of origin. For instance, a group leader referred to his dead grandfather telling him what to do and leading him to a career in science. While Daniel initially lacked a clear-cut origin based on a social interaction, he later on explained that he discovered how 'research was a thing' and something that one can actually do. To find out if research would be something for him, he explained that he took a year off before entering graduate school, which made him realise that he 'was born to do this' (science) while figuring out what he was not born to do. In his reasoning, he was able to tie his childhood interest in creative and explorative work to his later decision to follow a career in science.

Daniel's story is similar to other narratives that reference the urge to explore and discover the unknown. During her undergraduate studies in engineering, Abigail remembered that she was attracted to science while she was a part-time technician and doing maintenance lab work:

> It was always very exciting, everything was unknown, everything was new . . . You come to work with the feeling of 'ok today I might really figure out something that would not have been known before'. . . . You know, research is accumulation so you learn, you try, you fail, you go to other people, you discuss your ideas. And eventually after this process, you can come up with a solution or conclusion or knowledge that was not there before. Could be little change in our lives, it could be big change in our lives. So that's the drive that I felt, you know it was my calling if it is [laughs] a proper thing to say. (Abigail, postdoc)

Similar to Daniel, Abigail stated that she had always been inclined to discover the unknown and aimed to understand and develop unique innovative knowledge. But instead of tracing it back to her childhood, Abigail mentioned her encouraging supervisors in the time of her studies. She continued saying that science can be a frustrating experience because of the demands to publish and cases in which experiments do not work. Yet the memory of her former self as a graduate student and her enthusiasm for discovery provides her with a past that connects her 'calling' to her present self. The enthusiasm is still valid today and relates her emotional affection for scientific work to the profession she is working in. This is tangible as she is continuously fascinated with the ability to explore unknown frontiers and to feel the 'drive' towards this fascination, which she mentions as her calling.

---

[2] It is important to note that while the initial passion for scientific work or for a particular subject was often reflectively based on childhood experiences and interests, the decision to become a professional scientist was usually made later on during (under)graduate studies.

[3] All names of the quoted researchers are pseudonyms.

Another example of how scientists narrate their fascination with the profession is when they remember particular subjects or fields of study that serve 'as the primary sites' of their passion (Neumann 2006, p. 406). For example, Jonas, a department leader in Austria, started to tell his story as an intentional decision to avoid everything that would have been related to industry, an 'emotional' decision rather than a 'scientific' choice. By rejecting the corporate world in favour of basic research, he expressed his appreciation of and care for nature:

> I like nature and I have a close relation to nature and this is quite emotional. Not so scientific. I feel very good outdoors and I am able to calm down there. So I love nature. And this is what inspired me then in my studies to understand nature. (Jonas, department leader)

As Jonas narrated his relation to nature, his emotional experience became connected to his initial fascination with nature. Jonas, however, not only talked about this relationship when referencing the present but also when remembering his childhood. He continued to explain that as a child he liked to play at the beach and rescue fish and other sea animals from tide pools. While he stated that he had not been aware of his inclination to the natural sciences at first, he has always been fond of and inspired by nature. Jonas intentionally expressed his initial interest in fish, animals, or—more generally—nature as an emotional relationship to and interest in a subject. As a present day biologist, this interest has evolved throughout his career and now connects his childhood experience to his present life and success as a scientist. This appreciation of and passion for a subject was present especially in tales of biologists who aimed to avoid working with particular applications, such as those deployed in pharmaceutical companies.

Similarly, Sophia, a group leader, talked about her entry into science when she was 'just interested in soil science, in biology, in stuff'. She recalled that after she completed her bachelor's degree, she looked for jobs related to biology. While she stated that she was disappointed by the possibilities in the pharmaceutical sector, she also did not want to participate in 'selling agricultural chemicals', which she opposed as a scientist with environmental concerns. This reasoning was mentioned prominently in other tales of the past too, such as when a scientist stated his aversion to working as a pharmaceutical sales representative. Hence, Sophia and Jonas both expressed their passion for research on a subject, such as 'nature' or more specifically 'soil', and thereby tie their passion to earlier experiences as a child or as a graduate student when they first realised this inclination.

To conclude, the scientists remember childhood or graduate studies experiences as their entry into the profession and thereby create continuity as they follow their fascination with and motivation for pursuing a career in science. This creates a storyline that is situated in the past and related to the present, allowing for 'longer narratives of related lifetime experiences' (Neumann 2006, p. 411). Their passion for science reveals their identity as grounded in a past experience that reflects a strong inclination towards a subject or towards a practice that they have followed throughout their lives. Thus, the scientists build an identity that follows their initial fascination and 'calling', providing them with a consistent storyline. This identity becomes manifest when, for example, the group leaders and postdocs treated science first and foremost not as a job but as something that had always been there, that they had always felt passionate about, and that needed to be discovered. The calling represents a way for the researchers to distance themselves from an imagination that does not cohere with their expectations of working as a scientist (for some this is evident in the case of the pharmaceutical industry). This calling also coincides with an 'ideal scientist' who has always been fascinated with science and can trace passion back in time, which also means that the scientists conform to social and cultural expectations of the science field.

14.2.2   Committing to Science

Moving (internationally) from one laboratory to the other (e.g. Kerr and Lorenz-Meyer 2009) or being dedicated to scientific work around-the-clock (e.g. Garforth and Červinková 2009) are essential characteristics of the life of a scientist. As the researchers tell stories about their first experiences as professional scientists—such as during the late PhD phase, a junior postdoc, or the subsequent entry to the group leader phase—they account for their passion for a life in science through increasing temporal and spatial commitment to the profession. In this part, I focus on commitment as having a spatial dimension, when moving (internationally) from one laboratory to the other, and a temporal dimension, when working without temporal limitation.

Many scientists narrated moving as an essential element of their scientific career. Changing places most often represents a step forward in a career, such as when a postdoc position or professorship is offered at a new location. The researchers either moved with their research groups (as PhD students or postdocs) or relocated for a job to another university. As it is an integral part of their lives, the scientists spoke elaborately about moving.

For example, Sophia, a group leader in the US, regarded moving as 'an adventure'. She recalled that her journey started with different locations in the US and a postdoc in Stockholm and found a temporary ending upon returning to the US. She recounted her travels for scientific projects and meeting collaborators in 'exotic locations', such as when she was 'wining and dining' in a castle in Scotland while working 'really hard'. In short, and like many others, Sophia spent her early career moving. In so doing, she was rewarded with a wide scientific network across Europe and the US and enjoyed an exciting time. While moving internationally was an investment in her career, it also meant that she had to leave places that she had adapted to in order to seize opportunities somewhere else. Thus, an identity is expressed that builds on Sophia's commitment to move. In a similar vein, other scientists state the importance of moving for the sake of their CVs and giving up the spatial freedom of moving to a place of choice in exchange for relying on places that have a high reputation and visibility in their field.

The importance of different locations is also tangible when Camilla, an assistant professor, recalled her time as a graduate student. This period of her life was filled with abundant fieldwork trips to different areas, such as Barbados, the Caribbean, San Francisco, Santa Cruz, and Hawaii. Constantly moving led her to thank Fedex as part of her thesis—they shipped her samples from one place to the other and were 'fantastic because they were on time [laughs]'. She declared that she enjoyed travelling before having a family as 'it was one of those times'. This portrayal also calls to mind an ideal worker who is fully immersed in a scientific life, changing this habit only later after starting a family and having reached a relatively secure position as assistant professor. Hence, the spatial commitment allows for the creation of distinct phases in a professional career: a phase characterised by flexibility and a phase which allows for continuity. As part of the first phase, Camilla remembered a busy time full of enthusiasm for doing what she loves. Doing fieldwork or being enthusiastic about fieldwork is often acknowledged as a space of release and enjoyment in the biosciences in contrast to the exhausting and manual-labour intensive times in the laboratory. Similar to Sophia, Camilla's self-understanding as a biologist who is devoted to fieldwork emerged together with her commitment to move from one place to the other and conduct research in different areas of the world.

Moreover, Camilla's passion for fieldwork and for the subject of her study also relates to a specific set of skills that enabled her to be 'very adaptable to new situations'. Such skills are key for scientists as they have to show adaptability. For example, Abigail, a postdoc, explained that to 'function anywhere in any working culture' and to be flexible 'enough' to fit in is important for her. Many interviewees emphasised that their moving experience was exciting and spurred their enthusiasm for a life in science. At the same time, researchers also described these times in their early career phase as exhausting and without scope for respite or staying long in one place. Accordingly, flexibility and adaptability are part of the requirements for progressing in a career and thus demarcate characteristics of a powerful scientific ideal that enables a self-regime in which scientists are individually responsible for their commitment. Hence, in line with Gill's interpretation of passionate work (2010), expressing the enjoyment of moving also indicates an underlying tolerance of the demands of a life in science and bearing its uncertainties, such as not knowing where the next position will be found.

In addition to spatial commitments, scientists talked a lot about temporal dedication to their scientific work. The interviewed biologists commonly did not associate their work with an explicit 9-to-5 routine but rather with working late and on weekends. Their tales about time ranged from the PhD phase to professor level onwards and most often ascribed a lack of temporal boundaries to a specific time of the past—the early career stage:

> I mean I really cannot think of myself as I was working for my PhD, actually. I used to be in the lab 16 hours. No problem. I don't think that I have it in me anymore [laughs]. (Abigail, postdoc)

> So this postdoc time is still stuck in my bones. And I still remember [this time] now, also because you know that this is sort of the point where it is revealed, the next step [a group leader position]. (Marie, group leader)

Such stories about missing temporal boundaries are abundant and often similar, such as when talking about 'these times' when you are 'all in' (Jonas, department leader). Abigail and Marie described this early career period as graduate students or postdocs as time-intensive and physically exhausting. Marie added that when she worked at Stanford University, the laboratory was always busy, even in the middle of the night when she finished her work at 2 a.m. and on weekends. This time-intensive work is not limited to but characteristic of work in a laboratory since 'bench work' most often means gruelling manual labour on graduate and postdoctoral levels. Hence, I argue that the scientists' narratives of temporal dedication to their work in the early career phase provide an explanation for and legitimisation of their present position and career, which has an impact on their self-understanding as scientists in a specific position. Thereby, the next step—a group leader position—can be reached. This temporal commitment spurs an individualised self-regime in which the scientists are responsible for their career progress through demonstrating that they follow the demands of a powerful ideal: a scientist who never stops working.

In other narratives, I found explicit references to playful relationships and social interactions at work that were formative for the scientists' in the stages of their career. Through sharing fun and experiencing joy, the scientists become emotionally committed—not necessarily to research as a practice or as a subject but as a social endeavour. For example, Felix, a group leader, described playing volleyball in the afternoon before returning to work in the evening, as well as drinking gin and tonic together with colleagues when they finished work early:

> We did a lot ... there was practically no time, day or night, when there was nobody in the lab... Instead [of working at the bench] we were playing volleyball all together on the lawn in the afternoon. Or we sometimes, if something happened, then it was ... I don't know ... [we] finished in the early afternoon and everybody was drinking gin and tonic. (Felix, group leader)

In Felix's recollections, these earlier periods seem bound to a communality that he shared with others. He noted that everybody was experiencing similar circumstances as they would invest long hours in their work, which was specific but not limited to laboratory work. They would spend weekends and nights together as they worked on their experiments. In addition to dedicating time to work, they also shared playful engagements, such as drinking gin and tonic, playing volleyball before going back to work, or celebrating achievements together.

In his current position, Felix draws on these previous experiences when arranging 'social meetings' for his group, allowing members to devote time to informal interaction. While person-to-person relationships have been described as fuelling the scientists' passion in terms of knowledge production and ideas (Neumann 2006), the social interactions that I depict here demarcate a playful relationship with the environment in which scientists make themselves at home, thereby eliding the distinction between work and play or private and professional personae. This also means that interacting with others makes them more likely to tolerate long work hours and demanding labour.

In these stories, the group leaders and postdocs describe their commitment as a prominent characteristic in their early careers in science. In formulating specific demands for their lives, such as spatial and temporal dedication, they foster an emotional relationship between them and their profession. Aside from periods of hard work, they also recall times of enthusiasm and joy. These emotions provide the basis for their present self-identity as scientists who (successfully) established a career. The strong sense of commitment also provides the scientists with the potential to cope with insecurities (Parker and Hackett 2014), such as when they declare exhausting times to be 'fun' and exciting when moving from one place to another or when spending day and night in the laboratory. The related identity is fragile, however, as is evident when one is 'unable to be a good role model' any longer (Felix, group leader) because one cannot be in the lab day and night and start a family at the same time. Or when Emma, a staff scientist in the US, mentioned that she 'was never a 100 per cent scientist' as she could not simultaneously care for her child and take advantage of the time after her 'sexy paper' came out by working all night. This is also evidenced in studies that show how scientists who reduce their work hours are stigmatised as not fulfilling the characteristics of an ideal worker (e.g. Cech and Blair-Loy 2014). These deviations highlight what is imagined as the norm for an identity as a scientist: a 'drive' towards scientific work and life without any spatial or temporal boundaries.

14.2.3   Imagining a Future Generation of Scientists

Subsequent to the scientists' commitment in their early career stages, past experiences were also used to explicitly address the demands of the profession for students and nascent scientists. Thus, I now focus on how scientists position future scientists when reproducing the past (Garforth and Červinková 2009). In these attempts to imagine a future life in science, I stress how scientists formulate necessary requirements for students and how they articulate what makes a 'good scientist' (Traweek 1988) while discussing how both are entangled with their own stories of the past.

Some biologists regarded passion as an essential quality of a scientist and specified an explicit reference frame for this quality based on their own self-understanding as scientists. Jonas, for example, stated that the students would not necessarily have to abandon their life outside the lab. Yet he demands that they demonstrate their passion and enthusiasm for and commitment to the job:

> But . . . of course you can have a relationship, it is not about that. It is about the passion. I do have . . . of course the relationship is important . . . it doesn't mean that you have to be there 24 hours, 7 days a week . . . but this is your thing. Your passion. (Jonas, department leader)

Consequently, passion should be expressed in an appropriate time frame as the students should not waste their time on things that are 'not important'. Jonas continued with what he misses in his own students by reflecting on a specific spirit and time during which a scientist would have to give 'a hundred percent' in order to live and work in academia:

> But I think at the beginning I miss—not from everybody—I miss the absolute passion from many [students]. [Working] on weekends too, and so on. Not because I find it important to put them under pressure but as a sign that they will be head over heels for a while. It's only really fun if you are all in. Everything in life is. I have always done everything a hundred per cent. (Jonas, department leader)

This imagination of fun and excitement was entangled with his own past experience: Jonas remembered that he had always been inclined to experience everything to its fullest. When remembering his entry into science, Jonas stated that he was a 'punk' with long blue-dyed hair who was developing an ambition for challenge as he experienced his transformation from 'freak to wunderkind'. At the same time, Jonas missed an 'absolute passion' from his students, which he understood as a sign of devoting oneself to science and embracing its fun character. Accordingly, his self-understanding featured an assertiveness that distinguished him from others but also demonstrated a devotion to progressing his scientific self.

In a similar vein, another group leader explained her selection criteria for young scientists. For Sophia, students would not only need to be eager and intelligent but also show that they are dedicated to science by working all night:

> The main thing with students for me at that time was that they are really excited. And enthusiastic. Of course, intelligent and all that kind of stuff. But if they were burning for the science, then they would work all night, you know [laughs]. (Sophia, group leader)

She stated further that the students' excitement would lead to spin-off projects in which they would produce data for potential papers and work on their own reputation as scientists. As a group leader, Sophia profits from such behaviour as it might result in good publications and meet her aim to 'produce quality people that go out into the world'. Moreover, the 'burning' was often based on the group leaders' or senior postdocs' own experiences in which they immersed themselves into science and were 'all in' (Jonas, department leader). And while recalling their past experience in the lab, the scientists frequently bemoaned a lack of commitment and devotion in the younger generation. Consequently, I depict passion as becoming constitutive for a hard-working ethos of an excited scientist.

In addition to searching for students who would show a passionate relationship with scientific work, some group leaders and senior postdocs were concerned with how to select the most dedicated students. For example, Sophia, a group leader, was concerned with education quality and how students could be chosen on the basis of specific criteria. According to her, a screening mechanism similar to the European master's system would be worthwhile to 'see "is this really what you want to do or not?"' A mandatory master's programme and figuring out what you 'love' and need in order to be happy were common imaginations for a better method of student selection. By saying that 'not every person should get a PhD just because a person starts on (a career) track', Sophia is in company with others who declare that: 'It is not for everyone, obviously' (Abigail, postdoc). Some scientists would be good at analytical thinking, while others would either not be 'trained' properly or lack 'the natural ability to analyse what they see' (Abigail, postdoc). Furthermore, Abigail declared this path as not possible for everyone due to long work hours and time periods in which frustration would outweigh positive results. Hence, I find the imaginations of group leaders and senior postdocs about potential future students as closely entwined with how they envision selection criteria.

The researchers claim that the future generation needs to show excitement for and dedication to the job. Related to this are expectations assuring that these students have what it takes to accomplish a scientific career, such as enthusiasm about work and life in the laboratory and the willingness to work all night. Group leaders and senior postdocs also described a specific time *when* young academics would need to demonstrate this emotional trait. They positioned the younger generation as in need of working in a similar fashion and emotional state as they did during their own time in the lab. Formulating a prospective vision implies potential for selection, as the group leaders and postdocs envisioned possible qualifying gates for students that would make it easier to decide if a student would be dedicated to scientific work and if science would be something the students would like to do. Hence, while the narration of the group leaders' and senior postdocs' own past experiences signifies continuity in their life, which is 'tied to envisioned futures, or constructed retrospectively, as past events gain coherence and purpose through narrative' (Garforth and Červinková 2009, p. 175), the tales also imply an imagination of a future generation of scientists. This imagination creates a condition in which young scientists are urged to adapt to a powerful ideal—a condition that allows for a self-regime in which scientists are individually responsible for managing their emotional relationship with the profession.

14.3   Conclusions

In this chapter, I have shown that emotional relationships of scientists with their profession are having an impact on their self-perception and identity construction. As the scientists talk about their individual emotional states of being, they not only construct their career stage-dependent identities as researchers but also reinforce a quasi-romantic ideal of how researchers need to think and feel about a life in science according to a particular emotional scheme. In line with this, passionate tales provide some order in and for scientific communities and show that the scientists' emotional relationship with their profession is key to understanding the construction of their identities.

As the scientists narrated their passion—'drive'—for the profession, I have shown first that the group leaders and postdocs revisited their time of initiation and narrated their scientific self by invoking a past in which they had discovered their initial fascination with science as subject and practice. Second, when reflecting back on their early career phases, the scientists showed their commitment to their work by describing the spatial and temporal concessions they were willing to make and that have formed their self-understanding ever since. Finally, I have depicted which characteristics group leaders and senior postdocs looked for in the younger generation as they positioned the future generation by drawing on their own pasts in science. These stories not only inherit a quasi-nostalgic memory of the past (Holden 2015) but also comprise a passionate tale in which the scientists build on their 'drive', such as by remembering their initial fascination, enthusiasm, and commitment towards science. Being a scientist, then, means to be passionately driven and to demonstrate this in specific ways in the respective career stages.

I derive two main conclusions from these analyses: first, passionate tales provide order for the scientific community as they build an imaginative ideal of a passionate scientist through emotional schemes and behavioural patterns; second, the narratives provide the scientists with continuity by constructing an emotional track record.

Passionate tales about the past declare characteristics of a 'good scientist' (Traweek 1988) as they (re-)construct an ideal in which the fate of a scientist's soul depends on her/his work (Weber 1918/1946). And although the imagination of passion might not deviate much in its characteristics from that proposed a hundred years ago, it represents a cultural stability for science communities that are otherwise subject to change. It is crucial to note that the scientists' narratives were all similar to each other and lacked multiplicity in their storylines. Accordingly, these stories also imply marginalisation. This is evident, for example, when a scientist does not make use of the right timing to develop a career, such as when having a family and not being able to work all night, or in deciding that one does not want to pursue tenure but rather get a more stable position. Although such alternative ways of living a life in science exist, they are not dominant in the conceptualisation of a scientific identity. When the biologists refer to their shortcomings as not-ideal scientists, they support and manifest a norm of the hardworking, committed academic who is dedicated to scientific concerns, expanding the unknown frontier, and caring for the scientific community. In this sense, I argue that passionate tales also demarcate a regime of selection.

Scientists need to constantly demonstrate passion in order to correspond to an ideal imagination of a scientist. They need to work as experts of their emotions and their skills (Rose 1996). This is the case, for example, when researchers imagine a future generation working in accordance with temporal and spatial requirements. Yet, only those who conform to the demands of the profession can *be* scientists. Passion provides distinctiveness for those who *are* scientists and is recounted as intrinsic to everyone who *becomes* a scientist in the early career phase. While this brings forth an exclusive identity nourished by tales of the past and embedded in a common emotional culture, this also means that scientists have to work continuously towards an ideal that disciplines them into a 'neoliberal regime with ever-growing costs', not least to themselves (Gill 2010, p. 241).

Aside from creating selection criteria, I state that passionate tales introduce a profound stability for the scientists. The proposed stability is based on the storyline that the scientists create and that—at least temporarily—merges the initial detection of ones' interest, the early career stage, and the present. The group leaders and postdocs reflexively create a 'plot' in their lives that corresponds with their identity as passionate scientists. This plot also provides for consistency in their lives, as they have been and always will be passionate scientists. While this is an expression of their emotional relationship with the profession, their love and affection for what they do, it is also a sign of the inevitable need to embed one's identity in a temporary stable framing while committing to the demands of a life in science.

This stability is of high importance given the instabilities and insecurities that pervade work in the biosciences. These resemble the tenuous working conditions on a societal level outside academia. Work has transformed from a 'vocation or a life's mission' into a notion that is much more 'fragile and friable' (Bauman 2001, p. 45) with a regime of short-term work permeating today's society in which identity work is never stable and fixed, always appearing open and unfinished (Keupp 1994). This ephemeral nature of inhabited positions is significant for today's society—and for academia in particular—as traditional and contemporary ideals merge in researchers' narratives (Felt and Stöckelova 2009). In line with merging traditional and contemporary ideals is a passionate self that creates consistency in scientific communities and a steady reference frame for the scientists' identity construction and performance.

To conclude, focussing on scientists' pasts allows for a profound analysis of the imagined appropriate characteristics for a life in science. As scientists continuously produce and reproduce an imagination of passion while building on their own past in science, they create selection criteria for those to come. For instance, they determine that passion is mandatory in order to become a successful researcher. As such, passionate tales serve to seduce the scientists into a romantic logic while also underpinning a neoliberal regime in which scientists have to show extensive commitment. Aside from creating individual stability in the scientists' lives, such tales also seem to produce a normative ideal that promotes scientists who are fully committed through an emotional relationship to their profession and work. Passionate tales thereby nourish a self-regime that focusses on individual capabilities for managing behaviour as well as emotional responses. This condition raises urgent questions such as: who is able to conform to a powerful ideal of a passionate scientist and who is not? How does a self-regime that centres on the individual and her/his ability to manage an emotional relationship with the profession possibly change a future generation of scientists?

Based on these questions, I argue that there is an urgent need to trace stories that are different, those that resist a normative ideal. And we as academics need to take a critical stance on what 'passionate work' in the sciences means and how identity at work can be built and criticised. Only then can scientific identities be reflected upon in ways that do not disguise alternative forms but open up multiple possibilities for stories to be told.

Acknowledgements First of all, I would like to thank the biologists who generously agreed to talk to me and shared their stories. I would like to thank the editors of this volume, Susan

Molyneux-Hodgson and Karen Kastenhofer, for their continuous help with and care for this chapter. Many thanks are due to the reviewers for their constructive comments and help in re-writing this chapter. I would also like to thank Ulrike Felt for sharing her worthwhile comments when I first presented this work at the STS Austria workshop on 'Identity and Community' in February 2017 and Niki Vermeulen for her invaluable feedback on the manuscript. And finally, I would like to thank the University of Vienna for supporting me in the time of writing this chapter with a Dissertation Completion Fellowship.

References

Acker, J. 1990. Hierarchies, jobs, bodies: A theory of gendered organizations. *Gender & Society* 4 (2): 139–158.

Barbalet, J. 2002. Science and emotions. *The Sociological Review* 50 (S2): 132–150.

Bauman, Z. 2001. *Community: Seeking safety in an insecure world.* Oxford: Blackwell Publishers.

Beaufaÿs, S. 2003. *Wie werden Wissenschaftler gemacht?* Bielefeld: Transcript.

———. 2015. Die Freiheit arbeiten zu dürfen. Akademische Laufbahn und legitime Lebenspraxis. *Beiträge zur Hochschulforschung* 37 (3): 40–58.

Cech, E.A., and M. Blair-Loy. 2014. Consequences of flexibility stigma among academic scientists and engineers. *Work and Occupations* 41 (1): 86–110.

Charmaz, K. 2006. *Constructing grounded theory.* London: Sage.

Collins, R. 1998. *The sociology of philosophies.* Cambridge, MA: Harvard University Press.

Csikszentmihalyi, M. 1996. *Creativity: Flow and the psychology of discovery and invention.* New York: Harper Perennial.

Daston, L. 2003. Die wissenschaftliche Persona. Arbeit und Berufung. In *Zwischen Vorderbühne und Hinterbühne: Beiträge zum Wandel der Geschlechterbeziehungen in der Wissenschaft vom 17. Jahrhundert bis zur Gegenwart,* ed. T. Wobbe, 109–136. Transcript: Bielefeld.

Felt, U., and T. Stöckelova. 2009. Modes of ordering and boundaries that matter in academic knowledge production. In *Knowing and living in academic research. Convergence and heterogeneity in research cultures in the European context,* ed. U. Felt, 41–124. Prague: Institute of Sociology of the Academy of Sciences of the Czech Republic.

Fitzgerald, D. 2013. The affective labour of autism neuroscience: Entangling emotions, thoughts and feelings in a scientific research practice. *Subjectivity* 6 (2): 131–152.

Garforth, L., and A. Červinková. 2009. Times and trajectories in academic knowledge production. In *Knowing and living in academic research. Convergence and heterogeneity in research cultures in the European context,* ed. U. Felt, 169–226. Prague: Institute of Sociology of the Academy of Sciences of the Czech Republic.

Giddens, A. 1991. *Modernity and self-identity: Self and society in the late modern age.* Cambridge: Polity Press and Blackwell Publishing Ltd.

Gill, R. 2010. Breaking the silence: The hidden injuries of neo-liberal academia. In *Secrecy and silence in the research process: Feminist reflections,* ed. R. Flood and R.C. Gill, 228–244. New York: Routledge.

Hackett, E.J. 1990. Science as a vocation in the 1990s: The changing organizational culture of academic science. *The Journal of Higher Education* 61 (3): 241–279.

Hakala, J. 2009. The future of the academic calling? Junior researchers in the entrepreneurial university. *Higher Education* 57 (2): 173.

Holden, K. 2015. Lamenting the golden age: Love, labour and loss in the collective memory of scientists. *Science as Culture* 24 (1): 24–45.

Kerr, A., and D. Lorenz-Meyer. 2009. Working together apart. In *Knowing and living in academic research. Convergence and heterogeneity in research cultures in the European context,* ed. U. Felt, 127–167. Prague: Institute of Sociology of the Academy of Sciences of the Czech Republic.

Keupp, H. 1994. Ambivalenzen postmoderner Identität. In *Riskante Freiheiten. Individualisierung in modernen Gesellschaften*, ed. U. Beck and E. Beck-Gernsheim, 336–350. Frankfurt am Main: Suhrkamp.

Koppman, S., C.L. Cain, and E. Leahey. 2015. The joy of science: Disciplinary diversity in emotional accounts. *Science, Technology, & Human Values* 40 (1): 30–70.

Laufenberg, M., M. Erlemann, M. Norkus, and G. Petschick. 2018. Prekäre Gleichstellung – Eine Einleitung. In *Prekäre Gleichstellung*, 1–24. Wiesbaden: Springer.

Law, J. 1994. *Organizing modernity*. Oxford: Blackwell Publishers.

McRobbie, A. 2004. Making a living in London's small-scale creative sector. In *Cultural industries and the production of culture*, ed. D. Power and A.J. Scott, 130–144. London: Psychology Press.

Merton, R.K. 1973. *The sociology of science: Theoretical and empirical investigations*. Chicago: University of Chicago Press.

Mitroff, I.I. 1974. Norms and counter-norms in a selected group of Apollo moon scientists: A case study in the ambivalence of scientists. *American Sociological Review* 39: 579–595.

Mumby, D.K. 1987. The political function of narrative in organizations. *Communications Monographs* 54 (2): 113–127.

Neumann, A. 2006. Professing passion: Emotion in the scholarship of professors at research universities. *American Educational Research Journal* 43 (3): 381–424.

Parker, J.N., and E.J. Hackett. 2012. Hot spots and hot moments in scientific collaborations and social movements. *American Sociological Review* 77 (1): 21–44.

———. 2014. The sociology of science and emotions. In *Handbook of the Sociology of emotions: Volume II*, ed. J.E. Stets and J.H. Turner. Springer.

Pickersgill, M. 2012. The co-production of science, ethics, and emotion. *Science, Technology, & Human Values* 37 (6): 579–603.

Ricoeur, P. 1980. Narrative time. *Critical Inquiry* 7 (1): 169–190.

Rose, N. 1996. *Inventing ourselves. Psychology, power and regulation*. London: Sage.

Schönbauer, S. 2018. We are standing together in front. How scientists and research groups form identities in the life sciences. *Science & Technology Studies*: 1–21.

Sigl, L. 2016. On the tacit governance of research by uncertainty. How early stage researchers contribute to the governance of life science research. *Science, Technology & Human Values* 41: 347–374.

Slaughter, S., and L.L. Leslie. 1999. Academic capitalism. In *Politics, policies, and the entrepreneurial university*. Baltimore: Johns Hopkins University Press.

Traweek, S. 1988. *Beamtimes and lifetimes*. Cambridge: Harvard University Press.

Weber, M. 1930. *The Protestant ethic and the spirit of capitalism*. London: Allen & Unwin. (Original work published 1905).

———. 1946. Science as a vocation. In *From Max Weber: Essays in sociology*, ed. H. Gerth and C.W. Mills, 129–158. New York: Oxford. (Original work published 1918).

Ylijoki, O., and J. Ursin. 2015. High-flyers and underdogs: The polarisation of Finnish academic identities. In *Academic identities in higher education: The changing European landscape*, ed. L. Evans and J. Nixon, 187–202. London: Bloomsbury Academic.

Name Index

© The Author(s) 2021
K. Kastenhofer, S. Molyneux-Hodgson (eds.), *Community and Identity in
Contemporary Technosciences*, Sociology of the Sciences Yearbook 31,
https://doi.org/10.1007/978-3-030-61728-8

303

Subject Index

© The Author(s) 2021
K. Kastenhofer, S. Molyneux-Hodgson (eds.), *Community and Identity in Contemporary Technosciences*, Sociology of the Sciences Yearbook 31,
https://doi.org/10.1007/978-3-030-61728-8